Science and Applications of Nanoparticles

Science and Applications of Nanoparticles

edited by
Waqar Ahmed
Ehsan Nourafkan

JENNY STANFORD
PUBLISHING

Published by

Jenny Stanford Publishing Pte. Ltd.
101 Thomson Road
#06-01, United Square
Singapore 307591

Email: editorial@jennystanford.com
Web: www.jennystanford.com

British Library Cataloguing-in-Publication Data
A catalogue record for this book is available from the British Library.

Science and Applications of Nanoparticles

ISBN 978-981-4267-34-2 (Hardcover)
ISBN 978-1-003-28029-3 (eBook)

Contents

Preface

Nanoparticles are materials with overall dimensions in the nanoscale, i.e., under 100 nm. In recent years, these materials have emerged as important players in modern techniques, having applications from medicine and drug delivery to textile and agriculture. Nanoparticles are already integrated into our everyday lives in various forms, such as sunscreen lotions containing nanoparticles, dirt-resistant glass with nanocoatings, and nanodrugs. Much has already been written on the fundamental aspects of nanoparticles. This book is refreshing because it deals with recent studies and techniques to produce nanoparticles for practical application in diverse industrial fields. Although several aspects of nanoparticles have been dealt with in innumerable journal articles individually, sources covering this area comprehensively are limited, and this book tries to fill this gap. Some other interesting topics, including nanoencapsulation, nanoparticle formation, and surface functionalization of nanoparticles, have been covered in detail to make it a valuable resource for practical use.

Waqar Ahmed

Ehsan Nourafkan

Summer 2022

Chapter 1

Application of Nanoparticles in the Field of Photonics

Syed Asad Hussain
Department of Engineering Science, University of Oxford, Parks Road, Oxford, OX1 3PJ, UK
asad.hussain@outllook.fr

In this chapter, we have presented applications of nanoparticles in the field of photonics. The field of photonics is very vast, and nanoparticles have proved to be a valuable tool in this field. We have limited and focused our discussion by providing in-depth discussion about microscopy, laser, nonlinear optical conversion, solar cells, and therapy. Nanoparticles are very useful in all these areas due to the unique properties they offer in comparison to their bulk counterpart. Although these materials are entering the commercial markets of microscopy, laser, and therapy, further understanding is required to optimize the properties and make them superior.

Science and Applications of Nanoparticles
Edited by Waqar Ahmed and Ehsan Nourafkan

ISBN 978-981-4267-34-2 (Hardcover), 978-1-003-28029-3 (eBook)
www.jennystanford.com

1.1 Introduction

Due to reduction in size, nanoparticles with size in the range of 1 to 100 nm give rise to properties that are not observable from their bulk counterparts as we reach quantum confinement at the atom scale (Pavitra et al., 2019; Yanhong et al., 2004). Various properties of these nanoparticles are very important, which include high surface area, dedicated surface characteristic, electrical, optical, thermal, magnetic, and their interaction with the other systems. Intensive work has been done on their synthesis (Murphy et al., 2005; Panáček et al., 2006; Lu et al., 2007; Narayanan and Sakthivel, 2010; Tang et al., 2012). The landscape is rich in various methods, and each has its own advantages and disadvantages. This also includes various bottom-up or top-down routines. Due to the pressing climate issues and making them in an environmentally friendly manner, focus is now on making them without using chemicals. In this connection, using lasers to create organic or inorganic particles or making nanoparticles from natural elements has become a primary focus of the field (Eliezer et al., 2004; Hu et al., 2009; Amendola and Meneghetti, 2009; Bar et al., 2009; Iravani, 2011). Apart from environmental issues, these methods are important due to the cost involved in buying chemicals and deposing them. In this comparison, the capital cost of utilizing laser is one off, including its occasional maintenance.

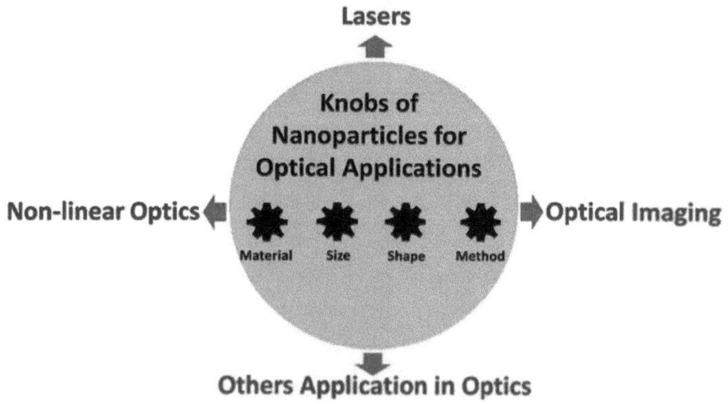

Figure 1.1 Various properties that need to be modified for the nanoparticles to be used for applications in optics.

Understanding the interaction and dynamics of different biological system is a key process to apply different strategies to find the cure of diseases. These processes are hard to understand by using conventional microscopes as the biological dynamics is not possible to image. Moreover, they are limited in terms of resolution. To overcome this problem, various nanoparticles have been made in the past that could be attached to different parts of microscopic biological systems. As an example, silicon, which is available in abundance and extensively used in the electronics industry, shows size-dependent photoluminescence at room temperature (Park et al., 2009; Pavitra et al., 2019). Once attached, they could then be used for imaging. Gold attracted a lot of attention in the past and has been used for many years. Not only silicon and gold, semiconductors and iron-based nanoparticles have also attracted attention. By using them inside the biological system, key imaging matrixes such as signal-to-noise ratio, spatial resolution, and selectivity go high.

In this chapter, we will mainly focus our attention on the application of nanoparticles in the field of optics. The chapter will highlight progress that has been made so far and will also provide knowledge about the science behind each application.

1.2 Nanoparticles for Bioimaging

Iron-based nanoparticles can provide magnetic properties. The properties of these nanoparticles could be tuned to provide functional properties that could be applied for magnetic resonance imaging (MRI) (Hadjipanayis et al., 2008; Ris et al., 2010; Wen et al., 2014). In the family of these nanoparticles, iron oxide is the most important material. One problem associated with this type of nanoparticles is stabilization, which could be overcome by using them inside composites (Masotti et al., 2009; Santra et al., 2012). However, this could affect the capabilities of these nanoparticles. Another factor is to make these types of nanoparticles with multifunctionality.

In MRI, a strong magnetic field is applied that helps to align the protons present in the tissue (Fig. 1.2). This helps to align the proton along the field direction. In the next step, a radiofrequency pulse is applied, which helps them to align in another direction. The radiofrequency is then turned off, due to which the protons relax back to their original orientation. In MRI, a unique image is obtained due to this relaxation. The magnetic nanoparticles could be attached to different types of tissues and, therefore, could interact with the relaxation process, making it faster or slower in comparison to other types of tissues in the surrounding. Once this is done, the MRI images generated from the setup have much better contrast in comparison to those without them.

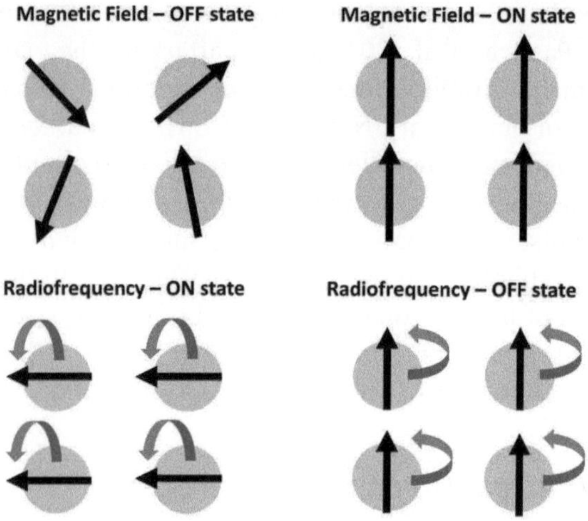

Figure 1.2 Dynamics of atoms in the presence of magnetic field and radiofrequency pulse. The atoms are aligned in one direction due to a magnetic field. The radiofrequency signal aligns them in another direction. By turning off the radiofrequency signal, the atom will relax back to its original aligned state emitted radiofrequency signal to be detected by the detector. This generated signal will be used to create an image.

As an example, Fig. 1.3 shows the results for $Zn_{0.2}Fe_{2.8}O_4$ nanoparticles with sizes 4, 7, and 10 nm for their magnetic properties (Ma et al., 2019). Figure 1.3(a–c) shows the sizes of these nanoparticles in transmission electron microscopy (TEM). When a similar magnetic field is applied to these nanoparticles, they show different behavior and depending upon the size give rise to magnetization different in aptitude. Figure 1.3d shows this phenomenon; in this case for the same magnitude of the magnetic field, the nanoparticles with size 10 nm attain higher magnetization and then 4 and 7 nm particles. Moreover, it was also shown that if a magnet is placed next to the solution of these nanoparticles, they will also be attracted toward the magnet due to the presence of iron.

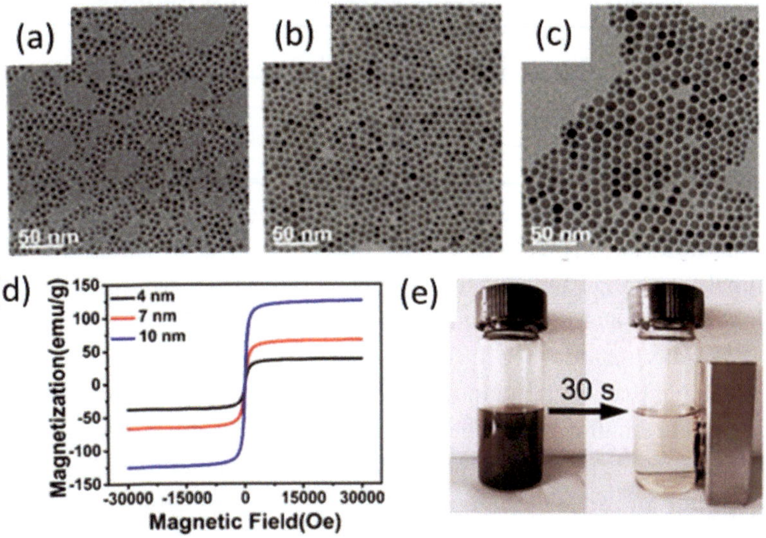

Figure 1.3 (a–c) $Zn_{0.2}Fe_{2.8}O_4$ nanoparticles of different sizes. From left to right, 4, 7, and 10 nm. (d) Magnetic field applied to these particles and magnetization obtained. (e) Nanoparticles attracted toward the magnet. Reproduced from Ma et al. (2019) with permission.

Although the technique based on MRI has proved to provide superior spatial resolution, the setup cost is expensive. To overcome this difficulty, we could use nanoparticles for optical microscopy. The optical properties are highly dependent on size and shape. Based on the size of these particles, fluorescence wavelength changes. They could be injected or delivered into the body where the imaging is required. However, they need to be connected to distinguish between different bodies. To do so, the surface is modified to allow the attachment of different molecules (McConnell et al., 2000; Cheng et al., 2006; Kango et al., 2013). These molecules act as a linker between the nanoparticles and the biological tissue. After these molecules are attached, they provide superior signal-to-noise ratio and are resistant to photobleaching in comparison to staining alone with organic or fluorescent dye molecules.

Let us now focus on different types of optical microscopes available. All microscopes normally consist of a light source that excites the nanoparticles, a detector to detect the light coming from the sample, an optics to direct the light toward the sample, a microscope objective to focus the light beam on the sample, and a set of optical filters that allow light beams of certain wavelengths and block light beams of other wavelengths. The microscope creates a spot diameter by using a microscope objective, which has a set of lenses packed inside a body to give better optical performance. The focused light creates a diameter of light known as the point spread function (PSF) (Fig. 1.4). We can see that the light has an intensity peak at the center and the decaying side lobes around it. This is because when the light, which is an electromagnetic field, is solved in three-dimensional space considering a microscope objective gives a mathematical expression known as the Airy pattern. This PSF could be made smaller by using a high numerical aperture (NA) objective lens (Fig. 1.4a). The inverse relation between the numerical aperture number and spot diameter can be simplified by using the following equation:

$$\text{Spot diameter} = 1.22\frac{\text{Wavelength of light}}{\text{Numerical aperture of microscope objective}}$$

On the other hand, the relation between the focusing angle (α) of the light and the numerical aperture and refractive index of the medium is given by

$$\begin{matrix}\text{Numerical aperture of} \\ \text{microscope objective}\end{matrix} = \text{Refractive index} \times \sin\alpha$$

When the light is focused inside the tissue and if within the region there are nanoparticles that could emit light, we will get more than one PSF originating from each nanoparticle. In this case, it will be possible to distinguish between each nanoparticle if the distance between them is at least the Rayleigh limit; otherwise, they will look one single quantity. Finally, in the lateral dimension, the PSF is a regular ring and an elongated ellipsoid in the other direction. To do the imaging of smallest molecules and to study their dynamics, we need to wisely select a microscope objective to get the smallest PSF. However, after a certain limit, it is not possible to go further down and we need to use clever optical techniques.

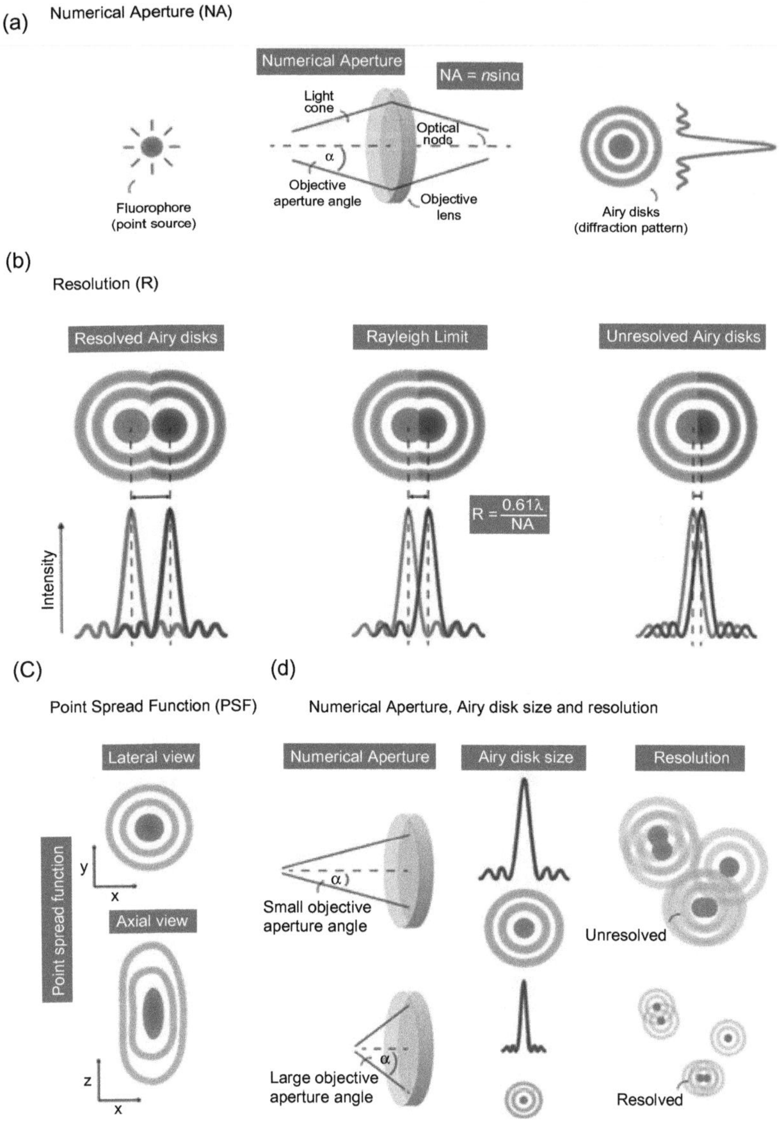

Figure 1.4 (a) Focusing of a light source or beam to a spot by using a lens or microscope objective. (b) Resolution of the microscope objective defined by different situations. Two spots will be distinguishable only when the distance between them is at least equal to the Rayleigh length. (c) The PSF of the microscope objective in lateral and axial view and (d) various situations when a large and small numerical aperture is used. Reproduced from Dunst and Tomancak (2019) under CC-BY 4.0 (http://creativecommons.org/licenses/by/4.0/).

Figure 1.5 shows different types of microscopes with which we can carry out many observations. Figure 1.5a shows a conventional microscope with white light in which wavelengths could be selected by using a filter in front of it. When a laser is used instead of white light, we achieve a smaller spot due to a narrow spectrum (Fig. 1.5b). Selected regions could be excited by using light sheet geometry by using a cylindrical lens or by using a pin hole (Fig. 1.5c,d). There are several types of super-resolution microscopes. All have the same target to go below the PSF limit to obtain the images. Two of them are shown here: One is structured illumination in which N images are captured for various orientations of the grating in front of the light source. This creates dark and bright regions for the PSF. By mathematical processing, we can obtain the final image that has contributions from each nanoparticle (Fig. 1.5d). Another type shown here is stimulated emission depletion (STED) in which by using a combination of excitation and deexcitation of the light beam, super-resolution images can be obtained (Fig. 1.5e).

Since there are several microscopes, nanoparticles, and tissues, the landscape of various studies that could be found in the literature is huge. By using interferometric scattering microscopy, it has been shown that it is possible to attach gold nanoparticles to giant unilamellar vesicles (GUVs) (Taylor and Sandoghdar, 2019) (Fig. 1.6a). By attaching these nanoparticles to them, they have achieved tracking of these in real time, which is useful to understand their behavior in the environment. On the other hand, Deka et al. presented the application of gold nanoparticles for deep tissue imaging. Figure 1.6b shows the images of these nanoparticles at different depths. When the depth is increased, the intensity of these nanoparticles goes down due to the scattering of the tissue. Still imaging at the depth of 200 μm has significant light, which is important for imaging in cells and small microbodies (Deka et al., 2018). Figure 1.6c shows the image of the HeLa cell with and without the two-photon photothermal microscope (Bijeesh et al., 2017). When the two-photon capability was not used, the images produced no significant information. However, the figure on the right shows that as soon as the two-photon capability was used, small features inside the cell became visible and clear.

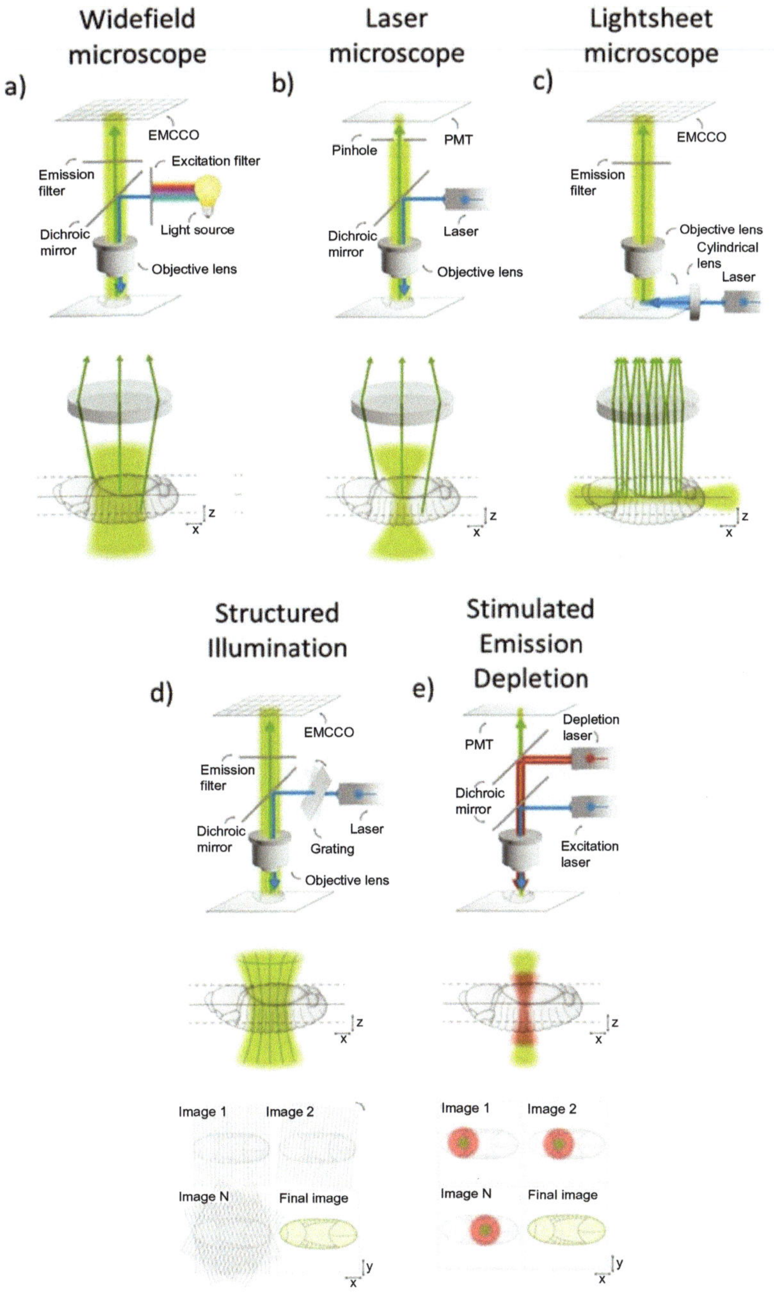

Figure 1.5 Examples of a few microscopes that could be found in the literature: (a) widefield microscope, (b) laser microscope, (c) light sheet microscope, (d) structured illuminations, and (e) stimulated emission depletion. Reproduced from Dunst and Tomancak (2019) under CC-BY 4.0 (http://creativecommons.org/licenses/by/4.0/).

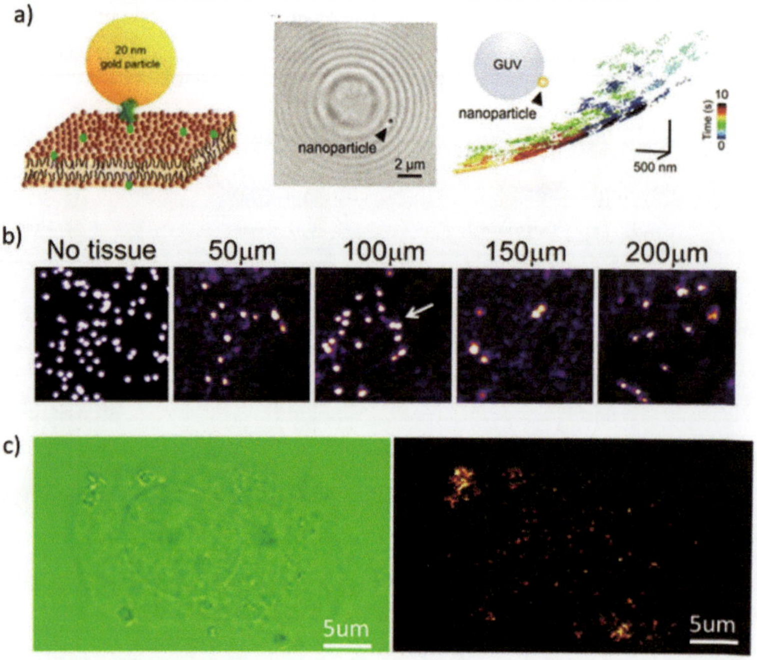

Figure 1.6 (a) Gold nanoparticles attached to giant unilamellar vesicles (GUVs) to show tracking inside the tissue. Gold nanoparticles could also be used for tracking deep inside the tissue. Reproduced from Taylor and Sandoghdar (2019) under CC-BY 4.0 (http://creativecommons.org/licenses/by/4.0/). The authors have managed to do imaging up to 200 μm. (b) This is significantly deep for most of the tissues in which biologist do study. Reproduced from Deka et al. (2018) under CC-BY 4.0 (http://creativecommons.org/licenses/by/4.0/). In the left, we can observe the HeLa cell in the widefield image; no useful information is available in this. On the other hand, the figure on the right shows the same structure with a two-photon photothermal microscope where small features of the cell are visible (c). Reproduced from Bijeesh et al. (2017) under CC-BY 4.0 (http://creativecommons.org/licenses/by/4.0/).

Applications in lasers: Laser is an acronym for Light Amplified by Stimulated Emission by Radiation. Since its discovery in 1960, it has provided many useful properties, including power as well as shorter pulse duration and wavelength. Depending upon the characteristic of the laser, one type of laser may be cost effective in one application but not the other (Kärtner and Keller, 1995; Hussain et al., 2012; Plamann et al., 2016; Coddington et al., 2016; Anastasiou et al., 2016; Hussain et al., 2017; Anastasiou et al., 2017; Singh et al., 2018;

Anastasiou et al., 2018; Fu et al., 2018; Hussain, 2019c; Hussain, 2019a; Hussain, 2019b; Hussain, 2020). Even after several decades, there are still areas that require advancement in different aspects. The laser consists of a gain medium that is sandwiched between two mirrors to provide optical feedback (Fig. 1.7a). The laser consists of an electrical or optical source that provides pumping from the ground state to a higher state and, therefore, providing lasing. The laser medium needs to be based on different energy levels that are different from the two-level system (Fig. 1.7b left). This is due to the fact that in a two-level system, the same photon will be absorbed and transmitted, making the system transparent (Silfvast, 2004; Svelto, 2010; Renk, 2017). In the laser system, we have three-level and four-level systems (Fig. 1.7b center and right). They are designed in such a way that the wavelength of the pumped photon is at the lower end of the spectrum and the emitted photon is at the higher end. Let us consider a three-level system. In this system, the atom will absorb the photon, and the electron will jump from the ground state to the excited state. Next it will decay in a non-radiative process followed by a radiative decay. For the system to act as a laser, we need an optical feedback system that could be achieved by using, in the simplest case, two mirrors. So far, we have talked about absorption and emission. This emission is known as spontaneous emission. In order to get a laser, we need gain. This gain is achieved through a stimulated process, also known as stimulated emission. In this process, the photon, which is already in the excited state, is stimulated to deexcite due to the presence of the same energy photon. These two photons will give rise to four photons, and these four photons will give eight photons. This will fill up the laser cavity quickly. To extract a useful amount of laser power from the cavity, one of the cavity mirrors is made semitransparent, which will give us the laser light (Fig. 1.7a).

To achieve lasers from nanoparticles, extensive work has been done on quantum dots and wells due to their compatibility with the current fabrication procedures in the silicon industry. The push on this kind of laser is achieved due to silicon being a non-direct bandgap material. Without going into details, non-direct bandgap materials cannot produce light when their electrons are excited. Although various optical components could be made from silicon, especially waveguides, it is not possible to get laser action from it

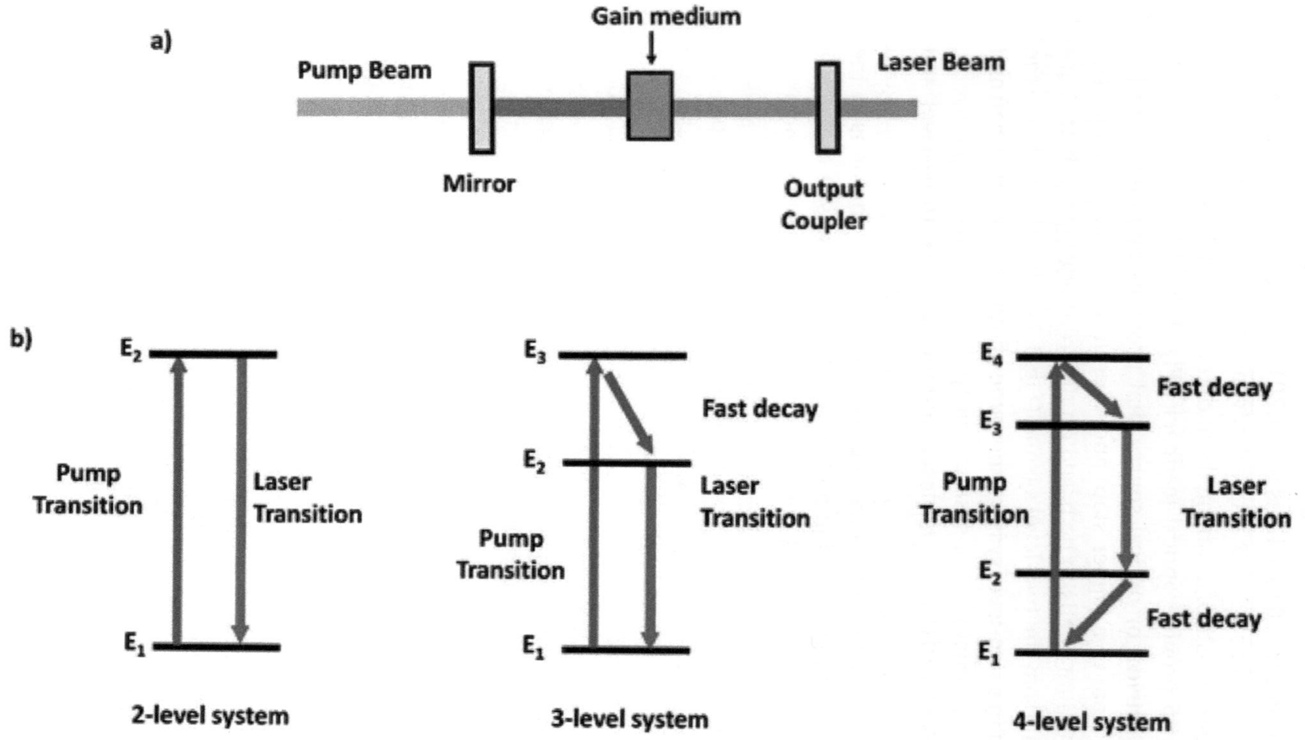

Figure 1.7 (a) The laser cavity consists of a laser gain material and two mirrors to provide the optical feedback. One of the mirrors is made semitransparent to take useful light out from the cavity. (b) Two three- and four-level systems. The laser will only work for three- and four-level systems.

due to this reason. Still researchers are trying to get laser from it, and a huge amount of investment is made every year. Therefore, the only possible route adapted by researchers is growing quantum dot structures on it to get lasing. There is always a strong competition between the quantum dot and well structures, and a lot of work has been done on it. Recently, it seems that the dot structure is winning the battle due to its low energy consumption for the same optical data transmission (Huang et al., 2020).

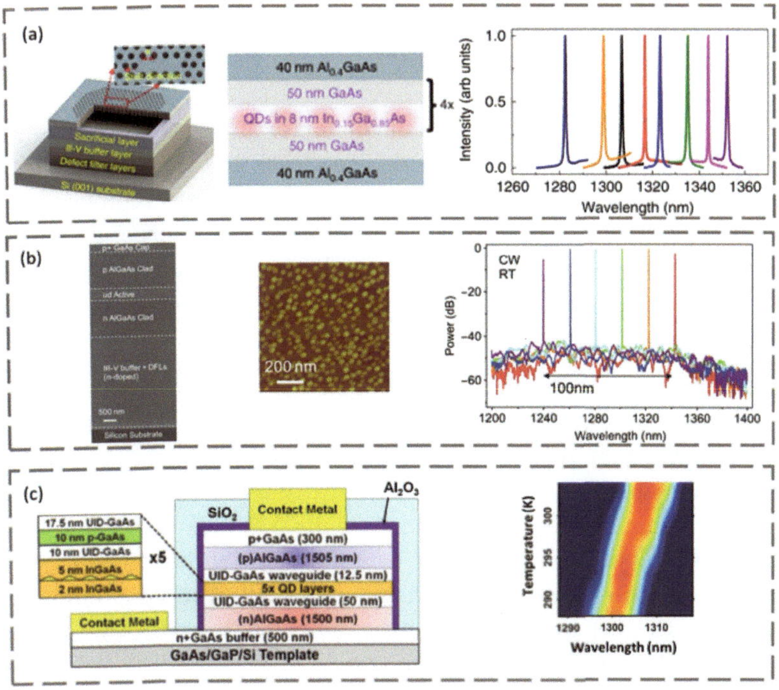

Figure 1.8 Various types of lasers that have been achieved by using the quantum nature of atoms. By using the quantum dots of InGaAs, near-infrared wavelengths were obtained (a). Reproduced from Zhou et al. (2020) under CC-BY 4.0 (http://creativecommons.org/licenses/by/4.0/). (b) In another study, 100 nm tunability was obtained (Wang et al., 2018). Reproduced from Wang et al. (2018) under CC-BY 4.0 (http://creativecommons.org/licenses/by/4.0/). Finally, a structure based on InGAs and GaAs was also capable of producing laser (Huang et al., 2020). Reproduced from Huang et al. (2020) under CC-BY 4.0 (http://creativecommons.org/licenses/by/4.0/).

Figure 1.8a–c shows structures for the fabrication of lasers based on these. As discussed before, the basic idea is to make them over silicon structures. On a silicon wafer, several layers are fabricated along with the gain medium. The gain medium is sandwiched between these layers. The task of these additional layers is to provide optical feedback and electrical contacts and to make the crystal structure compatible with silicon. The figure also shows the wavelength obtained in each case. In Fig. 1.8a, the wavelength could be tuned from 1280 to around 1350 nm (Zhou et al., 2020). Figure 1.8b presents a tuneability of 100 nm for this kind of laser (Wang et al., 2018). The same picture also shows an atomic force microscope with various active layers on it. Finally, Fig. 1.8c shows the change in wavelength at 1300 nm with temperature (Huang et al., 2020).

Apart from using the semiconductor nanostructures, other types of nanostructures are also used. Recently, $CsPbBr_3$ Perovskite quantum dot lasing was achieved in the green region of the electromagnetic spectrum (Huang et al., 2017). Fernandez-Bravo et al. developed a microlaser my enclosing Thulium nanoparticles inside polystyrene. This provided lasers at 800 and 450 nm wavelengths (Fernandez-Bravo et al., 2018).

Nonlinear optical processes based on nanoparticles: In this universe, substances around us are made of atoms arranged in such a way that gives us solid, liquid, or gas. When light, which is an electromagnetic field, is incident on them, it polarizes the material, which reemits the same electric field. This makes linear optical processes such as reflection or refraction. However, when the value of this electric field starts reaching 105–108 V/m, nonlinear processes start to happen, which cannot be described by using a linear relation and additional terms need to be considered (Murti and Vijayan, 2014; He, 2014; Boyd, 2019). In such a case, we need to write the equation in the following form:

$$P = \varepsilon_0 (\chi^{(1)}E^1 + \chi^{(2)}E^2 + \chi^{(3)}E^3)$$

where P is the polarization of the atoms, E is the electric field, χ is susceptibility, and ε_0 is the permittivity of free space. In this equation, the numbers on χ represent the order number. The first elment in the equation represents linear processes such as reflection and refraction. As we go higher in terms, other processes can be found such as second or third harmonics. We can also observe third-order processes such as saturable absorption, sum or difference frequency generation, optical parametric oscillation, optical parametric generation, and so on (Murti and Vijayan, 2014; He, 2014; Boyd, 2019). In this part of the chapter, we will discuss different nonlinear processes. Each process will be briefly introduced followed by applications of nanoparticles for this purpose.

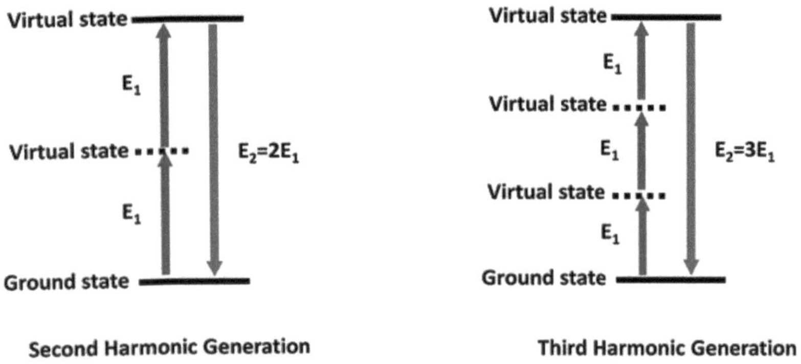

Figure 1.9 Energy level diagram for second-order (left) and third-order (right) effect.

Second and third harmonic generation: These two processes are widely applied in the industry and based upon $\chi^{(2)}$ (second harmonic generation) and $\chi^{(3)}$ (third harmonic generation) of the equation above. Considering the second harmonic generation, two photons are combined inside the nonlinear medium to provide a photon whose frequency is twice that of the incident photons. In this medium, the first photon excites the medium to an imaginary virtual state followed by another photon that excite it to another virtual state. After relaxation, the system will generate a photon whose frequency will be double (Fig. 1.9 left). In the third-order process, the same approximations are applied and, in this case, instead of two photons, three photons combine. For this situation, the frequency is thrice the incident photon frequency (Fig. 1.9 right). Their application in the industry and academia becomes obvious because one source of photons could give frequency at other parts of the electromagnetic spectrum. All we need is an efficient nonlinear medium whose conversion efficiency is high. The second and third harmonic generations are studied a lot in the past and a few systems have shown to be very promising, such as KDP (potassium dihydrogen phosphate), BBO (β-barium borate), lithium niobite, and

KTP (potassium titanyl phosphate) to name a few (Murti and Vijayan, 2014; He, 2014; Boyd, 2019). Still, nanoparticles can reach small places where these cannot reach. Therefore, the main application of nanoparticles is in the area where an extremely small footprint is required.

One of the simplest ways to show their usefulness is to mix them with the host material. Butet et al. used gold nanoparticles of 150 nm diameter, which were dispersed in gelatin to obtain SHG. In the study, they used femtosecond laser operating at 794 nm to achieve its second harmonic at 397 nm. Since the pumped laser provided femtosecond laser pulses, the obtained converted pulses could also be used in the applications where femtosecond pulses are required (Butet et al., 2010). Another paper presented an increase in the efficiency of the second harmonic signal due to the use of nanoparticles. When silicon nanoparticles were arranged in a manner that gives rise to Mie resonance, it has been shown that the efficiency of the SHG could be increased. In this case, a magnitude of at least two orders could be achieved (Makarov et al., 2017). A similar approach was applied to obtain THG (Shcherbakov et al., 2014). Gold nanoparticles, on the other hand, provided both SHG and THG (Lippitz et al., 2005).

Nanoparticles as saturable absorbers: A saturable absorber is a third-order nonlinear effect also observed in many materials, including nanomaterials. In a saturable absorber, the lower intensity light is blocked whereas high-intensity light can pass through the sample. This could be explained in terms of a two-level system (Fig. 1.10) when the light pulse contains a small number of photons and the number of electrons is enough to block the laser pulse. However, as the laser pulse starts to rise, the sample starts to become more and more transparent until the sample becomes completely transparent. This is because the number of photons is becoming higher and there are not enough electrons to block them. In some cases, the sample could be damaged before it could become completely transparent. Therefore, a saturable absorber sample can act like a switch that could be opaque for low-intensity pulse and transparent for high-intensity pulses. The experimental setup to test a saturable absorber normally has a light source that could produce nano to femtosecond laser pulses followed by a lens, the sample, and the detector to measure the power. Since a saturable absorber acts like a switch, such a switch is very useful for the creation of ultrashort laser pulses from the laser cavity. Apart from having a saturable absorber, another condition is to have a wide bandgap laser gain material inside the laser cavity. This supports longitudinal modes of the cavity. The saturable absorber will then provide modulation for the cavity. Another requirement is the absorption of the saturable absorber at the laser cavity wavelength. Moreover, the electron relaxation time should be faster than the cavity round trip time and the intracavity power should be high enough to make the absorber at least semitransparent.

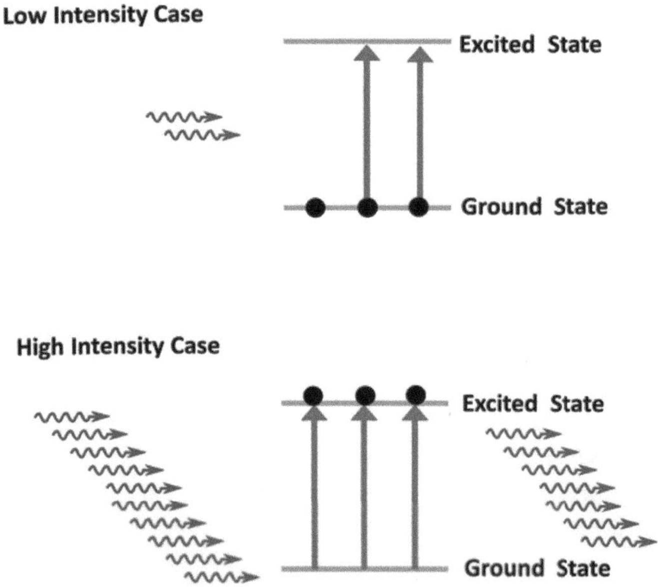

Figure 1.10 Energy level diagram to understand saturable absorption inside the material. For the low-intensity case, the light will be absorbed by the material. For the high-intensity case, the light will manage to pass through the material.

Various materials have been used in the literature as saturable absorbers inside the laser to obtain laser pulses from the laser cavity. They include semiconductors (Tilma et al., 2015), two-dimensional materials (Martinez and Sun, 2013), and other nanomaterials (Ahmad et al., 2016; Mohanraj et al., 2016; Song et al., 2016; Chu et al., 2018) that could be made by using different methods. Figure 1.11 shows the image of obtaining ultrashort laser pulses by using different materials. Figure 1.11a shows results that were obtained for carbon nanotubes. In this figure, the extreme left shows change in transmission as the power increases. In this case, the maximum change of 6.44% was obtained. Beyond 2000 MW/cm^2, the sample was damaged due to excessive power. By working with this saturable absorber below this power, we could use it inside the laser (Hussain, 2019a). Since carbon nanotubes have broad absorption, this absorber could be used for the tunability of wavelength. In this, we can see a tunability of 40 nm. For this laser, the pulse separation was found to be 59 ns. In another paper, by using a titanium disulfide saturable absorber, the authors showed pulsed operation at around 1530 nm. In this case, the optical transmission increased as they increased the power. The pulse separation in this case was found to be 291.3 ns (Shang et al., 2020). Finally, a saturable absorber based on tellurene could also be used for the same purpose (Xu et al., 2020). The same kind of observations were also made in previous papers. The sample showed an increase in transmission with the power. In this case, the pulse separation was 82.17 ns. Although we have shown examples of wavelengths around 1550 nm, several examples in the literature use different wavelengths. All these wavelengths depend on the availability of gain materials either for research or for commerce.

Nanoparticles for other miscellaneous applications: So far, we have shown applications of nanoparticles for microscopy, nonlinear conversion, and generation of laser pulses. The application is not just limited here. This is because of their quantum nature, and the dependence on size and shape could give so many applications that still need to be further discovered. Nanoparticles, for example, could be used for chemotherapy. For this application, the substance used, cisplatin, can be toxic for the tissue. Therefore, a platform is needed where this could be minimized while providing at least the same performance. Gold (Ajnai et al., 2014), carbon-based systems (Liu et al., 2011), polydopamine nanoparticles (Zhu and Su, 2017), and polymer also proved for such applications (Priya James et al., 2014). For this, Zhu et al. prepared the required nanoparticles by first modifying them with glycol followed by attaching cisplatin (Zhu and Su, 2017) (Fig. 1.12).

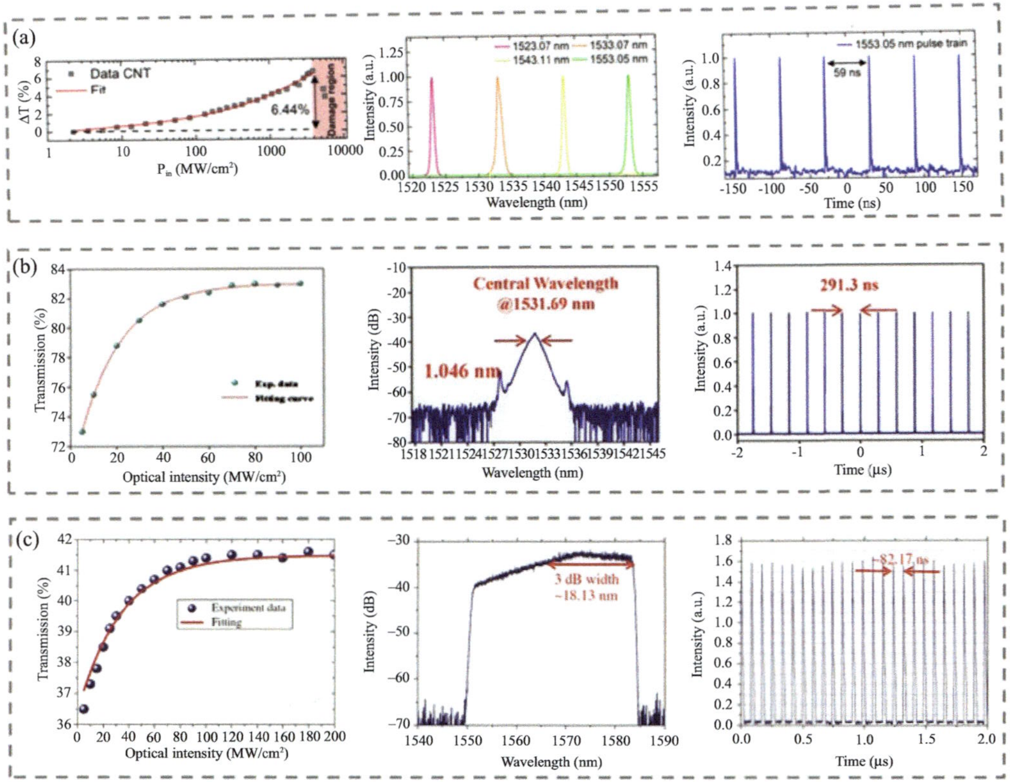

Figure 1.11 Use of various saturable absorbers to obtain laser pulses from the cavity. By using carbon nanotubes (a), titanium disulfide (b), and tellurene (c). Reproduced from Hussain (2019a), Xu et al. (2020), and Shang et al. (2020) under CC-BY 4.0 (http://creativecommons.org/licenses/by/4.0/).

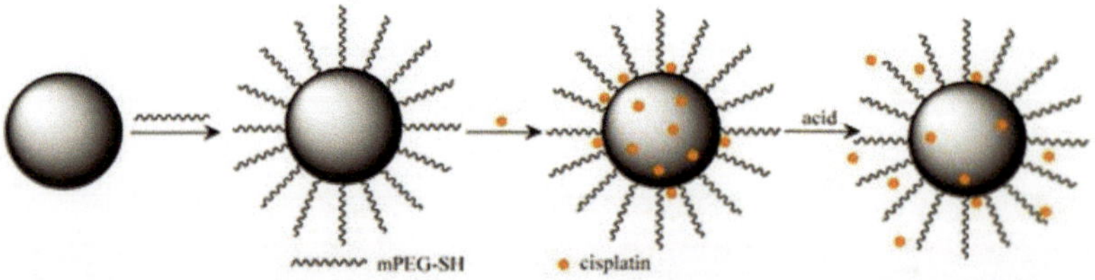

Figure 1.12 Modified polydopamine nanoparticles by attaching glycol and cisplatin for chemotherapy. Reproduced from Zhu and Su (2017) under CC-BY 4.0 (http://creativecommons.org/licenses/by/4.0/).

Another very hot area is using nanoparticles for energy applications. In this field, solar cells are one of the key areas as perovskite solar cells have shown better performance. These nanoparticles are used in the structure to make the performance superior. One example is to use plasmonic structures inside the solar cells to increase their absorption (Ghahremanirad et al., 2019). Ghahremanirad et al. (2019) used gold nanoparticles inside the structure (Fig. 1.13 left), which has proved to be very effective. The infrared absorption at 800 nm increased significantly (Fig. 1.13 right). We can see that the absorption increased with the radius of gold nanoparticles.

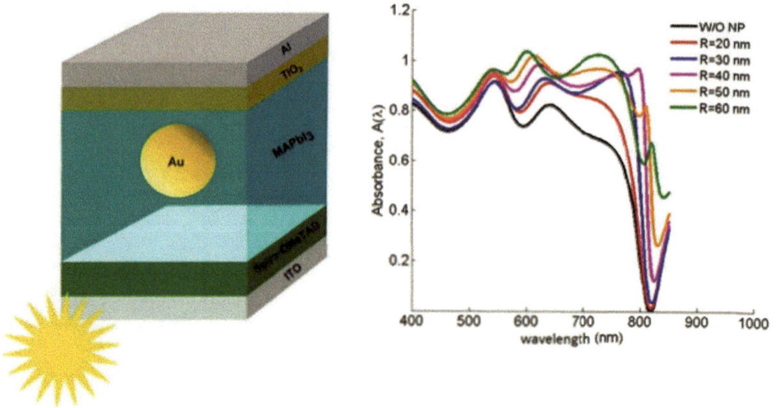

Figure 1.13 The structure used in solar cell application with gold nanoparticles (left). This showed high absorption with the increase in the radius of gold nanoparticles. Reproduced from Ghahremanirad et al. (2019) under CC-BY 4.0 (http://creativecommons.org/licenses/by/4.0/).

1.3 Conclusion

In this chapter, we discussed various applications of nanoparticles in optics. We presented various applications in microscopy, laser, nonlinear conversion, solar cells, and cancer therapy. Nanoparticles have been successfully applied in these areas. Due to their unique properties of size and shape, their applications are unlimited. Nanoparticles are being researched, where they must be fully applied in all commercial areas.

References

Ahmad, H., Ruslan, N. E., Ismail, M. A., Reduan, S. A., Lee, C. S. J., Sathiyan, S., Sivabalan, S., and Harun, S. W. (2016). Passively Q-switched erbium-doped fiber laser at C-band region based on WS_2 saturable absorber. *Applied Optics* **55**(5), p. 1001.

Ajnai, G., Chiu, A., Kan, T., Cheng, C.-C., Tsai, T.-H., and Chang, J. (2014). Trends of gold nanoparticle-based drug delivery system in cancer therapy. *Journal of Experimental & Clinical Medicine* **6**(6), pp. 172–178.

Amendola, V. and Meneghetti, M. (2009). Laser ablation synthesis in solution and size manipulation of noble metal nanoparticles. *Physical Chemistry Chemical Physics* **11**(20), p. 3805.

Anastasiou, A. D., Strafford, S., Posada-Estefan, O., Thomson, C. L., Hussain, S. A., Edwards, T. J., Malinowski, M., Hondow, N., Metzger, N. K., Brown, C. T. A., Routledge, M. N., Brown, A. P., Duggal, M. S., and Jha, A. (2017). β-pyrophosphate: A potential biomaterial for dental applications. *Materials Science and Engineering: C* **75**, pp. 885–894.

Anastasiou, A. D., Strafford, S., Thomson, C. L., Gardy, J., Edwards, T. J., Malinowski, M., Hussain, S. A., Metzger, N. K., Hassanpour, A., Brown, C. T. A., Brown, A. P., Duggal, M. S., and Jha, A. (2018). Exogenous mineralization of hard tissues using photo-absorptive minerals and femto-second lasers: The case of dental enamel. *Acta Biomaterialia* **71**, pp. 86–95.

Anastasiou, A. D., Thomson, C. L., Hussain, S. A., Edwards, T. J., Strafford, S., Malinowski, M., Mathieson, R., Brown, C. T. A., Brown, A. P., Duggal, M. S., and Jha, A. (2016). Sintering of calcium phosphates with a femtosecond pulsed laser for hard tissue engineering. *Materials & Design* **101**, pp. 346–354.

Bar, H., Bhui, D. K., Sahoo, G. P., Sarkar, P., De, S. P., and Misra, A. (2009). Green synthesis of silver nanoparticles using latex of *Jatropha curcas*. *Colloids and Surfaces A: Physicochemical and Engineering Aspects* **339**(1–3), pp. 134–139.

Bijeesh, M. M., Shakhi, P. K., Arunkarthick, S., Varier, G. K., and Nandakumar, P. (2017). Confocal imaging of single $BaTiO_3$ nanoparticles by two-photon photothermal microscopy. *Scientific Reports* **7**(1), p. 1643.

Boyd, R. W. (2019). *Nonlinear Optics*. Academic press.

Butet, J., Duboisset, J., Bachelier, G., Russier-Antoine, I., Benichou, E., Jonin, C., and Brevet, P.-F. (2010). Optical second harmonic generation of single metallic nanoparticles embedded in a homogeneous medium. *Nano Letters* **10**(5), pp. 1717–1721.

Cheng, Q., Li, C., Pavlinek, V., Saha, P., and Wang, H. (2006). Surface-modified antibacterial TiO_2/Ag^+ nanoparticles: Preparation and properties. *Applied Surface Science* **252**(12), pp. 4154–4160.

Chu, Z., Zhang, H., Wu, Y., Zhang, C., Liu, J., and Yang, J. (2018). Passively Q-switched laser based on gold nanobipyramids as saturable absorbers in the 1.3 μm region. *Optics Communications* **406**, pp. 209–213.

Coddington, I., Newbury, N., and Swann, W. (2016). Dual-comb spectroscopy. *Optica* **3**(4), p. 414.

Deka, G., Nishida, K., Mochizuki, K., and Fujita, K. (2018). Resolution enhancement in deep-tissue nanoparticle imaging based on plasmonic saturated excitation microscopy. *APL Photonics* **3**(3), p. 031301.

Dunst, S. and Tomancak, P. (2019). Imaging flies by fluorescence microscopy: Principles, technologies, and applications. *Genetics* **211**(1), pp. 15–34.

Eliezer, S., Eliaz, N., Grossman, E., Fisher, D., Gouzman, I., Henis, Z., Pecker, S., Horovitz, Y., Fraenkel, M., Maman, S., and Lereah, Y. (2004). Synthesis of nanoparticles with femtosecond laser pulses. *Physical Review B* **69**(14), p. 144119.

Fernandez-Bravo, A., Yao, K., Barnard, E. S., Borys, N. J., Levy, E. S., Tian, B., Tajon, C. A., Moretti, L., Altoe, M. V., Aloni, S., Beketayev, K., Scotognella, F., Cohen, B. E., Chan, E. M., and Schuck, P. J. (2018). Continuous-wave upconverting nanoparticle microlasers. *Nature Nanotechnology* **13**(7), pp. 572–577.

Fu, B., Popa, D., Zhao, Z., Hussain, S. A., Flahaut, E., Hasan, T., Soavi, G., and Ferrari, A. C. (2018). Wavelength tunable soliton rains in a nanotube-mode locked Tm-doped fiber laser. *Applied Physics Letters* **113**(19), p. 193102.

Ghahremanirad, E., Olyaee, S., and Hedayati, M. (2019). The influence of embedded plasmonic nanostructures on the optical absorption of perovskite solar cells. *Photonics* **6**(2), p. 37.

Hadjipanayis, C. G., Bonder, M. J., Balakrishnan, S., Wang, X., Mao, H., and Hadjipanayis, G. C. (2008). Metallic iron nanoparticles for MRI contrast enhancement and local hyperthermia. *Small* **4**(11), pp. 1925–1929.

He, G. S. (2014). *Nonlinear Optics and Photonics* [Online]. Oxford University Press. Available from: https://oxford.universitypressscholarship.com/view/10.1093/acprof:oso/9780198702764.001.0001/acprof-9780198702764.

Hu, S.-L., Niu, K.-Y., Sun, J., Yang, J., Zhao, N.-Q., and Du, X.-W. (2009). One-step synthesis of fluorescent carbon nanoparticles by laser irradiation. *Journal of Material Chemistry* **19**(4), pp. 484–488.

Huang, C.-Y., Zou, C., Mao, C., Corp, K. L., Yao, Y.-C., Lee, Y.-J., Schlenker, C. W., Jen, A. K. Y., and Lin, L. Y. (2017). $CsPbBr_3$ perovskite quantum dot vertical cavity lasers with low threshold and high stability. *ACS Photonics* **4**(9), pp. 2281–2289.

Huang, H., Duan, J., Dong, B., Norman, J., Jung, D., Bowers, J. E., and Grillot, F. (2020). Epitaxial quantum dot lasers on silicon with high thermal stability and strong resistance to optical feedback. *APL Photonics* **5**(1), p. 016103.

Hussain, S. A. (2019a). Comparison of graphene and carbon nanotube saturable absorbers for wavelength and pulse duration tunability. *Nature: Scientific Reports* **9**(1), p. 17282.

Hussain, S. A. (2019b). Discovery of several new families of saturable absorbers for ultrashort pulsed laser systems. *Nature: Scientific Reports* **9**(1), p. 19910.

Hussain, S. A. (2020). Tunable bound solitons from carbon nanotube saturable absorber. *Optik* **217**, p. 164733.

Hussain, S. A. (2019c). Ultrashort dual colour laser at 1.55 and 1.88 μm by using a carbon nanotube saturable absorber. *IEEE Photonics Technology Letters* **31**(12), pp. 990–993.

Hussain, S. A., Crotti, C., Deloison, F., Alahyane, F., Aptel, F., Kowalczuk, L., Legeais, J.-M., Peyrot, D. A., Savoldelli, M., and Plamann, K. (2012). Wavelength optimisation in femtosecond laser corneal surgery: Quantitative results on eye bank corneas. *Investigative Ophthalmology & Visual Science* **53**(14), p. 1506.

Hussain, S. A., Milián, C., Crotti, C., Kowalczuk, L., Alahyane, F., Essaïdi, Z., Couairon, A., Schanne-Klein, M.-C., and Plamann, K. (2017). Cell viability and shock wave amplitudes in the endothelium of porcine cornea exposed to ultrashort laser pulses. *Graefe's Archive for Clinical and Experimental Ophthalmology* **255**(5), pp. 945–953.

Iravani, S. (2011). Green synthesis of metal nanoparticles using plants. *Green Chemistry* **13**(10), p. 2638.

Kango, S., Kalia, S., Celli, A., Njuguna, J., Habibi, Y., and Kumar, R. (2013). Surface modification of inorganic nanoparticles for development of organic–inorganic nanocomposites: A review. *Progress in Polymer Science* **38**(8), pp. 1232–1261.

Kärtner, F. X. and Keller, U. (1995). Stabilization of solitonlike pulses with a slow saturable absorber. *Optics Letters* **20**(1), p. 16.

Lippitz, M., van Dijk, M. A., and Orrit, M. (2005). Third-harmonic generation from single gold nanoparticles. *Nano Letters* **5**(4), pp. 799–802.

Liu, Z., Robinson, J. T., Tabakman, S. M., Yang, K., and Dai, H. (2011). Carbon materials for drug delivery & cancer therapy. *Materials Today* **14**(7–8), pp. 316–323.

Lu, A.-H., Salabas, E. L., and Schüth, F. (2007). Magnetic nanoparticles: Synthesis, protection, functionalization, and application. *Angewandte Chemie International Edition* **46**(8), pp. 1222–1244.

Ma, Y., Xia, J., Yao, C., Yang, F., Stanciu, S. G., Li, P., Jin, Y., Chen, T., Zheng, J., Chen, G., Yang, H., Luo, L., and Wu, A. (2019). Precisely tuning the contrast properties of $Zn_xFe_{3-x}O_4$ nanoparticles in magnetic resonance imaging by controlling their doping content and size. *Chemistry of Materials* **31**(18), pp. 7255–7264.

Makarov, S. V., Petrov, M. I., Zywietz, U., Milichko, V., Zuev, D., Lopanitsyna, N., Kuksin, A., Mukhin, I., Zograf, G., Ubyivovk, E., Smirnova, D. A., Starikov, S., Chichkov, B. N., and Kivshar, Y. S. (2017). Efficient second-harmonic generation in nanocrystalline silicon nanoparticles. *Nano Letters* **17**(5), pp. 3047–3053.

Martinez, A. and Sun, Z. (2013). Nanotube and graphene saturable absorbers for fibre lasers. *Nature Photonics* **7**(11), pp. 842–845.

Masotti, A., Pitta, A., Ortaggi, G., Corti, M., Innocenti, C., Lascialfari, A., Marinone, M., Marzola, P., Daducci, A., Sbarbati, A., Micotti, E., Orsini, F., Poletti, G., and Sangregorio, C. (2009). Synthesis and characterization of polyethylenimine-based iron oxide composites as novel contrast agents for MRI. *Magnetic Resonance Materials in Physics, Biology and Medicine* **22**(2), pp. 77–87.

McConnell, W. P., Novak, J. P., Brousseau, L. C., Fuierer, R. R., Tenent, R. C., and Feldheim, D. L. (2000). Electronic and optical properties of chemically modified metal nanoparticles and molecularly bridged nanoparticle arrays. *The Journal of Physical Chemistry B* **104**(38), pp. 8925–8930.

Mohanraj, J., Velmurugan, V., and Sivabalan, S. (2016). Transition metal dichalcogenides based saturable absorbers for pulsed. *Optical Materials* **60**, pp. 601–617.

Murphy, C. J., Sau, T. K., Gole, A. M., Orendorff, C. J., Gao, J., Gou, L., Hunyadi, S. E., and Li, T. (2005). Anisotropic metal nanoparticles: Synthesis, assembly, and optical applications. *The Journal of Physical Chemistry B* **109**(29), pp. 13857–13870.

Murti, Y. and Vijayan, C. (2014). *Essentials of Nonlinear Optics* [Online]. Chichester, UK: John Wiley & Sons, Ltd. Available from: http://doi.wiley.com/10.1002/9781118902332.

Narayanan, K. B. and Sakthivel, N. (2010). Biological synthesis of metal nanoparticles by microbes. *Advances in Colloid and Interface Science* **156**(1–2), pp. 1–13.

Panáček, A., Kvítek, L., Prucek, R., Kolář, M., Večeřová, R., Pizúrová, N., Sharma, V.K., Nevěčná, T., and Zbořil, R. (2006). Silver colloid nanoparticles: Synthesis, characterization, and their antibacterial activity. *The Journal of Physical Chemistry B* **110**(33), pp. 16248–16253.

Park, J.-H., Gu, L., von Maltzahn, G., Ruoslahti, E., Bhatia, S. N., and Sailor, M. J. (2009). Biodegradable luminescent porous silicon nanoparticles for in vivo applications. *Nature Materials* **8**(4), pp. 331–336.

Pavitra, E., Dariya, B., Srivani, G., Kang, S.-M., Alam, A., Sudhir, P.-R., Kamal, M. A., Raju, G. S. R., Han, Y.-K., Lakkakula, B. V. K. S., Nagaraju, G. P., and Huh, Y. S. (2019). Engineered nanoparticles for imaging and drug delivery in colorectal cancer. *Seminars in Cancer Biology*. DOI: 10.1016/j.semcancer.2019.06.017.

Plamann, K., Hussain, S. A., Alahyane, F., Couairon, A., Crotti, C., Essaidi, Z., Kowalczuk, L., Enrique, C. M., and Schanne-Klein, M.-C. (2016). Cell viability and shock wave amplitudes in the endothelium of porcine cornea exposed to ultrashort laser pulses. *Investigative Ophthalmology & Visual Science* **57**(12), p. 1207.

Priya James, H., John, R., Alex, A., and Anoop, K. R. (2014). Smart polymers for the controlled delivery of drugs: A concise overview. *Acta Pharmaceutica Sinica B*. **4**(2), pp. 120–127.

Renk, K. F. (2017). *Basics of Laser Physics* [Online]. Cham: Springer International Publishing. Available from: http://link.springer.com/10.1007/978-3-319-50651-7.

Ris, F., Lepetit-Coiffe, M., Meda, P., Crowe, L. A., Toso, C., Armanet, M., Niclauss, N., Parnaud, G., Giovannoni, L., Bosco, D., Morel, P., Vallee, J.-P., and Berney, T. (2010). Assessment of human islet labeling with clinical grade iron nanoparticles prior to transplantation for graft monitoring by MRI. *Cell Transplantation* **19**(12), pp. 1573–1585.

Santra, S., Jativa, S. D., Kaittanis, C., Normand, G., Grimm, J., and Perez, J. M. (2012). Gadolinium-encapsulating iron oxide nanoprobe as activatable NMR/MRI contrast agent. *ACS Nano* **6**(8), pp. 7281–7294.

Shang, X., Guo, L., Zhang, H., Li, D., and Yue, Q. (2020). Titanium disulfide based saturable absorber for generating passively mode-locked and Q-switched ultra-fast fiber lasers. *Nanomaterials* **10**(10), p. 1922.

Shcherbakov, M. R., Neshev, D. N., Hopkins, B., Shorokhov, A. S., Staude, I., Melik-Gaykazyan, E. V., Decker, M., Ezhov, A. A., Miroshnichenko, A. E., Brener, I., Fedyanin, A. A., and Kivshar, Y. S. (2014). Enhanced third-harmonic generation in silicon nanoparticles driven by magnetic response. *Nano Letters* **14**(11), pp. 6488–6492.

Silfvast, W. T. (2004). *Laser Fundamentals* [Online]. Cambridge University Press. Available from: https://www.cambridge.org/core/product/identifier/9780511616426/type/book.

Singh, N., Xin, M., Vermeulen, D., Shtyrkova, K., Li, N., Callahan, P. T., Magden, E. S., Ruocco, A., Fahrenkopf, N., Baiocco, C., Kuo, B.P.-P., Radic, S., Ippen, E., Kärtner, F. X., and Watts, M. R. (2018). Octave-spanning coherent supercontinuum generation in silicon on insulator from 1.06 μm to beyond 2.4 μm. *Light: Science & Applications* **7**(1), p. 17131.

Song, Y., Chen, S., Zhang, Q., Li, L., Zhao, L., Zhang, H., and Tang, D. (2016). Vector soliton fiber laser passively mode locked by few layer black phosphorus-based optical saturable absorber. *Optics Express* **24**(23), p. 25933.

Svelto, O. (2010). *Principles of Lasers* [Online]. Boston, MA: Springer US. Available from: http://link.springer.com/10.1007/978-1-4419-1302-9.

Tang, F., Li, L., and Chen, D. (2012). Mesoporous silica nanoparticles: Synthesis, biocompatibility and drug delivery. *Advanced Materials* **24**(12), pp. 1504–1534.

Taylor, R. W. and Sandoghdar, V. (2019). Interferometric scattering microscopy: Seeing single nanoparticles and molecules via Rayleigh scattering. *Nano Letters* **19**(8), pp. 4827–4835.

Tilma, B. W., Mangold, M., Zaugg, C. A., Link, S. M., Waldburger, D., Klenner, A., Mayer, A. S., Gini, E., Golling, M., and Keller, U. (2015). Recent advances in ultrafast semiconductor disk lasers. *Light: Science & Applications* **4**(7), p. e310.

Wang, Y., Chen, S., Yu, Y., Zhou, L., Liu, L., Yang, C., Liao, M., Tang, M., Liu, Z., Wu, J., Li, W., Ross, I., Seeds, A. J., Liu, H., and Yu, S. (2018). Monolithic quantum-dot distributed feedback laser array on silicon. *Optica* **5**(5), p. 528.

Wen, S., Liu, D.-F., Cui, Y., Harris, S. S., Chen, Y., Li, K. C., Ju, S. and Teng, G.-J. (2014). In vivo MRI detection of carotid atherosclerotic lesions and kidney inflammation in ApoE-deficient mice by using LOX-1 targeted iron nanoparticles. *Nanomedicine: Nanotechnology, Biology and Medicine* **10**(3), pp. 639–649.

Xu, N., Ma, P., Fu, S., Shang, X., Jiang, S., Wang, S., Li, D., and Zhang, H. (2020). Tellurene-based saturable absorber to demonstrate large-energy dissipative soliton and noise-like pulse generations. *Nanophotonics* **9**(9), pp. 2783–2795.

Yanhong, L., Dejun, W., Qidong, Z., Min, Y., and Qinglin, Z. (2004). A study of quantum confinement properties of photogenerated charges in ZnO nanoparticles by surface photovoltage spectroscopy. *The Journal of Physical Chemistry B* **108**(10), pp. 3202–3206.

Zhou, T., Tang, M., Xiang, G., Xiang, B., Hark, S., Martin, M., Baron, T., Pan, S., Park, J.-S., Liu, Z., Chen, S., Zhang, Z., and Liu, H. (2020). Continuous-wave quantum dot photonic crystal lasers grown on on-axis Si (001). *Nature Communications* **11**(1), p. 977.

Zhu, Z. and Su, M. (2017). Polydopamine nanoparticles for combined chemo- and photothermal cancer therapy. *Nanomaterials* **7**(7), p. 160.

Chapter 2

What We Need to Know about Quantum Dots Nanoparticles

Shiv Prakash Singh,[a] Sudheer Kumar Yadav,[b] and Kunal Mondal[c]

[a]*Centre for Ceramic Processing, International Advanced Research Centre for Powder Metallurgy and New Materials (ARCI), 500005 Hyderabad, India*
[b]*Eduard-Zintl-Institut für Anorganische und Physikalische Chemie, Technische Universität Darmstadt, 64287 Darmstadt, Germany*
[c]*Materials Science and Engineering Department, Idaho National Laboratory, Idaho Falls, ID 83415, USA*
spsingh67@gmail.com

Quantum dots (QDs) have engrossed huge research interest over the last 10 years. QDs are within the range of a few nanometers (<10 nm), and to be part of photonic structures, they are usually entrenched in thin films or other micro- and nanostructures. After briefly reviewing the various fabrication methods of QDs, examples of results concerning the characterization of these nanostructures were discussed. Properties of QDs containing CdSe/ZnS, TiO_2, and other QDs are presented describing their optics, optoelectronics, absorption, and photoluminescence, including different ways of

Science and Applications of Nanoparticles
Edited by Waqar Ahmed and Ehsan Nourafkan

ISBN 978-981-4267-34-2 (Hardcover), 978-1-003-28029-3 (eBook)
www.jennystanford.com

characterization of these nanostructures. Finally, we discussed applications of QDs for drug delivery, energy materials, catalysis, and electronics with brief future prospects. It is important to enhance the understanding of QDs to explore the possibility of new properties and applications toward the advancement of technology as well as scientific discoveries.

2.1 Introduction

We are dealing with nanotechnology in our daily life by using different materials such as oxides, semiconductors, metals, polymers, etc. In all these cases, atoms play an important role. A group of atoms of semiconductors have revolutionized the whole world of materials science from simple home lights to modern computers and photovoltaic solar cells, several types of sensors, biomedical applications, and many in more advanced technologies. These groups of atoms of semiconductors are known as QDs [1–3]. It is essential to understand the importance of these QDs and their functions. The QDs are extremely small in diameter that they are effectively concentrated into a single point or we can say into a zero-dimension structure [4, 5]. Due to their size effect, the particles inside them are trapped and have well-defined energy levels according to the quantum theory. The QDs are nanocrystalline, with sizes within a few nanometers meaning a few dozens of atoms. These QDs are made up of semiconductors such as silicon, germanium, etc. and are nanocrystalline but behave as a single atom. Hence, QDs are also called artificial atoms [6, 7]. Recent developments in the processing of semiconductors allowed the artificial design of quantum confinement of only a few electrons [1, 8, 9]. This type of finite fermion structure has many similarities with atoms; however, they are artificial in structure, design, and development in the laboratory. The properties of QDs can be altered in a precise way by electrostatic gates, modifications in the dot geometry, induction of magnetic fields, etc. Their technological understanding provided the opportunity for quantum effects in a finite low-dimensional system that was generally unexplored. If the particle size of a semiconductor is below the Bohr radius of both the holes and electrons, strong quantum confinement or quantum effect will occur. The quantum confinement will lead to the breaking of continuous energy levels into discrete energy levels. The bandgap and the spacing between these energy levels increase with a decrease in the particle size of the semiconductor. Thus, the optical absorption peak shifts toward higher energies as the particle size decreases [10]. Figure 2.1a shows a schematic presentation of energy bands of a bulk semiconductor showing the continuity of conduction and valence bands separated by a bandgap, E_g. For QDs, the continuity of energy bands is broken into discrete atom-like energy levels whose energy is determined by the size of the quantum dot [11].

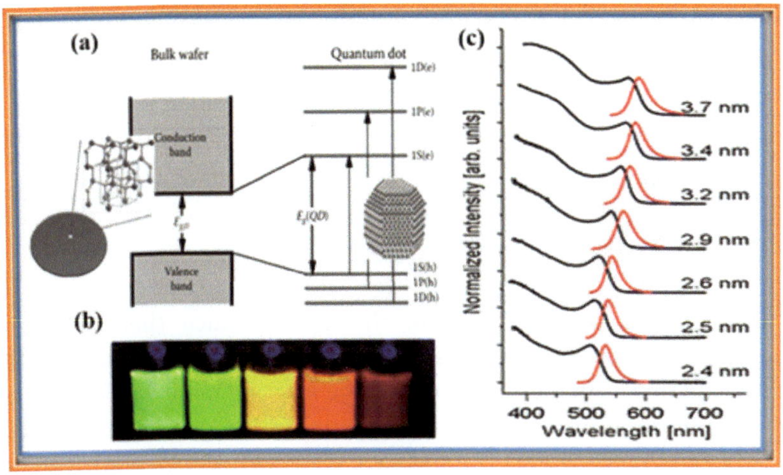

Figure 2.1 (a) Energy bands of bulk semiconductor (E_g) and for a quantum dot (QD) [11], (b) visual images of samples excited by a broadband UV lamp, and (c) UV-visible absorption (black) and PL emission (red) spectra (excitation at 350 nm) of CdSe quantum dots with different sizes [12]. Reproduced from Refs. [11] and [12] with permission from the Royal Society of Chemistry.

In recent developments, some QDs have attracted immense attention from researchers. Carbon QDs are one of them, which have inspired many studies for their potential technological applications. Some other types of QDs are known as III–V and II–VI semiconductors. The III–V semiconductors are compounds of the III and V groups of elements in the periodic table. One best-known example of this category of semiconductors is gallium arsenide (GaAs) used as an amplification medium in lasers. The II–VI semiconductors are compounds of elements that belong to groups II and V of the periodic table. Cadmium selenide (CdSe) and cadmium telluride (CdTe) are examples of the II–VI group of semiconductors. Figures 2.1b,c show the visual images of CdSe QDs excited by a broadband UV lamp and UV-visible absorption (black) and PL emission (red) spectra of CdSe QDs with different sizes, respectively [12].

These are extensively used in electronic, photonic, and photovoltaic applications. Silicon is the most known and widely used semiconductor in photovoltaic and electronic applications because of its abundancy and low cost. In the recent time, graphene QDs are another fascinating material for many cutting-edge technologies. These QDs have demonstrated many attractive properties such as tunable luminescence, biocompatibility, nontoxicity, etc. Moreover, it has been found that graphene QDs have superior properties over the conventional semiconductor QDs, which make them a potential for many applications such as photovoltaic solar cells, energy storage materials, optical sensing, bioimaging, etc.

2.2 Synthesis Techniques

QDs were first discovered in the 1980s. Since then, different techniques have been adopted to synthesize different types of QDs. The synthesis method depends on the starting materials as well as its application. Some of the most popular techniques are as follows.

2.2.1 Chemical Route Synthesis

The synthesis of QDs through the chemical rout is considered a conventional way of synthesis. Several categories of QDs such as II–VI and III–V semiconductors and graphene are synthesized by the chemical route. Different types of precursors are taken as starting materials, and different types of surfactants are added to organic or aqueous solvents. Depending on the requirement, at different temperatures and stirring rates, the size and shape of the QDs are controlled. The final product was analyzed for its purity using different characterization tools. For the practical applications, these QDs are embedded into some matrix either in liquid or solid states. Common examples of chemical synthesis are colloidal, thermal decomposition, and hot injection method.

2.2.2 Physical Vapor Deposition

QDs can be synthesized by physical vapor deposition techniques. In this case, the starting materials are in either solid or gaseous state. The starting solid materials for the synthesis of QDs can be evaporated by high temperature, laser ablation, or using a sputtering technique to remove the elements atom by atom and deposit on the substrate. The size and shape of the QDs can be controlled with the rates of evaporation, sputtering, or laser ablation. The other factors that control the size and shape of QDs are the microstructure, temperature, and rotation of the substrate on which the QDs are deposited. In this method, materials are vaporized and form a cluster of atoms, which are later on deposited on the specific substrate.

2.2.3 Chemical Vapor Deposition

Chemical vapor deposition is the technique where chemical reactions take place in the gaseous phase and the product deposits on the substrate. The synthesis process involves the gaseous precursors of the semiconductors, which react at elevated temperatures depending on the nature of semiconductors.

2.3 Characterization

The synthesized QDs are characterized for their purity, size, shape, and crystal phases. Some characterization techniques are listed in Fig. 2.2.

2.3.1 X-Ray Diffraction

X-ray diffraction (XRD) is the primary tool for the characterization of the purity of the phase and the crystal structure of the synthesized QDs. The sample for the XRD measurement can be powder, thin films, or in the bulk form. In most of the XRD measurements, the wavelength of X-rays is fixed. In general, Cu K_{α} source is used for X-ray in most of the XRD measurements. But in some cases, other sources such as Mo and Ag are also used. The angle where constructive interference from X-rays scattering by parallel planes of atoms is determined by Bragg's law [13]. Moreover, we can calculate the size of the QDs using Scherrer's formula [14]. For the detailed analysis of the crystal structure, lattice parameters, and phase, Rietveld refinement was performed using the XRD data. Figure 2.3 shows the XRD patterns of ZnO QDs.

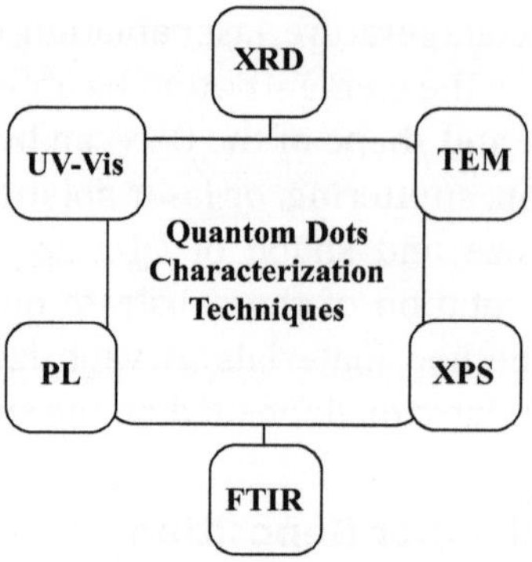

Figure 2.2 Schematic representation of different techniques used for the characterization of quantum dots.

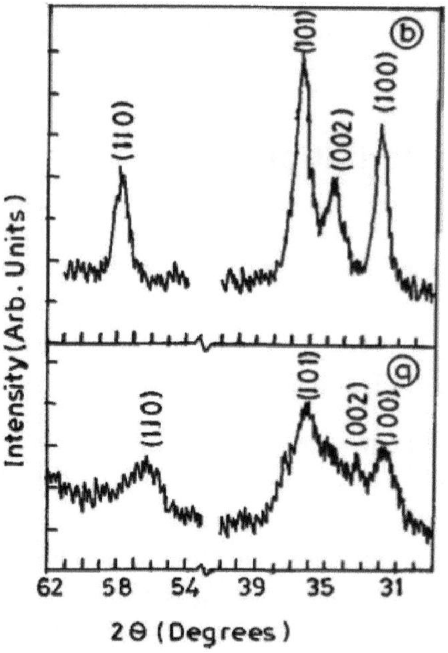

Figure 2.3 X-ray diffraction patterns of ZnO quantum dots showing the wurtzite phase. The average diameters determined using the Scherrer formula are (a) 4.0 nm and (b) 8.5 nm. Reproduced from Ref. [14] with permission from the publisher.

2.3.2 Transmission Electron Microscopy

Transmission electron microscopy (TEM) is a microscopy characterization tool used to see the two-dimensional (2D) image of QDs. In the TEM experiment, the electron beam transmits through the specimen of the QDs and forms the 2D image. The specimens for the TEM are in the form of either thin films or dispersed powders. The specimen is placed on the grid, which is made up of commonly carbon-coated copper or molybdenum depending on the nature of the QDs. Using the high-resolution TEM, we can see the image of QDs down to the atomic resolution. Moreover, TEM will give information about the size and shape of the QDs. The selected area electron diffraction (SAED) technique attached to the TEM will provide the information about the crystal phase and the structure of QDs. The SAED technique provides the projection of reciprocal lattices represented by the bright spots. The SAED results can be compared with the XRD to confirm the crystallographic information of the QDs. A representative TEM and a high-resolution TEM (HRTEM) of InAs nanocrystal QDs are provided in Fig. 2.4 [15].

2.3.3 X-Ray Photoelectron Spectroscopy

The X-ray photoelectron spectroscopy (XPS) is a surface-sensitive spectroscopy technique that provides quantitative information about the chemical nature (oxidation states) of the QDs. This is the most popular characterization technique for the film of QDs. The XPS technique helps to reveal the chemical environment of the QDs. Generally, two types of XPS are used for analysis: laboratory-based and synchrotron-based XPS. In the laboratory-based XPS, the source of the beam used is either Al K_{α} or Mg K_{α}. The synchrotron-based XPS produces high brilliance and a high flux photon beam. Borgohain et al. [16] have synthesized cupric oxide nanocrystals by the electrochemical method and reported the quantum size effects of diameters of about 4 and 6 nm. They have investigated the formation of CuO by XPS. They have dispersed the nanoparticles onto the graphite to avoid the effect of charging and have used the Al K_{α} ($h\upsilon$ = 51486.6 eV) as the incident radiation source. The XPS spectrometer functioned at the resolution of about 1 eV. The Au $4f_{7/2}$ line was used to define the Fermi level, and the C 1s line acts as an internal standard. The corresponding XPS graph is displayed in Fig. 2.5.

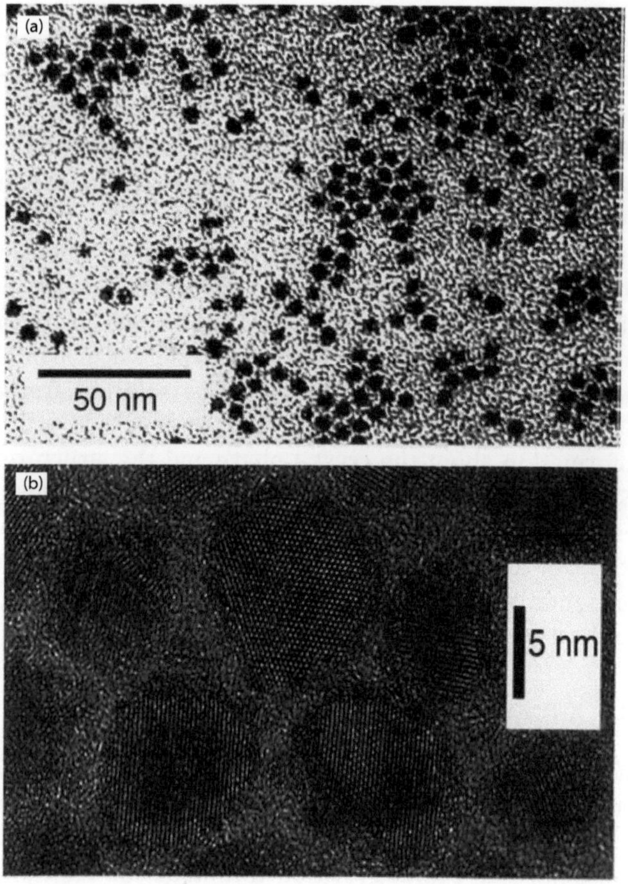

Figure 2.4 (a) TEM and its (b) higher resolution TEM images of InAs nanocrystals. Reproduced from Ref. [15] with permission from the publisher.

2.3.4 Fourier Transform Infrared Spectroscopy

Fourier transform infrared (FTIR) spectroscopy is the primary tool to confirm the chemical environment of QDs. For practical applications, QDs are dispersed in the matrix of a solid or liquid. In this case, it is important to know the nature of the chemical bonding and chemical environment surrounding the QDs. The effect of the matrix on the shape and size of QDs has been reported. The FTIR spectra of the QDs of CuO were studied by Borgohain et al. [16] at room temperature.

They implanted the sample in the host media of nujol and scanned the spectra from 4000 to 450 cm^{-1}. The FTIR transmission spectra of CuO-I and CuO-II QDs are recorded and displayed in Fig. 2.6. They reported the three characteristic vibrational modes at 468, 529, and 590 cm^{-1} assigned to CuO-I QDs. Moreover, broad peaks centered at about 454, 526, and 587 cm^{-1} are due to CuO-II QDs. The weak band positioned at about 540 cm^{-1} in both the spectra is due to the capping agent TOAB. The high-frequency bands at about 590 and 530 cm^{-1} are due to the Cu-O stretching vibration. The characteristic bands due to Cu_2O at 610 and 147 cm^{-1} are not observed in the FTIR spectra. This observation suggested that the QDs contain purely the CuO phase, without the presence of any impurity of Cu_2O.

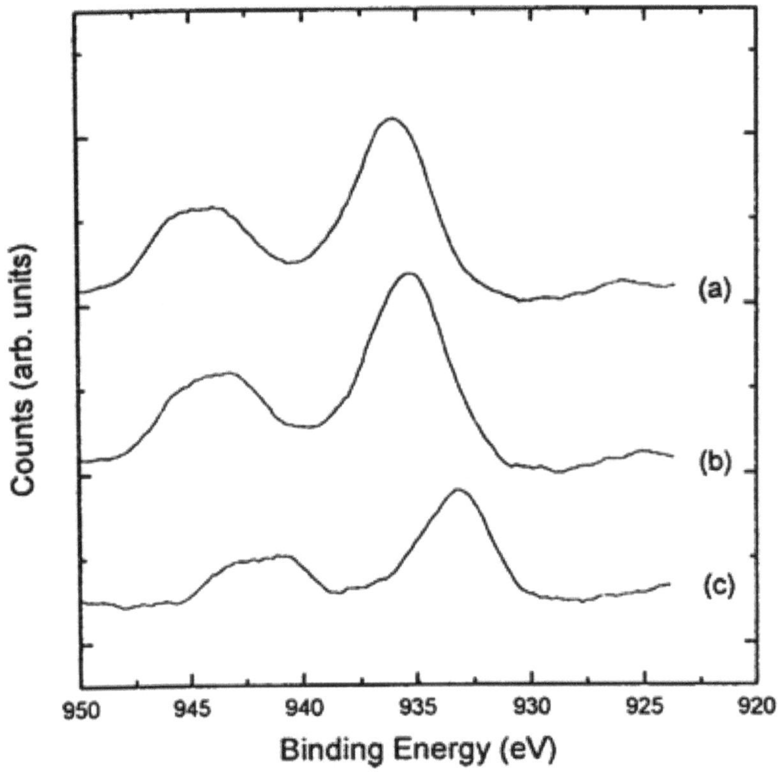

Figure 2.5 XPS of Cu 2p 3/2 level of (a) CuO-I, (b) CuO-II, and (c) CuO bulk powder. The spectra show the main peak along with the satellite feature positioned at higher binding energy at about 8 eV. Reproduced from Ref. [16] with permission from the publisher.

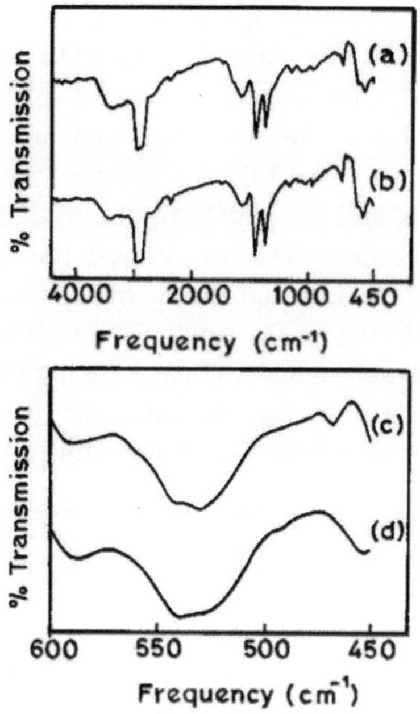

Figure 2.6 FTIR spectra of CuO-I and CuO-II nanocrystals. The absence of the Cu_2O phase revealed in (a) and (b). Spectra (c) and (d) illustrate the various CuO vibrational modes, at about 468, 529, and 590 cm^{-1}. The spectra (a) and (c) refer to CuO-I, and spectra (b) and (d) refer to CuO-II. The figure is reproduced from Ref. [16] with permission from the publisher.

2.3.5 Photoluminescence

Photoluminescence spectroscopy is used to study the optical properties of QDs. Photoluminescence spectra is the emission of light by QDs after the absorption of electromagnetic radiation. Photoluminescence spectroscopy works on the principle that on the absorption of electromagnetic radiation, the electrons get excited to a higher energy level and when the electrons return to the ground-state energy level, they emit radiation of a characteristic wavelength. For example, the normalized photoluminescence spectra for the ethanolic solution of ZnO QDs are shown in Fig. 2.7a [17]. We can see that with a decrease in ZnO QDs size, the visible emission peaks blue-shift toward the shorter wavelength. The direct photos captured under UV illumination light can vibrantly show the different color emission for the different sizes of ZnO QDs (Fig. 2.7b).

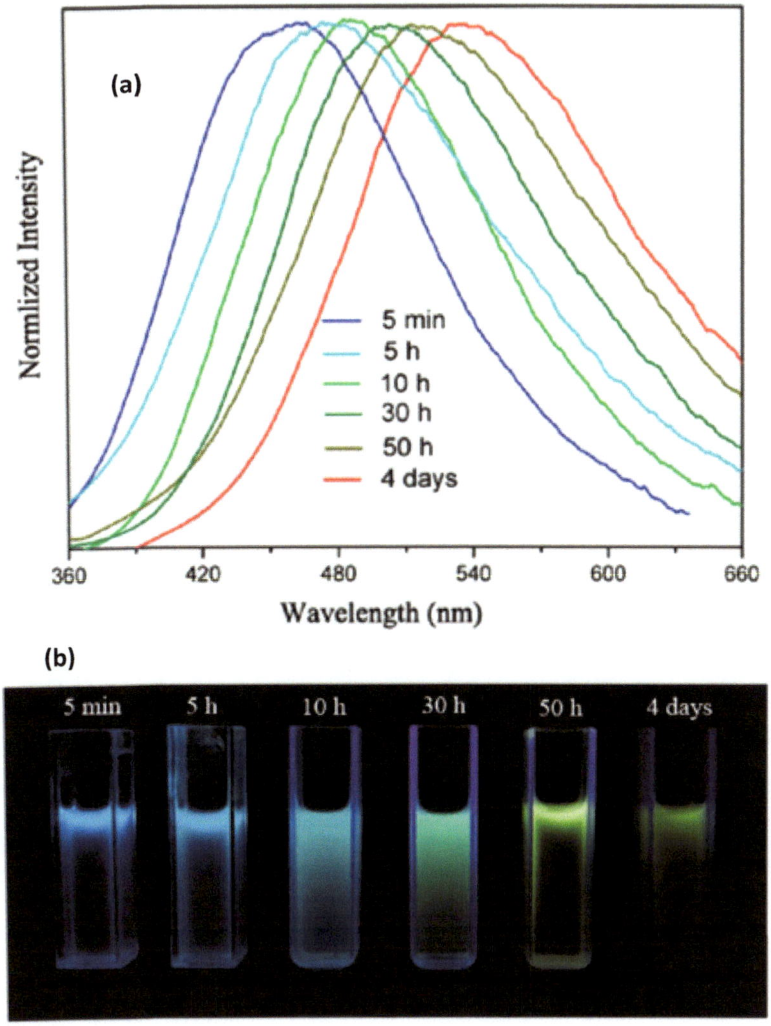

Figure 2.7 (a) Normalized photoluminescence emission spectra of QDs ZnO ethanolic suspensions and (b) direct photo of ethanolic solutions of ZnO QDs captured at different time intervals under the illumination of a 254 nm UV lamp. Reprinted with permission [17]. Copyright 2010, American Chemical Society.

2.3.6 UV-Visible Spectroscopy

UV-visible spectroscopy measurement is an important tool to determine the bandgap absorption for QDs. It helps to determine the absorption edge of the materials and also the variation in optical properties depending on QDs size distribution. Hens et al. [18] recorded the absorption spectra for PbS QDs, size varying from 3 to 6 nm, as shown in Fig. 2.8. Solid lines represent QDs synthesized using tri-n-octylphosphine (TOP) and the dotted lines for QDs synthesized without TOP.

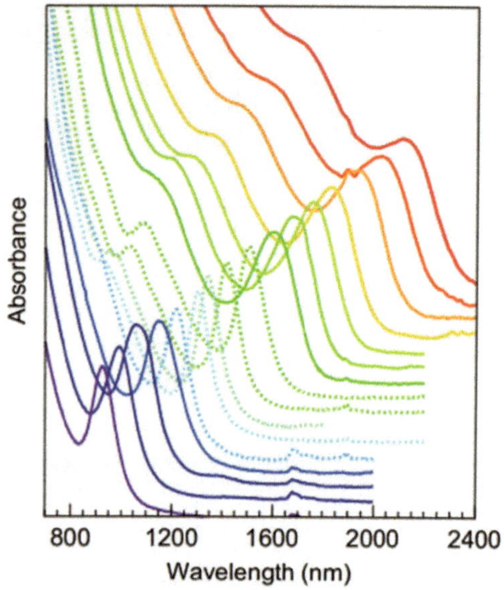

Figure 2.8 Characteristic absorbance spectra of PbS QDs suspensions. Solid lines are QDs prepared in the presence of TOP and dotted lines in the absence of TOP. Reprinted with permission [18]. Copyright 2011, American Chemical Society.

2.4 Applications

Several exciting and cutting-edge applications have been reported for QDs. Due to their small size and enhanced electronic, optical, and magnetic properties, QDs are widely used in different applications. Some of these applications are discussed and presented schematically in Fig. 2.9.

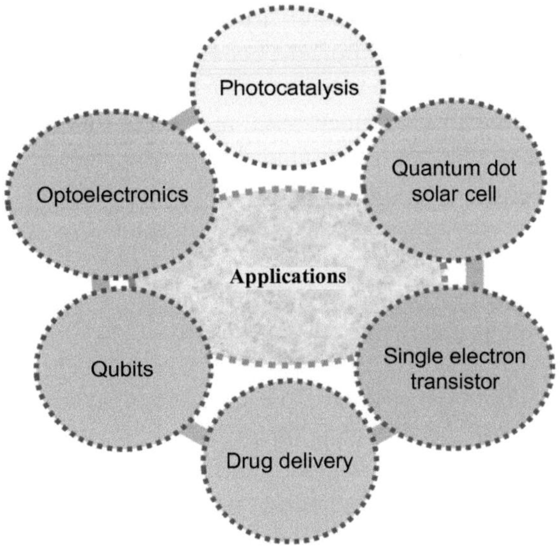

Figure 2.9 Schematic representation of different applications of quantum dots.

2.4.1 Qubits

QDs have a special interest in the field of quantum technology such as the development of quibits. A qubit is derived from the combination of two words—quantum and bit. It is the quantum mechanical equivalent of a classical bit used in classical computing. In classical computing, the data is programmed in bits. Each bit can have the binary value of either zero or one. However, in quantum computing, the data is encrypted in qubits. A qubit is a two-level quantum arrangement where the two basic qubit states are typically inscribed as $|0\rangle$ and $|1\rangle$. A qubit can be in the state of $|0\rangle$, $|1\rangle$ or in a linear combination of both states. In recent years, spin-orbit coupling is a striking factor in spin-based quantum information technologies. The spin of an electron is a likely candidate for a solid-state qubit. Most of the experimental developments have been accomplished in recent years using QDs fashioned using semiconductor-based technology. The extensive attention in quantum computing has encouraged the progress of conceptual designs across the various disciplines. Qubits are defined by the states of the spin of electrons in the semiconductor

QDs. Moreover, it should fit in an area of less than 5×5 mm^2. Small constituents can afford important benefits in terms of scalability. In addition to it, an effort to exhibit the physical operation has already been completed in the understanding of high-fidelity single-qubit rotations and small quantum algorithms. However, at present, the qubit technology needs control lines for each qubit, and a significant task is to circumvent an interconnect bottleneck for full regulator over a big qubit grid.

For the suitability of quantum technologies, electron-spin qubits need to have a long coherence time. Although the reports available to date have yielded short coherence times (i.e., 0.1 to 1 μs). Kobayashi et al. have demonstrated the spin-orbit coupled $J = 3/2$ systems that can have 10 ms coherence times resembling the best $S = ½$ and $S = 1$ structures in silicon and diamond, respectively [19]. It has been claimed that the results competing for the best electron-spin qubits are 104 to 105 longer than the earlier spin-orbit qubits. In another study, Li et al. have presented a design based on shared control and an accessible number of lines to increase the number of qubits to the thousands or millions for real-world quantum information with the use of silicon QDs [20]. Here, the control lines pronounce the qubit grid in a way that there is no need for local components. The fabrication is based on a three-layer structure to outline the qubit and the tunnel barrier gates. They have demonstrated the double stripline on the top of the configuration that can determine high-fidelity single-qubit rotations. A magnetic field, self-aligned and inhomogeneous, prompted by direct currents through superconducting gates allows qubit addressability and readout. The qubit coupling is dealt with the exchange interaction, and it has illustrated that the parallel two-qubit gates can be accomplished at the detuning-noise insensitive point. Recently, carbon-related materials can increase the coherence times of spin qubits and are considered among the advanced solid-state qubits. The research group of Burkard et al. has demonstrated the mechanism of forming the spin qubits in graphene QDs [21]. They have found the states of QDs where the valley degeneracy in bulk graphene is elevated, which are based on ribbons of graphene with armchair boundaries. In these proposed spin qubits, any two of them can likely be coupled in an array of several qubits through the Heisenberg exchange with the others being decoupled by detuning. Recently, there is immense interest in the development of solid-state qubits using different QDs.

2.4.2 Quantum Dot Solar Cell

Nowadays, a number of bulk materials are used in the fabrication of solar cells, such as silicon, gallium selenide, cadmium telluride, etc. However, there is a growing research interest in replacing these bulk materials with QDs. These QDs act as absorbing photovoltaic materials. The advantage with QDs is the alteration of bandgaps in a comprehensive range of energy levels by changing their sizes. This property is useful for developing the multi-junction solar cell to utilize the whole spectrum of the sun to increase its efficiency.

A QD acts basically as a three-dimensional potential well for electrons. Incorporation of a single-sized QD into an orderly arrangement within the intrinsic region of a p-i-n (here, i stands for insulating medium) solar cell resulted in a theoretical efficiency of 63% [22]. The increase in the efficiency of the solar cell by a QD intermediate followed a rudimentary principle—inclusion of the dots having discrete states of energies that allow the absorption of sub-bandgap energy states. This approach of the inclusion of QDs can enhance the efficiency of the solar cell than an ordinary dual-junction or multi-junction solar cell devices. This enhancement in efficiency occurs due to the extraction of current, which is controlled by the host bandgap, but not the independent photon energies. On the other hand, in a dual-junction solar cell, the current must resemble between the two junctions. An equal amount of current should be passed through both the junctions. Hence, the overall device efficiency in a dual-junction cell is limited by the current-generating capability of the weaker junction of the two. However, enormous successes have been achieved toward the development of multi-junction solar cells. But, the above limitations of current generation will be compounded as the number of junctions increases. In the case of QDs, the energy is influenced by multiple quantum wells. It is known that the energy states of each QD are inversely proportional to its size. If QDs are arranged in an ordered array within an insulating matrix, then the wave functions related to the discrete electronic states of the QDs will overlap. Such overlapping creates "mini-bands" within the insulating junction. It has been observed that the lowest empty mini-band energy level should be about one-third of the bandgap energy of the semiconductor above the valence band energy to maximize the device efficiency [23]. QD solar cells have materialized as a potential competitor for making transformative modifications. Three types of configurations have emerged as feasible candidates: solid-state heterojunction, polymer semiconductor hybrid, and liquid-junction QD-sensitized solar cells [24]. The schematic presentation is displayed in Fig. 2.10.

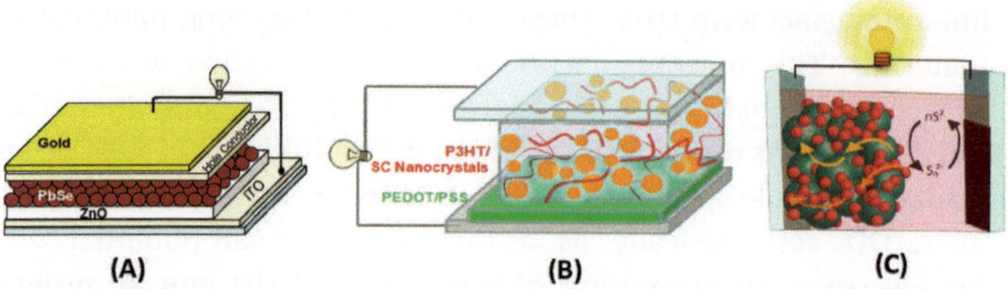

Figure 2.10 Schematic design of solar cells with semiconductor nanocrystals: (A) semiconductor heterojunction; (B) polymer semiconductor hybrid solar cells; and (C) liquid-junction solar cells. Reproduced from Ref. [24] with permission from the publisher.

2.4.3 Optoelectronics

QDs have active optical properties and can be tuned in a wide range by changing their size. In the past decades, silicon-based semiconductors are most popular for different photonic (optoelectronic) applications [25–28]. Over the last 20 years, the development of Si microelectronics by integrating photonic components has been a thrust area of research for many applications. However, Si-based light-emitting technologies have been explored broadly. The Si-based laser has been considered one of the prolific applications in the Si photonics domain. Both Si and Ge have an indirect band structure, which means the radiative recombination actions do not happen often [29]. The III–V compounds of direct bandgap have vigorous photonic properties that can be used for many photonics applications. The mingling of III–V photonic semiconductor components with Si microelectronics would provide the best solution for Si photonics. To date, the most effective methodology to realize the III–V lasers on Si has been hybrid integration using wafer bonding [30, 31]. However, there is a challenge that exists in the implementation of the direct monolithic integration of III–V lasers on Si substrates. The most critical difficulty in III–V on Si assimilation is the creation of high-density threading dislocations due to the lattice mismatch between them. To avoid such dislocations, for example, alternatively, the direct growth of GaAs on Si substrate via an intermediate epitaxial layer produces a close lattice constant to GaAs with fewer defects. The lattice constant of Ge is nearly lattice-matched to GaAs. Hence, Ge-coated Si layers have been extensively used as a model substrate for successive GaAs growth. Thus, the main task for integrating III–V photonic components into prospect Si microelectronics is to produce III–V layers directly on Ge/Si substrate [32].

2.4.4 Drug Delivery

The prerequisite of extraordinary specific and sensitive probes that defy the limitations of organic dyes attached to fluorescent proteins is of considerable interest in many areas of research, from cellular biology to medical imaging and diagnostics. Semiconductor QDs can covalently adhere to bio-recognition units, for example, peptides, nucleic acids, antibodies, or small-chain ligands for use as fluorescent probes [27–29]. Due to their wide excitation outlines and narrow emission spectra, the high-quality QDs are appropriate for optical mixing, in which numerous colors and intensities are united to encode proteins, genes, and small molecule libraries [30, 31]. Bagalkot et al. [33] have reported the development of CdSe/ZnS core–shell fluorescent (QD)-aptamer(Apt)-doxorubicin (Dox) conjugate [QD-Apt(Dox)] as a directed cancer imaging, therapy, and sensing system (Fig. 2.11). The QD surface was functionalized with an A10 RNA aptamer, which recognizes the extracellular domain of the prostate-specific membrane antigen (PSMA). The introduction of Dox, which is a broadly castoff antineoplastic anthracycline drug having fluorescent characteristics, in the stem of the A10 aptamer leads to a directed QD-Apt(Dox) conjugate with alterable self-quenching properties centered on a Bi-fluorescence resonance energy transfer (FRET) mechanism. A donor-acceptor model between QD and Dox through FRET mechanism, and a similar donor-quencher model between Dox and aptamer form when Dox intercalated within the A10 aptamer. The multifunctional nanoparticle arrangement can deliver Dox to the directed prostate cancer cells and sense the delivery of Dox by activating the fluorescence of QD, which concurrently images the cancer cells [33].

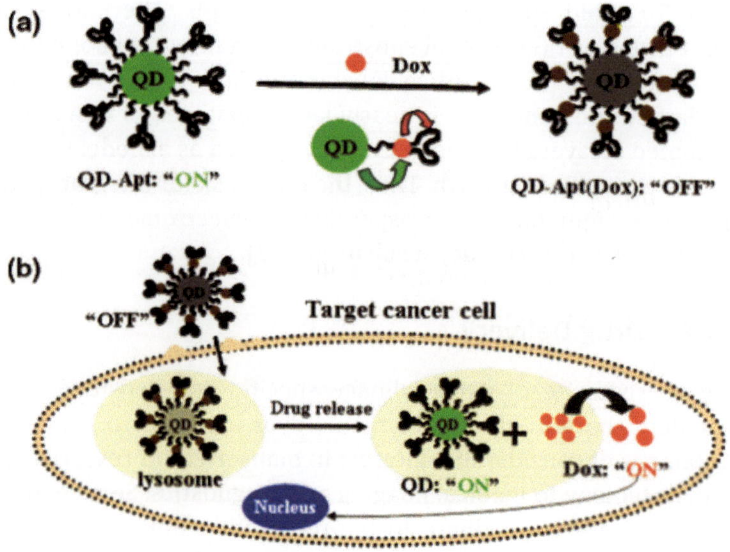

Figure 2.11 Diagrammatic representation of the QD-Apt (Dox) Bi-FRET system. The first step includes CdSe/ZnS QDs surface functionalization with the A10 PSMA aptamer. The insertion of Dox in the A10 PSMA aptamer on the surface of QDs results in the formation of the QD-Apt(Dox) conjugate and quenching of both QD and Dox fluorescence through a Bi-FRET mechanism: the fluorescence of the QD is appeased by Dox while concurrently the fluorescence of Dox is appeased by intercalation within the A10 PSMA aptamer resulting in the "OFF" state. (b) Diagrammatic representation of precise uptake of QD-Apt(Dox) conjugates into directed cancer cell through PSMA facilitate endocytosis. The release of Dox from the conjugates encourages the regaining of fluorescence from both QD and Dox ("ON" state), thus sensing the intracellular distribution of Dox and enabling the synchronous fluorescent localization and killing of cancer cells. Reprinted with permission [33], Copyright 2007, American Chemical Society.

2.4.5 Single-Electron Transistor

Colloidal synthesis allows routine production of semiconductor QDs dimensionally far smaller than those produced using lithography techniques. Detailed studies of both optical and electrical properties of individual QD show noticeably that a single excess charge on a QD can influence its properties markedly. McEuen et. al. [34] reported an electron transport study on a single-electron transistor made of colloidal CdSe QD. The device construction allows tuning of the amount of charge carriers, which in turn permit estimation of the energy needed for adding successive charge carriers. Such estimations are vital in realizing the energy-level spectra of nanoelectronic systems.

2.4.6 Photocatalysis

The construction of efficient and simple photocatalysts exploiting visible light (~43% of the sunlight spectrum) as contrast to UV light (~4% of the sunlight spectrum) is of prodigious importance for real-world application. Different semiconductor QDs, for example, CdS [35], CdSe [36], CdTe [37], PbS [38], and $CuInS_2$ [39], have been attached to photocatalysts to improve their photoactivity in the visible spectrum. Semiconductors QDs also help in efficient charge separation over the illuminated surface. For example, Peng et al. [40] reported a hydrogen evolution rate of 2200 μmol/hW in the presence of CdS QD-sensitized TiO_2 photocatalyst under visible light irradiation. Semiconductor QDs with high surface area and strong visible light absorption can also do better alone as a photocatalyst; for example, Osterloh et al. [40] reported photocatalytic H_2 evolution at the semiconductor (CdSe QDs)-solution interface. Precisely, the H_2 evolution rate from the irradiation of well-dispersed CdSe QDs in aqueous sodium sulfite solution is determined by the varying sizes of QDs (Fig. 2.12).

Singh et al. [41] reported that the hollow and porous TiO_2/CdS photocatalyst enhances the absorption of UV and visible light owing to the presence of CdS QDs on the nanofiber surfaces (shown in Fig. 2.13) and thus exhibits excellent photocatalytic activity as proven with the degradation of 4-NP (nitrophenols) dye in an aqueous medium. The effect of the deposition cycles of CdS QDs on dye-degradation activity was also demonstrated for TiO_2/CdS nanofibers. TiO_2/CdS photocatalyst for three successive ion layer adsorption and reaction (SILAR) deposition cycles was proven to be ~3 times more useful than hollow TiO_2 nanofibers and ~8 times effective than the pristine solid nanofibers. Following a similar pattern, semiconductor QDs were also used in the photocatalytic conversion of CO_2 into carbonate and degradation of harmful industrial pollutants such as dyes, chemicals, and mineral ions [42–45].

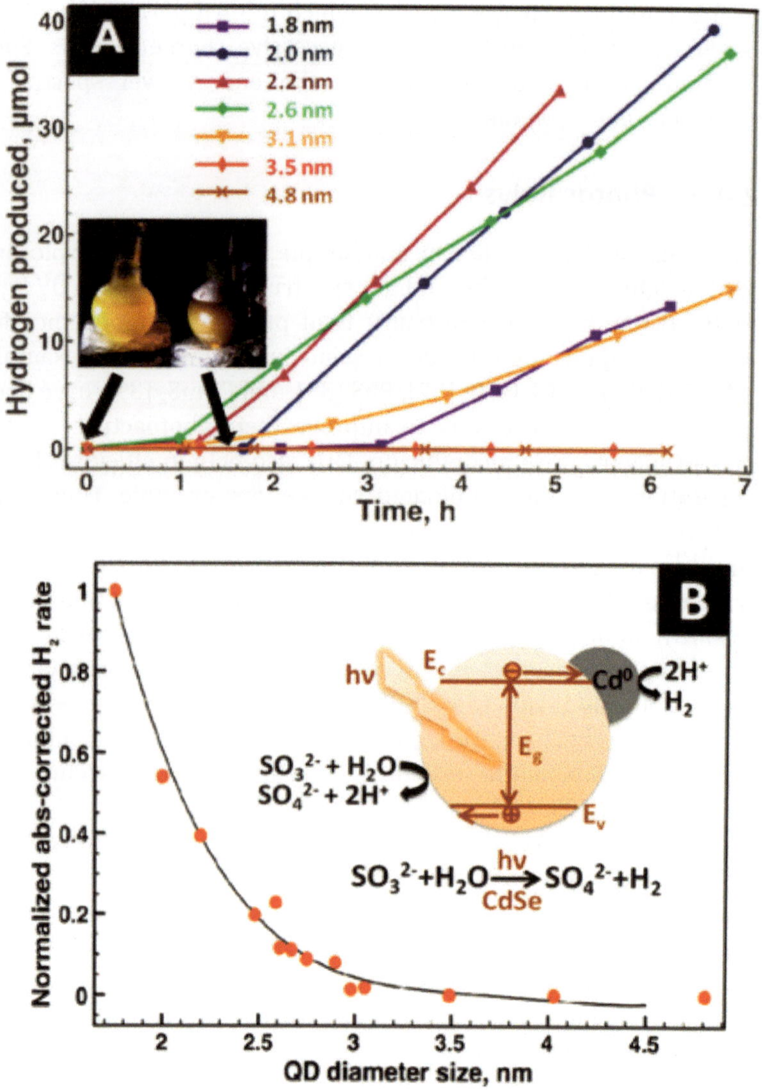

Figure 2.12 (A) H_2 evolution results in the presence of CdSe QDs dispersed in 0.10 M Na_2SO_3 aqueous solution (pH 9.2). Inset: Photographs depicting color alteration of CdSe QD solution without (left) and with (right) illumination. (B) Relative rates for H_2 evolution normalized with respect to catalyst concentration and engrossed photons schemed vs QD size. Inset: Schematic representation of H_2 evolution in the presence of CdSe QDs in sodium sulfite. Reprinted with permission [40], Copyright 2013, American Chemical Society.

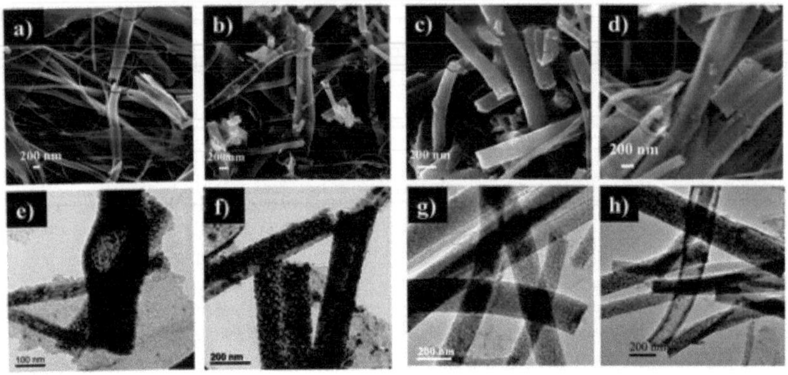

Figure 2.13 Field emission scanning electron microscopy (FESEM) images (a, b, c, and d) of CdS-loaded TiO_2 hollow nanofibers after 1, 2, 3, and 5 successive ion layer adsorption and reaction (SILAR) cycles, respectively (encircled portions show hollow morphology of nanofibers); transmission electron microscopy (TEM) micrographs (e, f, g, and h) of CdS-loaded TiO_2 hollow nanofibers after 1, 2, 3, and 5 SILAR cycles, respectively. Reproduced from Ref. [41] with permission from the Royal Society of Chemistry.

2.5 Conclusion and Future Scopes

QDs are semiconductors that have revolutionized the materials science in different ways. The size of QDs is within the range of a few nanometers (<10 nm). They have been prepared by several methods such as the wet-chemical route as well as physical, and chemical vapor depositions. Based on the synthesis methods and environment, the size and shape of the QDs are tuned. They can be synthesized in spherical to different non-spherical shapes. The property can be tuned in a wide range by altering the size and shape of the QDs. Different experimental tools are being used for the characterization of QDs such as different types of spectroscopies and microscopies. The application of QDs is widely spread in many areas of advanced technology. QDs are used in drug delivery, energy materials, catalysis, electronics, etc. Moreover, several studies are going on to explore new properties and applications of QDs. Future scope lies in the advancement of novel, easier, and economical methods for the fabrication of new QD materials for increasing performance efficiency toward various areas. This can be reached by the chemical modification of the materials and surfaces, exploration of high surface nanomaterials as well as manipulating the bandgap of QDs by doping metal centers to the semiconductors.

References

1. Manikandan, A., Chen, Y. Z., Shen, C. C., Sher, C. W., Kuo, H. C., and Chueh, Y. L. (2019). A critical review on two-dimensional quantum dots (2D QDs): From synthesis toward applications in energy and optoelectronics, *Prog. Quantum Electron.*, **68**, pp. 100226-1–28.
2. Alivisatos, A. P. (1996). Semiconductor clusters, nanocrystals, and quantum dots, *Science*, **271**, pp. 933–937.
3. Bednarek, S., Szafran, B., and Adamowski, J. (1999). Many-electron artificial atoms, *Phys. Rev. B*, **59**, pp. 13036–13042.
4. Ashoori, R. C. (1996). Electrons in artificial atoms, *Nature*, **379**, pp. 413–419.
5. Gammon, D. (2000). Electrons in artificial atoms, *Nature*, **405**, pp. 899–900.
6. Kastner, M. A. (1993). Artificial atoms, *Phys. Today*, **46**, pp. 24–31.
7. Mceuen, P. L. (1997). Artificial atoms: New boxes for electrons, *Science*, 278, pp. 1729–1730.
8. Burkard, G., Loss, D., and DiVincenzo, D. P. (1999). Coupled quantum dots as quantum gates, *Phys. Rev. B*, **59**, pp. 2070–2078.
9. Cronenwett, S. M., Oosterkamp, T. H., and Kouwenhoven, L. P. (1998). A tunable kondo effect in quantum dots, *Science*, **281**, pp. 540–544.
10. Su, H., Xie, Y., Gao, P., Xiong, Y., and Qian, Y. (2001). Synthesis of MS/TiO_2 (M = Pb, Zn, Cd) nanocomposites through a mild sol-gel process, *J. Mater. Chem.*, **11**, pp. 684–686.
11. Debnath, R., Bakr, O., and Sargent, E. H. (2011). Solution-processed colloidal quantum dot photovoltaics: A perspective, *Energy Environ. Sci.*, **4**, pp. 4870–4881.
12. Siy, J. T., Brauser, E. M., and Bartl, M. H. (2011). Low-temperature synthesis of CdSe nanocrystal quantum dots, *Chem. Commun.*, **47**, pp. 364–366.
13. M. Lee, *X-ray Diffraction for Materials Research: From Fundamentals to Applications*, 13:978-1-7, Apple Academic Press, New Jersey, United States, 2016.
14. Mahamuni, S., Borgohain, K., Bendre, B. S., Leppert, V. J., and Risbud, S. H. (1999). Spectroscopic and structural characterization of

electrochemically grown ZnO quantum dots, *J. Appl. Phys.*, **85**, pp. 2861–2865.

15. Guzelian, A. A., Banin, U., Kadavanich, A. V., Peng, X., and Alivisatos, A. P. (1996). Colloidal chemical synthesis and characterization of InAs nanocrystal quantum dots, *Appl. Phys. Lett.*, **69**, pp. 1432–1434.
16. Borgohain, K., Singh, J. B., Rao, M. V. R., Shripathi, T., and Mahamuni, S. (2000). Quantum size effects in CuO nanoparticles, *Phys. Rev. B.*, **61**, pp. 11093–11096.
17. Zhang, L., Yin, L., Wang, C., Lun, N., Qi, Y., and Xiang, D. (2010). Origin of visible photoluminescence of ZnO quantum dots: Defect-dependent and size-dependent, *J. Phys. Chem. C.*, **114**, pp. 9651–9658.
18. Moreels, I., Justo, Y., De Geyter, B., Haustraete, K., Martins, J. C., and Hens, Z. (2011). Size-tunable, bright, and stable PbS quantum dots: A surface chemistry study, *ACS Nano.*, **5**, pp. 2004–2012.
19. Kobayashi, T., Salfi, J., Chua, C., Van Der Heijden, J., House, M. G., Culcer, D., Hutchison, W. D., Johnson, B. C., Mccallum, J. C., Riemann, H., Abrosimov, N. V., Becker, P., Pohl, H., Simmons, M. Y., and Rogge, S. (2021). Engineering long spin coherence times of spin-orbit qubits in silicon, *Nat. Mater.*, **20**, pp. 38–42.
20. Li, R., Petit, L., Franke, D. P., Dehollain, J. P., Helsen, J., Steudtner, M., Thomas, N. K., Yoscovits, Z. R., Singh, K. J., Wehner, S., Vandersypen, L. M. K., Clarke, J. S., and Veldhorst, M. (2018). A crossbar network for silicon quantum dot qubits, *Sci. Adv.*, **4**, pp. 1–11.
21. Trauzettel, B., Bulaev, D. V., Loss, D., and Burkard, G. (2007). Spin qubits in graphene quantum dots, *Nat. Phys.*, **3**, pp. 192–196.
22. Luque, A. and Martí, A. (1997). Increasing the efficiency of ideal solar cells by photon induced transitions at intermediate levels, *Phys. Rev. Lett.*, **78**, pp. 5014–5017.
23. Raffaelle, R. P., Castro, S. L., Hepp, A. F., and Bailey, S. G. (2002). Quantum dot solar cells, *Prog. Photovoltaics Res. Appl.*, **10**, pp. 433–439.
24. Kamat, P. V. (2012). Boosting the efficiency of quantum dot sensitized solar cells through modulation of interfacial charge transfer, *Acc. Chem. Res.*, **45**, pp. 1906–1915.
25. Liang, D. and Bowers, J. E. (2010). Recent progress in lasers on silicon, *Nat. Photonics*, **4**, pp. 511–517.
26. Paniccia, M. (2010). Integrating silicon photonics, *Nat. Photonics.*, **4**, pp. 498–499.
27. Hochberg, M. and Baehr-Jones, T. (2010). Towards fabless silicon photonics, *Nat. Photonics*, **4**, pp. 492–494.

28. Soref, R. (2006). The past, present, and future of silicon photonics, *IEEE J. Sel. Top. Quantum Electron.*, **12**, pp. 1678–1687.

29. Soref, R. (2010). Mid-infrared photonics in silicon and germanium, *Nat. Photonics*, **4**, pp. 495–497.

30. Liu, H., Wang, T., Jiang, Q., Hogg, R., Tutu, F., Pozzi, F., and Seeds, A. (2011). Long-wavelength InAs/GaAs quantum-dot laser diode monolithically grown on Ge substrate, *Nat. Photonics*, **5**, pp. 416–419.

31. Liu, A., Jones, R., Liao, L., Samara-Rubio, D., Rubin, D., Cohen, O., Nicolaescu, R., and Paniccia, M. (2004). A high-speed silicon optical modulator based on a metal-oxide-semiconductor capacitor, *Nature*, **427**, pp. 615–618.

32. Brammertz, G., Caymax, M., Meuris, M., Heyns, M., Mols, Y., Degroote, S., and Leys, M. (2008). GaAs on Ge for CMOS, *Thin Solid Films*, **517**, pp. 148–151.

33. Bagalkot, V., Zhang, L., Levy-Nissenbaum, E., Jon, S., Kantoff, P. W., Langery, R., and Farokhzad, O. C. (2007). Quantum dot-aptamer conjugates for synchronous cancer imaging, therapy, and sensing of drug delivery based on Bi-fluorescence resonance energy transfer, *Nano Lett.*, **7**, pp. 3065–3070.

34. Klein, D. L., Rotht, R., Lim, A. K. L., Alivisatosti, A. P., and McEuen, P. L. (1997). A single-electron transistor made from a cadmium selenide nanocrystal, *Nature*, **389**, pp. 699–701.

35. Robel, I., Subramanian, V., Kuno, M., and Kamat, P. V. (2006). Quantum dot solar cells. Harvesting light energy with CdSe nanocrystals molecularly linked to mesoscopic TiO_2 films, *J. Am. Chem. Soc.*, **128**, pp. 2385–2393.

36. Lee, Y. L., Huang, B. M., and Chien, H. T. (2008). Highly efficient CdSe-sensitized TiO_2 photoelectrode for quantum-dot-sensitized solar cell applications, *Chem. Mater.*, **20**, pp. 6903–6905.

37. Gao, X. F., Li, H. B., Sun, W. T., Chen, Q., Tang, F. Q., and Peng, L. M. (2009). CdTe quantum dots-sensitized TiO_2 nanotube array photoelectrodes, *J. Phys. Chem. C.*, **113**, pp. 7531–7535.

38. Sambur, J. B., Novet, T., and Parkinson, B. A. (2010). Multiple exciton collection in a sensitized photovoltaic system, *Science*, **330**, pp. 63–66.

39. Li, T. L. and Teng, H. (2010). Solution synthesis of high-quality $CuInS_2$ quantum dots as sensitizers for TiO_2 photoelectrodes, *J. Mater. Chem.*, **20**, pp. 3656–3664.

40. J. Zhao, M. A., Holmes, F., and Osterloh, E. (2013). Quantum confinement controls photocatalysis: A free energy analysis for photocatalytic proton reduction at CdSe nanocrystals, *ACS Nano.*, **7**, pp. 4316–4325.

41. Singh, N., Mondal, K., Misra, M., Sharma, A., and Gupta, R. K. (2016). Quantum dot sensitized electrospun mesoporous titanium dioxide hollow nanofibers for photocatalytic applications, *RSC Adv.*, **6**, pp. 48109–48119.

42. Kuehnel, M. F., Orchard, K. L., Dalle, K. E., and Reisner, E. (2017). Selective photocatalytic CO_2 reduction in water through anchoring of a molecular Ni catalyst on CdS nanocrystals, *J. Am. Chem. Soc.*, **139**, pp. 7217–7223.

43. Hou, J., Yang, C., Wang, Z., Jiao, S., and Zhu, H. (2013). Bi_2O_3 quantum dots decorated anatase TiO_2 nanocrystals with exposed {001} facets on graphene sheets for enhanced visible-light photocatalytic performance, *Appl. Catal. B Environ.*, **129**, pp. 333–341.

44. Yadav, S. K. and Jeevanandam, P. (2015). Synthesis of Ag_2S-TiO_2 nanocomposites and their catalytic activity towards rhodamine B photodegradation, *J. Alloys Compd.*, **649**, pp. 483–490.

45. Yadav S. K. and Jeevanandam, P. (2015). Thermal decomposition approach for the synthesis of CdS-TiO_2 nanocomposites and their catalytic activity towards degradation of rhodamine B and reduction of Cr(VI), *Ceram. Int.*, **41**, pp. 2160–2179.

Chapter 3

Micro-/Nano-Encapsulation and Encapsulation Applications

Hayrunnisa Nadaroglu,[a,b] M. Kaan Yeşilyurt,[c] and Ömer Çomakli[d]

[a]*Department of Food Technology, Vocational College of Technical Sciences, Ataturk University, Turkey*
[b]*Department of Nano-Science and Nano-Engineering, Institute of Science and Technology, Ataturk University, Turkey*
[c]*Department of Machinery and Metal Technologies, Vocational College of Technical Sciences, Ataturk University, Turkey*
[d]*Department of Mechanical Engineering, Faculty of Engineering, Ataturk University, Turkey*
hnisa25@atauni.edu.tr

Micro-/nano-encapsulation involves coating of a gaseous, liquid, or solid material (i.e., core or active substance) with a secondary active material in capsules in a range of about 1 nm to 2000 μm. It protects the encapsulated active material at the desired place and time until it is released when appropriate or desired conditions occur or they have been mobilized to a desired location. The aim of micro-/nano-encapsulation is to give the active ingredient in the core of the

Science and Applications of Nanoparticles
Edited by Waqar Ahmed and Ehsan Nourafkan

ISBN 978-981-4267-34-2 (Hardcover), 978-1-003-28029-3 (eBook)
www.jennystanford.com

capsule superior properties compared to its pure form. Basically, realizing all micro-/nano-encapsulation requires multi-disciplinary processing steps, all of the selected ingredient, shell material, formulation, and use. In this chapter, the methods, properties, and usage areas of micro-/nano-encapsulation are discussed.

3.1 Introduction

A micro-/nanocapsule (M/N-Cap) is a structure formed from liquid/solid material covered by a polymeric wall. The wall material of the capsule is inert to the substance it covers. Micro-/nano-encapsulation, in simpler terms, is the system of confining active target ingredients in micro- or nano-sized capsules (Yeşilyurt et al., 2019). The capsule size varies from millimeters to micrometers and even to nanometers mainly depending on the materials and methods used in the capsule production. Capsules having a size between 1 and 1000 μm are called microcapsules, whereas capsules larger than 1000 μm are referred to as macrocapsules. Now capsules even smaller than 1 μm are nanocapsules (Gökmen et al., 2012) (Fig. 3.1).

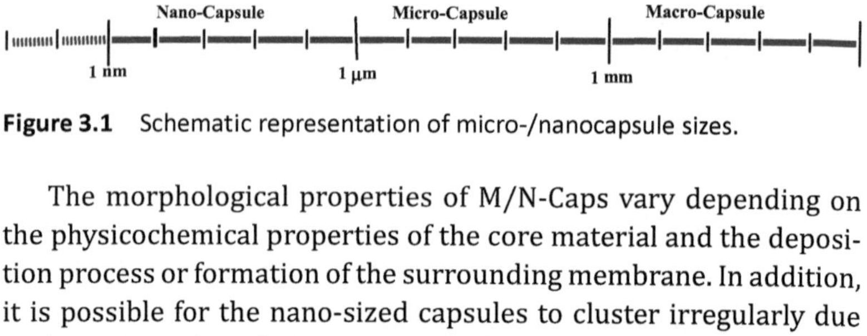

Figure 3.1 Schematic representation of micro-/nanocapsule sizes.

The morphological properties of M/N-Caps vary depending on the physicochemical properties of the core material and the deposition process or formation of the surrounding membrane. In addition, it is possible for the nano-sized capsules to cluster irregularly due to the aggregation phenomenon, which is encountered across most nanostructured molecules. Therefore, M/N-Caps can have regular or irregular shapes. The basis of the morphology of M/N-Caps is that they may have either single or multiple cores or the shell structure can consist of a single layer or multiple layers (Ghosh, 2006; Thies, 1996).

The technique for microencapsulation has been known for decades, and a large number of different encapsulation methods can be seen reported in the literature and patents with regards to the synthesis of the M/N-Cap. Although the core and shell materials could be different, the basic stages through which M/N-Caps are synthesized are common and consist of three steps. These synthesis steps, in order, are preservation of the core component, formation, and hardening. Such processes were almost too expensive for wider practical application. However, the development of new synthesis methods and the improvement in the existing ones, by taking the advantage of new techniques and technological possibilities, paved the way for economic mass production, as a result of which micro-/nano-encapsulation has found numerous application areas and new ones are being added to these application areas day by day.

Most capsulation processes are based on the principle that the starting components are added to the reaction medium as a monomer or prepolymer whereafter they polymerize and form the shell material of the capsule in the reaction medium. In the process of capsule formation, the purpose of chemical techniques such as suspension polymerization or multiple condensation is to achieve the entrapment of the core material and to set off a reaction that will ensure the shell formation all around the target core material. While physical features, cost, toxicity, etc. are important in the selection of production methods for M/N-Cap, the main concern should be the properties of the active substance and the reaction conditions of encapsulation (Andrade et al., 2015). A successful synthesis of M/N-Cap depends on the morphology as well as the chemical and physical properties of the polymeric shell material (Yadav et al., 1990). In the decision making of the method to be employed in the synthesis of M/N-Cap, several variables, such as the structure and concentration of polymer (El Bahri and Taverdet, 2007), concentration of emulsifiers (Alexandridou, 1994), the stirring rate (Maa and Hsu, 1996), temperature (Constantin et al., 2007), and the ratio of the dispersed volume to the continuous phase (Yadav et al., 1990), are chosen as parameters along with many others. Thus, the encapsulation efficiency and size distribution of the M/N-Cap will be achieved (Bleiel et al., 2017).

Encapsulation methods can be classified into categories such as physical, chemical, and physicochemical methods, which involve the use of physical and chemical procedures in combination. Some mechanical methods may also be used in physical methods (see Table 3.1).

Table 3.1 Classification of encapsulation processes (Al Shannaq and Farid, 2015)

	Physical Processes	
Chemical Process	**Physicochemical Processes**	**Physical–Mechanical Processes**
• Suspension, dispersion, and emulsion polymerization • Polycondensation	• Coacervation • Layer-by-layer assembly • Sol–gel encapsulation • Supercritical CO_2-assisted microencapsulation	• Spray drying • Multiple nozzle spraying • Fluid bed coating • Vacuum encapsulation • Electrostatic encapsulation

The number of methods is around 25. However, among these methods, the most applied method group, despite the fact that they can be harmful to health and environment, is chemical. Because physical methods, while posing little to no negative impacts, are not capable of meeting requirements with regard to capsule diameter in particular (Zhao and Zhang, 2011; Chen et al., 2014). Among the chemical methods, *in situ* polymerization is the method that is highly developed and commercialized (Duan, 2015) and is the most applied method (Zondag et al., 2016) (Table 3.2).

Table 3.2 Various encapsulation techniques as given by Mishra (2015)

Various Encapsulation Processes	
Annular jet process	Layer-by-layer deposition
Centrifugal suspension separation	Matrix encapsulation
Coacervation and phase separation	Melt-dispersion
Cocrystallization	Microfluidic encapsulation
Controlled precipitation	Molecular encapsulation
Dispersion/suspension polymerization	Organogels
Dripping and jet break-up	Pan coating technology
Electrospraying	Phase inversion/precipitation
Emulsification/emulsion polymerization	Polyelectrolyte complexation
Emulsion-solvent removal	Self-assembly

Various Encapsulation Processes	
Fluidized bed coating technology	Sol–gel process
Freeze drying	Solvent evaporation
Granulation	Spinning disk technology
Hot-melt extrusion process	Spray chilling/cooling/ congealing
In situ polymerization	Spray drying
Interfacial crosslinking/reaction	Supercritical fluid technology
Interfacial polymerization	Templating
Ionotropic gelation	Vapor phase deposition

3.2 Chemical Methods

Chemical processes can be initiated by free radicals attacking on monomers, triggered by a number of factors such as polymerization, heating, or pH change. Most chemical processes involve polymerization techniques, such as suspension polymerization, emulsion polymerization, interface polymerization, *in situ* polymerization, and condensation polymerization. In some studies, a combination of multiple chemical techniques was used to achieve microencapsulation (Alva et al., 2017).

More details on some widely used chemical synthesis methods are provided in the following paragraphs.

3.2.1 Interfacial Polymerization

The interfacial polymerization method is used if there are two or more monomers associated with the shell material (Fig. 3.2). The solvent medium is generally water. PCM and hydrophobic monomers form the emulsion medium in the reaction medium. Since hydrophilic monomers are soluble in the medium, they allow the capsule to form as a very thin layer in the later stages of the reaction. This method is mostly used for organic shell materials such as polyurea and polyurethane. The essence of this method is as follows: Polyurea is based on the reaction of an isocyanate with a polyfunctional amine, and polyurethane is based on the reaction of an isocyanate with a polyol. Briefly, first, the isocyanate PCM is prepared in an oil phase in an organic solvent environment (chloroform, etc.) and added to the water environment to prepare the oil-in-water emulsion and slowly heated. Then polymerization is carried out at the interface, and the microcapsules obtained are washed and dried (Alva et al., 2017).

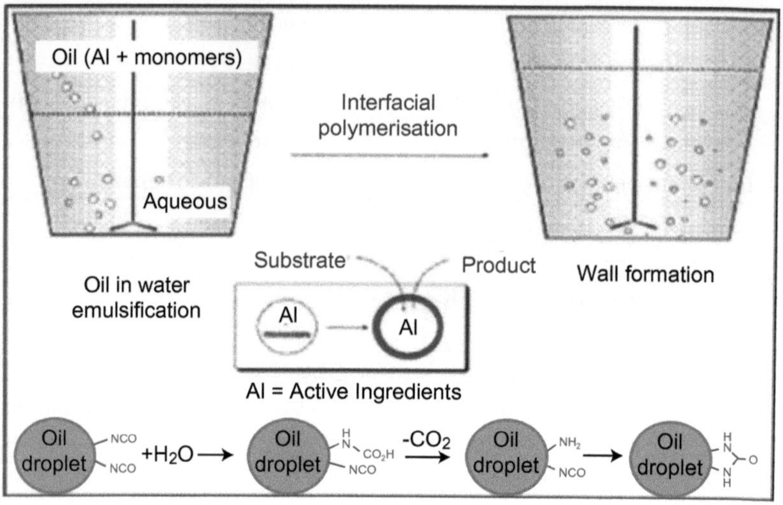

Figure 3.2 Synthesis process in interfacial polymerization (Chen et al., 2014).

The interface polymerization method is a suitable method for encapsulating both oil- and water-soluble capsules (Liu et al., 2015). The substance to be encapsulated is dispersed in an aqueous phase created as an oil-in-water or water-in-oil emulsion with appropriate surfactants, where needed. As a result of rapid polymerization of monomers that starts at the surface of core droplets, a polymer shell that constitutes the capsule wall is formed on the surface of the droplets or particles of the core material upon the addition of a reactant to the monomer dispersed in the aqueous phase. Once the process is completed, the capsules are separated from the oil or water phase (Erdoğan, 2017; Mishra, 2015).

Altering the constituents of capsules, namely using different monomers or utilizing additives, or changing the conditions in the reaction medium will help obtain capsules in different sizes or having different properties. The process, however, can be adjusted to produce small microcapsules in the range of 20 to 30 μm or larger (Mishra, 2015).

In the process of preparing phase change capsules through the interfacial polymerization method, the core material is emulsified first, after the formation of the oil/water or water/oil emulsion by using an appropriate emulsifier. Then the polymer as capsule is formed on the surface of the core by the polymerization of the monomers. Finally, the capsule is separated from the oil phase or the water phase. When the method for preparing nano PCM is used, the core must be added to the syringe with a capillary tube. It requires the setting of high voltage and direct current. Moreover, the distance between the needle of the syringe and the liquid level of monomer solution needs to be as close as possible.

Nanoencapsulated PCMs have been produced by the interfacial polymerization method using paraffin as core and polyurea as shell. Park et al. (2014) synthesized the nanoencapsulated PCM whose core and shell were paraffin and polyurea, respectively (Liu et al., 2015).

3.2.2 Suspension Polymerization

Suspension polymerization is the method used to form a thermoplastic polymer. The polymerization reaction is carried out in the form of large droplets or stabilized polymerization with gum material dissolved in water. Peroxide initiators used in the reaction medium are used to generate radicals in droplets. A solvent is used to dissolve the monomer in high doses. The suspension polymerization method is preferred for the polymerization of monomers alone, and the initiators are used in the insoluble form or slightly soluble form in the aqueous phase solvent (Fig. 3.3). Also, water is generally preferred as the solvent medium. The essence of this method is that during emulsification, the core, monomers, and initiators form the oil phase and are suspended in the continuous aqueous phase as dispersed droplets. Emulsion can be formed by adding surfactants to the reaction medium and by vigorous mechanical stirring. 2,2-Azobisisobutyronitrile (AIBN) and benzoyl peroxides, which are poorly soluble in water, are among the radical initiators that are used in suspension polymerization (Alva et al., 2017). Many common commercial resins such as polyvinyl chloride (PVC), polystyrene, and (poly) methyl methacrylate have been produced using the suspension polymerization method. Over time, suspension polymerization has been extensively used for the synthesis of functional microspheres and more recently successfully in the manufacture of microcapsules having phase change materials as core (Al Shannaq and Farid, 2015).

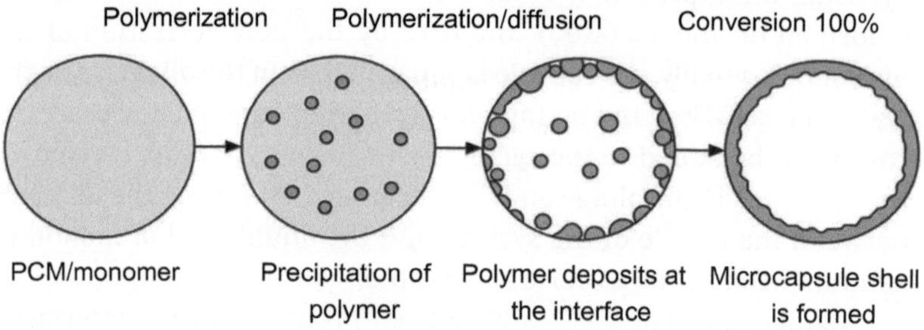

Figure 3.3 Schematic cross-sectional diagram showing the mechanism of capsule formation using non-aqueous precipitation polymerization within a droplet (suspension polymerization) (Smith, 2009).

3.2.3 Emulsification/Emulsion Polymerization

The emulsion polymerization method is one of the most widely used methods for producing organically structured encapsulation. Emulsions can be used as a delivery vehicle for the water-soluble and/or lipophilic active ingredient. In emulsion polymerization, the surfactant is dissolved in water until the critical micelle concentration is reached. The inner side of the micelle medium provides the necessary core formation for polymerization. As polymerization progresses, these cores gradually grow and simultaneously trap the core material to form the final microcapsules. The most common type of emulsion polymerization is an oil-in-water emulsion formed by emulsifying monomer oil droplets in a continuous water phase using surfactants. The water-soluble active ingredients can be encapsulated in water-in-oil (w/o) emulsions or double emulsions of the water-in-oil-in-water (w/o/w) type. Likewise, lyophilic active substances can be encapsulated by forming oil-in-water emulsions (Appelqvist et al., 2007). Among typical initiators used in suspension polymerization are ferrous sulfate, ammonium persulfate, sodium thiosulfate, which are easily soluble in water. The emulsion polymerization method is also a suitable method for the encapsulation of liquid PCM. This method is typically used to encapsulate alkanes with organic shell materials such as PMMA and polystyrene. Many researchers have successfully performed microencapsulation using this technique (Alva et al., 2017; Sari et al., 2014). Alkan et al. (2009) microencapsulated docosane with polymethylmethacrylate (PMMA) as phase change material for thermal energy storage using emulsion polymerization. Cao and Yang (2014) synthesized octadecane microcapsules using the emulsion polymerization method.

Nanocapsule production by emulsification consists of the following steps. First, the monomer, which is insoluble in the solvent, is homogeneously dispersed with the emulsifying agent by means of a mechanical stirrer. The initiator is then added to initiate the polymerization reaction, and the polymer shell is formed on the surface of the core (Erdoğan, 2017). Chemical microencapsulation methods are shown in Fig. 3.4.

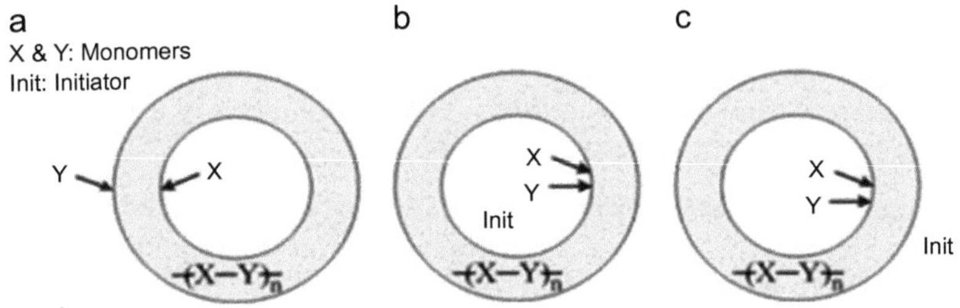

Figure 3.4 Schematic diagrams that illustrate the difference between the chemical microencapsulation methods: (a) interfacial polymerization, (b) suspension polymerization, and (c) emulsion polymerization (Jamekhorshid et al., 2014).

3.2.4 Miniemulsion Polymerization

The miniemulsion polymerization method is the most widely used method for the production of nanoencapsulated PCMs. The essence of this method is that polymerization is carried out in very small emulsion drops. To state briefly, nanoencapsulated PCMs are synthesized by forming nano-sized droplets under the influences of water, monomer, initiator effect of polymer in the reaction medium together with the emulsifier added into the reaction medium under a high mechanical stirrer (Erdoğan, 2017). The monomer preferred at the beginning gives some information about the polymeric structure of the nanocapsule to be formed. Using the miniemulsion method, the encapsulation of alkane materials is carried out by using mostly polystyrene, polyurea, styrene-butyl acrylate (SBA), styrene (S)-methyl methacrylate (MMA) copolymer as the shell material in the encapsulation.

Chen et al. (2012) synthesized nano PCMs made of n-dodecanol and styrene-butyl acrylate (SBA) copolymer in the encapsulation they performed using the miniencapsulation method. Again Fang et al. (2008, 2009, 2013, 2014) synthesized spherical capsules of n-tetradecane, n-octadecane, and dotriacontan by using polystyrene as the shell material. They determined that the sizes of the nanoencapsulated PCM they obtained were 132 nm, 124 nm, and 168.2 nm. Similarly, Tumirah et al. (2014) produced and characterized nanocapsules using the miniemulsion method with styrene-methylmethacrylate shell material and n-octadecane core material.

The size and morphology of the polymer emulsion are the same as that of the droplets initially formed. Compared to emulsion polymerization, this method has the advantages of less energy use and high stability (Erdoğan, 2017).

3.2.5 *In situ* Polymerization

The *in situ* polymerization method usually combines two immiscible liquids, such as water and organic solvents, which contain complementary, direct acting, organic intermediates that will react with each other to form a precondensate. The whole polymerization process takes place in a continuous phase, as in interfacial polymerization. *In situ* polymerization is distinguished in that there is no reactant in the core material (Qiu et al., 2019). Numerous encapsulation studies have been conducted using (Zhao and Zhang, 2011) the *in situ* polymerization method, in which the monomer and catalyst dissolved in the continuous phase are separated from the core material, and the prepolymer is gradually formed on the surface of the core material during the polymerization reaction (Erdoğan, 2017).

It is possible to produce the best quality microcapsules in terms of the diffusion tightness of the walls and the capability to produce capsules in sizes ranging from 5 to 100 μm by the *in situ* polymerization method. Therefore, *in situ* polymerization methods are widely used in the manufacture of microencapsulated PCM (Qiu et al., 2019).

Although the processes can slightly differ depending on the starting materials and formulations, the basic encapsulation steps of an oil immiscible with water, as outlined by Duan (2015), are as follows:

- Forming a stable emulsion in the desired oil droplet size by dispersing it in the oil-in-water phase containing reactants (i.e., melamine-formaldehyde or urea-formaldehyde resin prepolymer).
- Polymerization from water phase and precipitation of monomer/oligomer on oil phase.
- Shell formation and curing.

A more detailed and understandable process example is given by Koehler as seen in Fig. 3.5.

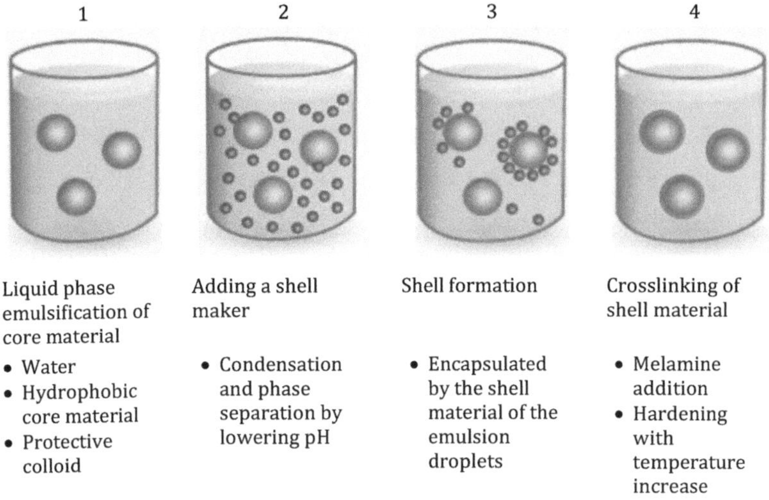

Figure 3.5 General scheme of the microencapsulation process by *in situ* polymerization (Koehler).

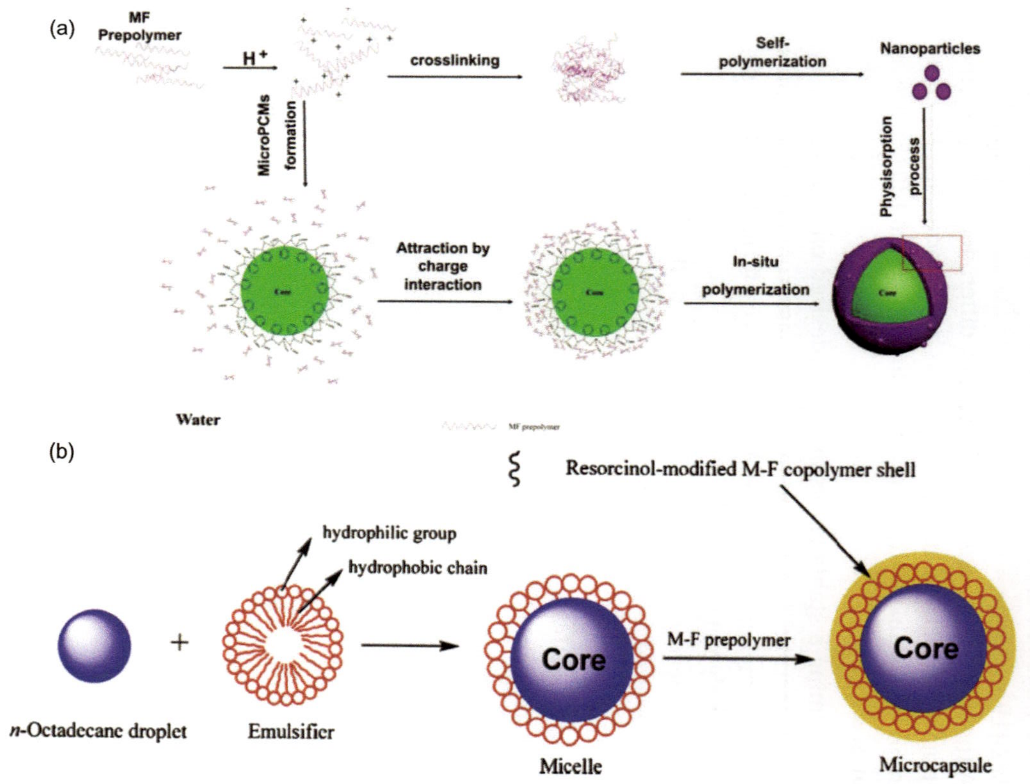

Figure 3.6 Schematic formation of encapsulation by *in situ* polymerization. (a) (Zhang et al., 2017) and (b) (Chen et al., 2014; Qiu et al., 2019; Alva et al., 2017).

In the first stage, oil-in-water or water-in-oil emulsions are prepared according to the structure of the core material. After the shell material is added to this mixture, the pH value of the environment is lowered and condensation and phase separation are achieved. After the shell material begins to bond to the core surface, melamine is added to the environment and the shell material is hardened by increasing the temperature. Thus, the core material gets encapsulated (Boh et al., 2005).

For a clearer illustration of the chemical process, Fig. 3.6 presents two schematics for *in situ* polymerization.

During the shell formation phase, the dropping rate and the stirring rate of the emulsion have a determining effect on the capsule size to be obtained as well as the sphericity and physical structure of the capsules. In addition, the temperature of the reaction medium is also among the important factor parameters. Figure 3.7 shows the effect of stirring rate on capsule size during the microencapsulation process.

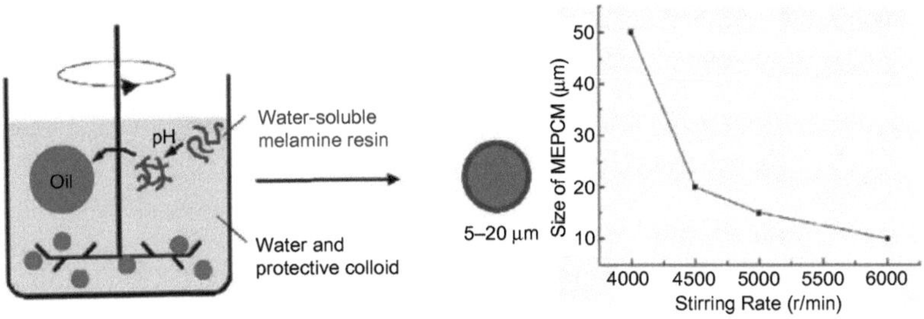

Figure 3.7 Microencapsulation process in *in situ* polymerization and effect of stirring rate on capsule size (Zondag et al., 2016).

The steps of a typical procedure can briefly be given as follows. The core material and surfactant, if necessary, are dissolved in distilled water being stirred in a magnetic stirrer. In a different flask, the shell material is dissolved in distilled water, whereafter the mixture is added with a pH regulator to prepare the prepolymer, which will then be added into the oil emulsion through the dropping funnel at an appropriate dropping rate. The pH of the medium is then lowered by adding regulatory solutions to allow shell formation. The resulting capsules are washed and dried in an oven at 40°C for 24 h.

A schematic presentation of the stepwise preparation of M-PCMs is given in Fig. 3.8.

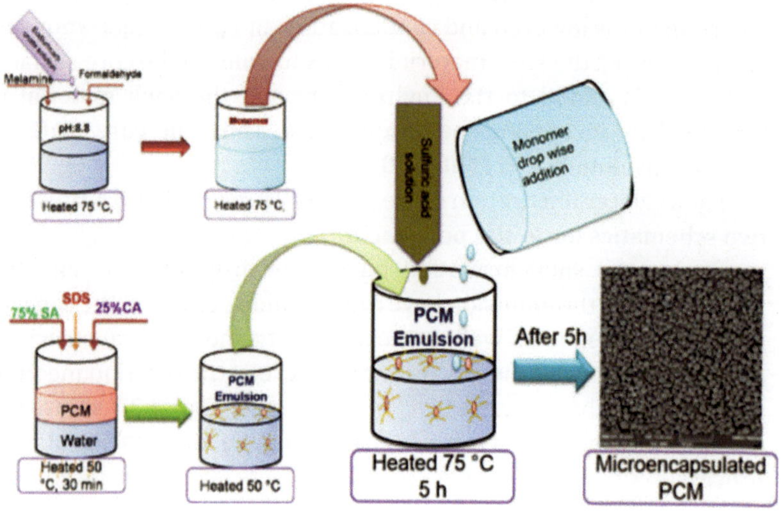

Figure 3.8 Schematic presentation of stepwise preparation of M-PCMs (Srinivasaraonaik et al., 2020).

3.2.6 Condensation Polymerization

The condensation polymerization method is based on the polymerization of monomers with functional groups (–OH (alcohol), $-NH_2$ (amino), –COOH (carboxylic group), etc.) by a condensation reaction, shown in Fig. 3.9. Condensation polymerization is generally a part of synthetic methods like sol–gel, *in situ* polymerization, and interfacial polymerization (Alva et al., 2017).

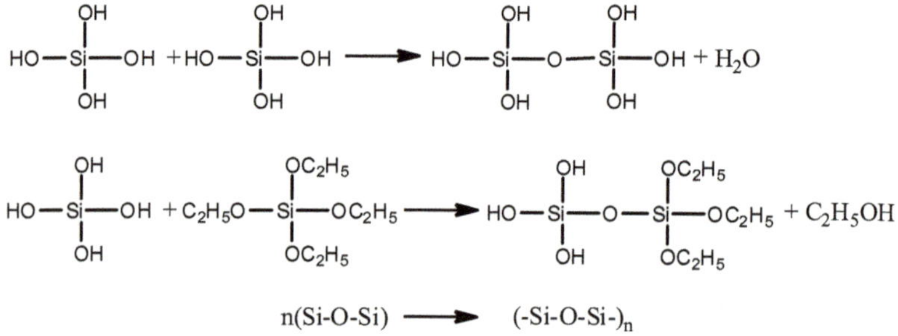

Figure 3.9 Condensation polymerization (Alva et al., 2017).

3.2.7 Solvent Evaporation

The solvent evaporation method is one of the most widely used microencapsulation methods. In this method, a water-immiscible polymer prepared in an organic solvent environment under strong mechanical stirring is added with an oil emulsion. In order to create multiple emulsifiers (like water-in-oil-in-water), poly (vinyl alcohol) or poly (vinyl pyrrolidone) is added dropwise to the reaction medium. Microcapsule formation can then be achieved in the medium by the evaporation of solvent. The microcapsules obtained in the last stage of the reaction are washed and dried (Mishra, 2015).

3.2.8 Emulsification-Diffusion

Many different methods used in the synthesis of nanoparticles are based on the quick removal of solvent as a result of which tiny droplets or particles are formed in the reaction medium. Nanoparticles are thus produced in similar synthesis methods. Monodispersed particles can be obtained without any emulsifying agent. In this kind of synthesis, polymers are formed by two-stage process steps based on the formation of an emulsion by a dilution process that allows the droplets of the shell material to accumulate around the droplets of the core material, which results in the formation of nanocapsules. This method involves the use of a partially water-soluble solvent previously saturated with water in order to achieve an initial thermodynamic equilibrium between the two liquids (i.e., water and solvent) (Piñón-Segundo et al., 2018).

The polymer material is dissolved in the organic solvent and then emulsified in an aqueous solution with a stabilizing agent under vigorous mechanical stirring. The smaller the size of the emulsion droplets formed, the smaller the size of the nanocapsules to be obtained (see Fig. 3.10). Adding water to the environment together with mechanical stirring provides solvent diffusion, allowing movement to the external phase and accumulation of nanocapsules at the interface.

The organic solvent in the environment can be removed using distillation or ultrafiltration techniques. However, the size range cannot be controlled in this method and this constitutes the most prominent disadvantage of the method. The emulsification-diffusion method has been developed by using the emulsification stage in synthesis in order to overcome this disadvantage. Then rapid solvent diffusion occurs, which transforms the droplets into particles. The emulsification-diffusion technique, patented in 1999 by Quintanar-Guerrero et al., is used to synthesize nanocapsules based on preformed polymers (Fig. 3.10) (Puel et al., 2005).

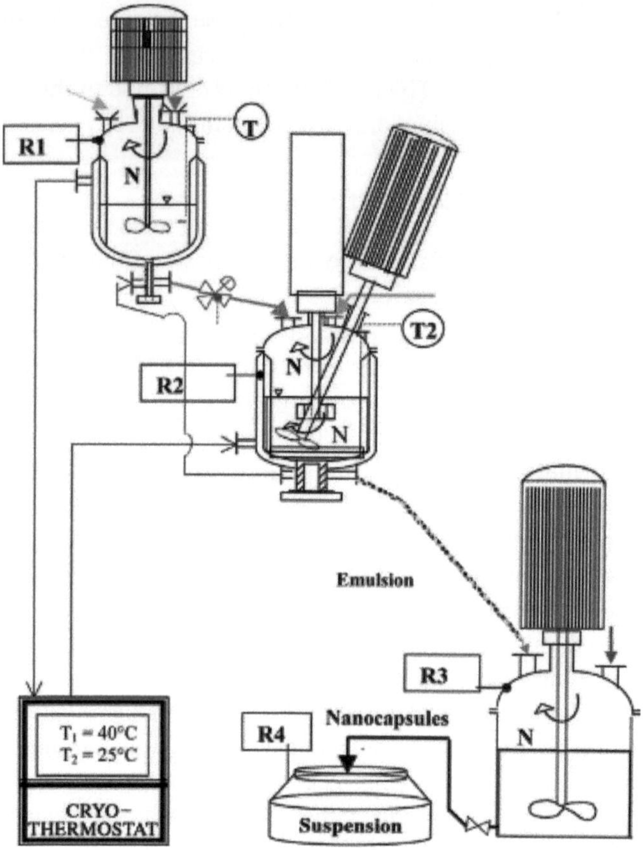

Figure 3.10 A sample setup schema for emulsification-diffusion (Puel et al., 2005).

3.2.9 Molecular Encapsulation/Inclusion Complexation

Molecular encapsulation can be defined as the confinement/entrapping of a biomolecule in a room within another molecular structure. The supramolecules that emerged in the last five decades being used as a cage system constitute the basis of this method of microencapsulation (Fig. 3.11). The molecular capsulation method, which is also called the host–guest complexation or inclusion complexation method, has wide use in many industries, especially in the food and pharmaceutical industry. It is accepted that the beginning of molecular encapsulation started with the discovery of cyclodextrins and crown ethers. In general, cyclodextrin units (six, seven, and eight glucopyranose units), which are carbohydrate units with a cyclic structure, have hydrophilic exterior and hydrophobic interior. Therefore, compounds in apolar structure can be trapped inside the cyclodextrin structure. However, no tight encapsulation is possible in molecular capsulation as the biomolecular agent placed in the lattice molecule cannot be covered completely (Fujita, 2014).

The most important advantage of the method is that it does not contain any toxic material. Cyclodextrins (cyclicoligasaccharides) come first among the supramolecules used in molecular encapsulation. Cyclodextrins used in encapsulation are hydrophobic in character and provide molecular encapsulation of hydrophobic bioagents (Szejtli, 1998). Compounds known as crown ethers are cyclic oligomers of ethylene oxide and are involved in the molecular encapsulation of cation structures by forming complexes. Apart from these structures, cryptanes, cryptophanes, calycarenes, carbohydrates, and cucurbituriles have also been used in the molecular encapsulation of biological agents, thanks to their cage-like molecular structure (Neta and Evans, 1981; Rebek Jr., 2000; Warmuth and Yoon, 2001). The encapsulation area created during encapsulation was only sufficient for organic and inorganic compounds smaller than 1 nm in initial studies. There was no way that larger bioagents could be encapsulated. This disadvantage was not easy to overcome. New techniques had to be developed in order to encapsulate large molecules. This difficulty could be overcome with a practical solution. In the new technique, the lattice structure is expanded with multi-component joining. Thus, less synthetic reaction steps were applied and larger caged molecules were formed. In a study, a large lattice structure was formed with dimeric structures using coordination bonds with Pd-pyridine, and thus they achieved encapsulation into the large lattice structure (Kawase and Kurata, 2006; Yoshizawa et al., 2009).

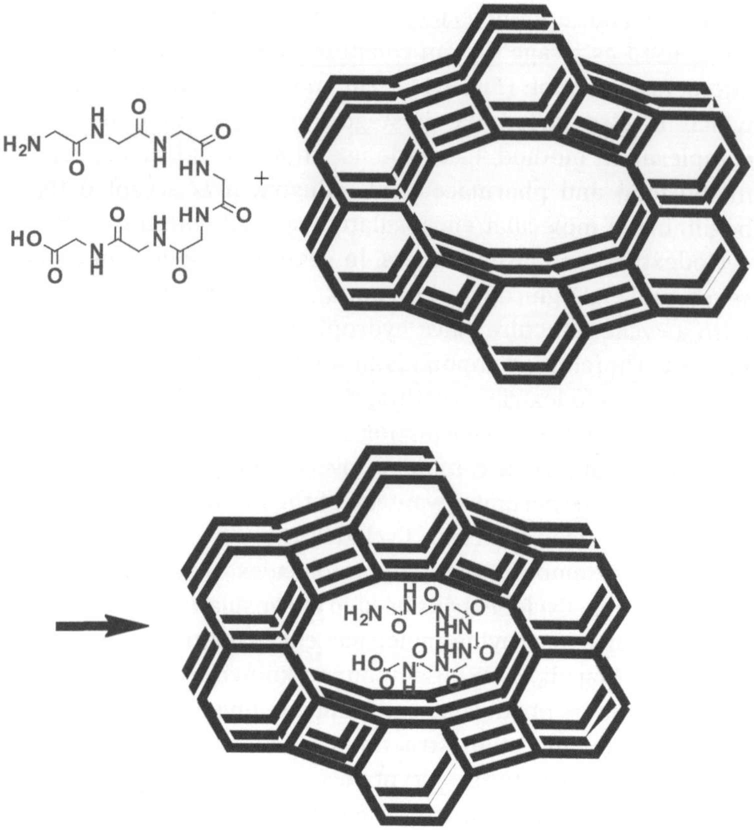

Figure 3.11 An example of molecular encapsulation formation reaction.

Phenolic compounds found in fruits and plants can be encapsulated in cyclodextrin structures. Gallic acid, caffeic acid, p-coumaric acid, ferulic acid, hydroxycinnamic acid, and elaidic acid, which comprise the phenolic compounds and have antioxidative activity, were encapsulated using the molecular encapsulation method in some studies (Aytac et al., 2016; Kfoury et al., 2016; Kurkov and Loftsson, 2013). In addition, essential oils (thymol, geraniol, aneholethole, linalool, citral, menthol, citronellal, etc.) found in some plant and fruit (orange, lemon, garlic, onion, mustard, bergamot, etc.) structures are also encapsulated in cyclodextrin structures and have gained structural stability (Marquis et al., 2015).

3.3 Physical Methods

In the encapsulation technology, there exist physical methods, in which the constituents that form the shell of the micro-/nanocapsules do not chemically interact with each other or with the core material, and retain their initial chemical composition throughout the encapsulation process. The shell formation is achieved as a result of physical processes, such as binding as a result of drying, or adhesion. Pure mechanical methods, like pan coating, also fall under physical methods (Alva et al., 2017).

Among several physical methods, spray drying is the most widely applied method. Another common drying method for the preparation of microencapsulation is freeze-sublimation drying (Mahdavi et al., 2014).

3.3.1 Spray Drying

Spray drying is a fully developed method suitable for the encapsulation of volatile, sensitive, and functional constituents. Sensitive core materials can be successfully stabilized in a shell having forming a closed matrix by transforming them from a liquid medium into a dry powder. Spray drying is a unit process that involves the continuous conversion of the feed from a liquid state to a dried microparticle by spraying the synthesized capsule material into a hot drying medium (Reineccius, 2004; Fang and Bhandari, 2010). The capsule core is equally distributed within the shell matrix; therefore, the resulting capsules are also called matrix capsules.

With spray drying, it is also possible to protect the core content from interacting with other substances as well as from hostile ambient conditions, by making use of a carbohydrate-based matrix. The most crucial problem that needs to be tackled can be said to be the preservation of the physical properties and physical stability of emulsion during and after drying. Another advantage of spray drying is that it is a low-cost method (1/30 cheaper than freeze drying) (Drusch and Diekmann, 2015).

The basis of the method can be briefly stated as follows: After the bioactive material is dissolved in an aqueous dispersing medium, it is atomized with a high-pressure nozzle or centrifuge wheel and dried by spraying on the hot air stream. The resulting droplets are dehydrated in the heated chamber of the spray dryer and transformed into capsules of 10 to 200 μm in diameter and then collected from the lower chamber of the dryer (Fig. 3.12). The procedure can be summarized in three basic steps as follows (Bonaui et al., 1996; Cal and Sollohub, 2010):

1. Atomizing a liquid feed into fine droplets;
2. Mixing these spray droplets with a heated gas stream, which enables the liquid to evaporate and the dried solids separate;
3. Separation and collection of the dried powder from the gas stream.

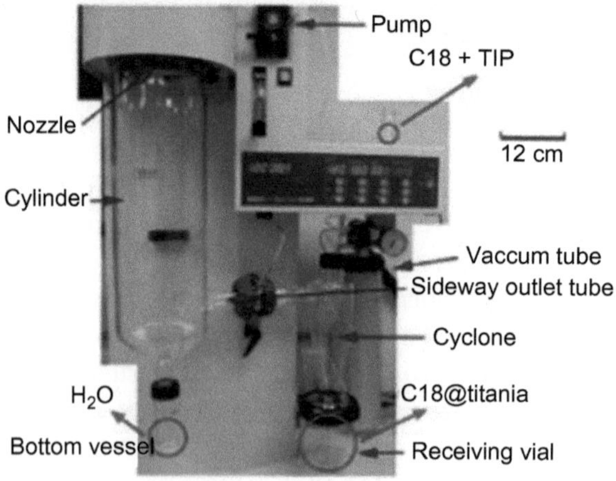

Figure 3.12 A spray drying equipment (Alva et al., 2017).

Spray-dried microcapsules have important properties such as uniform particle size, almost uniform spherical particle shape, improved fluidity, particle compressibility, reduction in mass density, increased solubility, reduced moisture content, and increased thermal stability (Jafari et al., 2008; Kandansamy and Somasundaram, 2012; Reineccius, 2004).

Approximately 90% of the microcapsules in the food industry are produced using this method. Hydrolytic carbohydrates and proteins (Gum Arabic, octyl derivatives, and modified starch) and starch hydrolysates, milk, casein, and soy proteins are used in the encapsulation applied in the food industry (Zhao and Tang, 2016).

When making the wall formulation of encapsulated components, the use of other colloidal compounds is often limited due to that they restrict solubility and viscosity. Instead of these, lactose and maltodextrins with small molecular weight are used. Although high-molecular-weight hydrocolloids are generally hydrophilic as wall materials, the bioactive compound in the core material can be hydrophilic or hydrophobic. Hydrophobic molecules are mostly emulsified with surfactants and then spray dried (Aghbashlo et al., 2013).

The encapsulation of foodstuff, such as fructose syrup and glucose, with spray drying may result in a sticky mass as the bioactive material in the capsule core, sensitive to heat, may degrade upon exposure to high temperature. Another disadvantage is that the capsules clump together in an uneven size and shape during spray drying (Ghorbanzade et al., 2017).

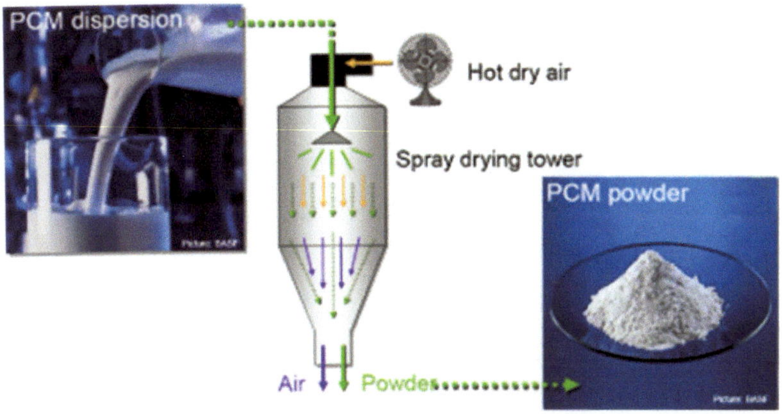

Figure 3.13 Spray drying process applied by BASF (Delgado et al., 2012).

In this method, a polyanion emulsion solution such as gelatin is mixed at a speed of 10,000 rpm and added dropwise into a polycation solution such as Gum Arabic. The emulsion solution obtained is delivered to a sprayer. The moisture in the droplets coming into the drying chamber by being dragged with hot air or nitrogen is dried. Thus, a shell is formed around the core. The retrieval of dried capsules is done in cyclones, filter bags, or electrostatic precipitate bags. With the appropriate design of the atomizer in the process, it is possible to increase and decrease the capsule size. Figures 3.12 and

3.13 show an example spray drying equipment. In a study conducted by Hawaldar et al., microcapsules of gelatin and Gum Arabic were obtained by the paraffin-wax spraying technique. The resulting capsules were reported to be spherical with smooth surfaces and have uniform size distribution (Alva et al., 2017).

3.3.2 Spray Chilling/Spray Congealing

The spray chilling method is similar to the spray drying method. In this technique, instead of applying heat to remove the solvent of the reaction medium, the carrier sprayed into a cooling chamber is solidified by drawing heat. The device for spray chilling, therefore, consists of two compartments: a cooling chamber and an atomizer—a rotating wheel, a high-pressure nozzle, a fountain nozzle, a vibrating nozzle, and a pair of liquid or rotating disk atomizer. A schematic diagram of the process is shown in Fig. 3.14.

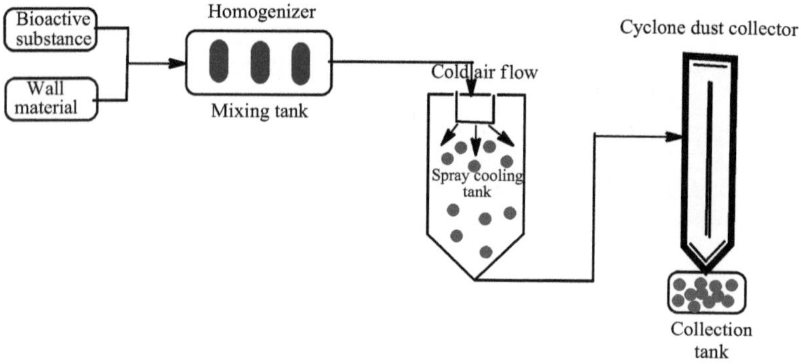

Figure 3.14 A scheme of the encapsulation process with spray chilling method.

In the spray chilling method, a stock solution, consisting of emulsion solutions, drug solutions prepared in a carrier fluid, and suspended solutions, is sprayed into the chilling chamber. A hot mixture of core and shell materials is atomized into the cooled/cold ambience, and the shell material is solidified around the core, covering it all around, hence forming micro-/nanocapsules. The cooling of the ambience is provided by liquid nitrogen, by which the carrier material sprayed into the environment solidifies, and microspheres are formed.

With spray chilling, capsules varying in size from a few microns to millimeters can be obtained by the congealing of the liquid sprayed into the environment with simultaneous or counter-current cold air flow. In this method, after the bioactive component is added to an oil carrier, dispersing the suspended solid particles or the emulsion particles, the particles are encapsulated by spraying. The spray cooling method can be solvent-free encapsulation during the encapsulation of bioactive substances sensitive to water (Lakkis, 2016).

Three steps are generally followed in the encapsulation procedure of the spray cooling method: the dispersion of the bioactive component to be coated in the matrix material, the atomization of the resulting mixture, and the solidification of the matrix material by spray chilling (Garti and McClements, 2012).

In order for the process steps to be successful, the liquid content should be well atomized and solidified. In the atomization step, the content is fragmented into spherical droplets in the form of a spray, and the process is completed with capsule formation as a result of the cooling. Cordeiro et al. (2013) outlined three rules that define the conditions for spray chilling encapsulation:

1. The temperature of the melt must be 10°C above the melting temperature of the carrier.
2. The feed viscosity is recommended to be less than 500 mPas. Higher viscosities may be possible, but only with proper nozzle design.
3. Loads for microencapsulation are generally limited to 50%.

The most important advantages of spray chilling are the absence of high heat application, low operating cost, and solvent-free operation. Whereas the main disadvantages of the method are that it is the most expensive encapsulation method and that the resulting capsules show a weak barrier to oxygen permeability (Matos-Jr. et al., 2017).

Another problem experienced during the encapsulation phase is insufficient cooling of the droplets. If the process is insufficiently cooled, the morphological structure of the capsules deteriorates, and it also causes clumping on the wall surface of the cooling chamber. The process design should be adjusted to eliminate these disadvantages. This can be achieved by optimizing the particle size, particle size distribution, specific heat of the synthesized capsule, heating and delivery rate of the cooling system, and optimization of the time required for solidification to occur.

The ideal coating material used in the spray cooling method:

1. Should have good rheological properties at high concentration and can be easily processed during the encapsulation process.
2. Must have emulsion and dispersion properties and must have high emulsion stability.
3. Should not react in a way that impairs the property of the core material during the coating process and/or during storage.
4. Should be able to cover the core material and protect it stably during both processing and storage (Varhan and Koç, 2018).

Vegetable oils and their derivatives are used as coating materials. Fats generally have a melting point in the range of 32–42°C and can be either functional or hydrogenated. The capsules intended to be prepared have acquired a hydrophobic character and the material inside can be hydrophilic with water solubility.

The spray chilling method is widely used in the encapsulation of mineral substances, water-soluble vitamins, enzymes, vitamins, flavor components as food ingredients or additives (Garti and McClements, 2012; Rokka and Rantamäki, 2010). It has been reported that these microcapsules synthesized in the literature have been used for many different purposes. These include masking of ambient odors, increasing the resistance to environmental conditions (pH, temperature, humidity, oxygen, light, etc.) by using the enzyme as filling. The spray chilling method has been used successfully in the encapsulation of ginger, vitamin C, ascorbic acid, vitamin D3, vitamin A, gallic acid, and lycopene (Consoli et al., 2016; Matos-Jr. et al., 2017; Matos-Jr. et al., 2015; Oriani et al., 2016; Paucar et al., 2016; Pelissari et al., 2016; Sartori et al., 2015; Wegmüller et al., 2006).

Capsules produced by using the spray chilling method with the determination of the appropriate polymer material can be used in drug-delivery systems by providing diffusion and erosion control or can be used successfully in changing the dissolution behavior of the drug. In addition, it is possible to successfully encapsulate the active compounds that are desired to be carried in the cosmetic industry (Mishra, 2015; Favaro-Trindade et al., 2015).

3.3.3 Freeze Drying

Freeze drying is a simple method and is very convenient provided that a correct carrier shell material is selected for certain ingredients, such as drug, enzyme, protein, etc. However, it is costly. Freeze drying costs about 30 times more than spray drying. In addition, capsules produced by freeze drying can provide relatively weak protection to the core content, since the wall has a structure with too many pores, hence a greater oxygen or moisture permeability. Therefore, it provides weak protection in the prolonged release of the bioactive component in the core structure (Pikal et al., 2002).

In the freeze drying process, sublimation, also known as lyophilization, is applied and covered (Ezhilarasi et al., 2013). Freeze drying encapsulation is the most commonly used method of encapsulation for heat-sensitive food and pharmaceutical bioactive ingredients. The basic principle of forming capsules by the freeze drying method is based on the lyophilization process after the core bioactive component is homogenized in the reaction medium. In the lyophilization process, after being frozen at a temperature (between −90°C and −40°C) selected according to the homogenized capsule sample, it is sublimated at a lower temperature. Freeze capsulation allows the bioactive substance, especially when it is an enzyme, to not lose its activity by preserving its molecular structure, as well as enabling the raw material to keep color, smell, and taste (Young et al., 1993).

In freeze drying encapsulation, adding a small amount of a cryoprotectant (e.g., disaccharides such as sorbitol, trehalose, maltodextrins) to the solution together with other wall materials to stabilize the freeze-sensitive bioagents can eliminate inconveniences. The most commonly used wall materials are Gum Arabic, maltodextrins, modified starches or other polysaccharides, and whey proteins (Shahidi and Han, 1993).

3.3.4 Fluid Bed Coating

The encapsulation by fluid bed coating is based on the principle of forming capsules in the form of layers by spraying the coating material through the spray nozzle onto the substances to be coated in a fluid bed. Core particles suspending in a blown stream of air get covered with a spray of atomized liquid shell materials in the thermally controlled bed (Mishra, 2015). The fluid bed coating method can be applied in three different ways: top, bottom, and tangential spray (Desai and Park, 2005). First developed by D. E. Wurster in the 1950s, the method has been used in the pharmaceutical industry since then for coating of drugs with a film layer (Fig. 3.15) (Hall and Pondell, 1980). Compared to the spray drying method, it is much more expensive and, therefore, it is not a widely preferred method. However, since the film quality varies depending on processing conditions, it started to be used in recent years as a good alternative for coating foodstuff and pharmaceuticals, as it provides a high level of film homogeneity (Koç et al., 2010).

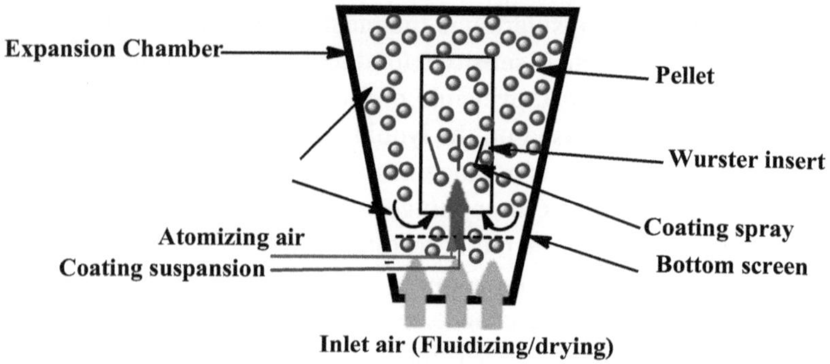

Figure 3.15 Schematic diagram of the Wurster coating process in the fluidized bed chamber.

The method requires spherical and dense particles that are almost identical in size and have good flowability. The core particles are relatively large and are around 100 μm in size. The sizes of the capsules, however, can be altered by making some changes in the process conditions that govern during encapsulation (temperature, atomization air flow, solution concentration, etc.). The shell thickness and hence the weight of capsules, on the other hand, can be adjusted by cycling through the stages of suspending, spraying, and cooling over and over again. In this regard, compared to pan coating, fluid bed coating offers greater control and flexibility on the encapsulation process (Khandbahale, 2020).

Coating materials need to be capable of forming a layer over the surface of the core material. Generally, a thermally stable polymer material having a suitable viscosity for pumping and atomization is used as the coating material. The most used materials are cellulosic esters, acrylic polymers, and copolymers. Common aspects of the polymers used are their good solubility, viscosity, and impermeability properties. Coating can be done by using multiple polymers in combination to provide these properties (Werner et al., 2007).

Using this method, B vitamin derivatives, a mixture of various vitamins and minerals, KCl, iron sulfate, and iron fumarate, were encapsulated (Havaić et al., 2018). In addition, various food acids, encapsulated using this method in order to improve the color and flavor, have been used in the meat industry (DeZarn, 1995).

As a result, fluid bed coating has been widely used to encapsulate micronutrients (some vitamins and minerals), ingredients with antimicrobial properties (sorbates, propionates, lactic acid, etc.), processing aids (acids, leavening agents), and salts. It has also been applied to encapsulate some plant extracts, essential oils of herbs, and carotenoids. Capsules obtained with this method were reported to have improved physicochemical properties and increased stability in the food industry as well as in cosmetic and pharmacological applications (Benelli and Oliveira, 2016; Coronel-Aguilera and San Martín-González, 2015).

3.3.5 Air Suspension Coating

In the coating technique known as spray coating or fluid bed processing, it is carried out by applying heating or cooling of the core material with an upward moving air flow.

3.3.6 Melt Extrusion

In the extrusion method, a coating material covers a core material after a series of dies into dehydrating liquid bath, whereby the encapsulation of aromatic components is achieved. The shell formation occurs as a result of molten carbohydrate, which entraps the core material, hardening on coming into contact with liquid. In general, the extrusion process is carried out under 100 psi pressure and at about 115°C. When the coating material comes into contact with the liquids in the reaction medium, it hardens, covering the core material, thus forming the encapsulation. It can then be separated by precipitating it to the bottom of the solution, followed by washing and drying (Mishra, 2015; Shahidi and Han, 1993).

3.3.7 Annular Jet (Vibrational Nozzle)

The annular jet technology is based on coating the core with the shell material using two concentric jets. The interior of the jets contains the core material, and the exterior generally contains the shell material that solidifies as it exits the jet. Core–shell encapsulation is accomplished by applying additional vibration as it passes through the nozzle. With the application of vibration, it is also possible to form submicron droplets. Thus, the capsule size is kept under control. Simultaneously, the core material and the shell material can be cured by applying vibration, cooling, crosslinking, or evaporation processes, after the flow is ensured by applying vibration from the nozzles. By using this method and optimizing the system, it is ensured that more uniform and smaller encapsulations are formed (Mishra, 2015).

3.3.8 Entrapment in Liposome Vesicles

Liposomes are phospholipid capsules formed in a laboratory environment that can only be seen through an electron microscope. They can be considered sacs with a kind of a lipid structure (Fig. 3.16). The liposome structures are self-assembled spherical vesicles (25 nm to 1 μm in diameter) with one or more phospholipid bilayers that separate the inner aqueous environment (liquid core) from the external aqueous environment. Liposomes have begun to be used in cosmetics as well as in pharmacology. Especially skincare products make use of collagen and elastin loaded on liposomes because liposomes can cross the cell membrane and carry any desired substance to the lowest layer of the skin. In the cosmetic industry, these capsules are often loaded with collagen and elastin as the main ingredients, as well as various herbal extracts and vitamins. A wide variety of bioactive compounds can be encapsulated through liposome entrapment, due to the amphiphilicity of the phospholipid encapsulation material (da Silva Malheiros et al., 2010; Singh et al., 2012; Zuidam and Shimoni, 2010).

Nano-sized liposomes, also known as nano-liposomes or nanosomes, can easily penetrate the targeted cells thanks to being one millionth of a millimeter in size and for their outer coatings made of phospholipids being identical to the cell membrane. Nanoliposomes act at the cellular level and, therefore, can yield results much faster and more powerfully than conventional products. Active substances placed in nanosomes can reach the lower layers of the skin and target hair follicle cells very quickly, and can be much more effective than liposomes and uncoated molecules. Some other uses of nanoliposomes include the placement of oxidized vitamin K into nanosomes in some haircare products and removal of bruises after aesthetic surgery (Hallaj-Nezhadi and Hassan, 2013).

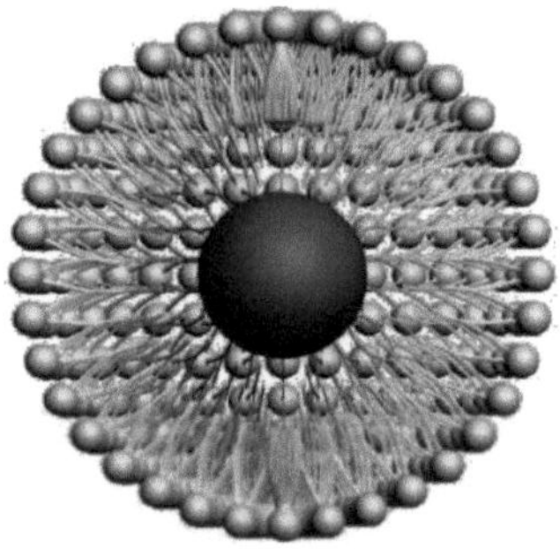

Figure 3.16 Liposome vesicle structures.

Liposome vesicles can be synthesized in different sizes as having a single-layered or multilayered structure. The resulting liposomes can be found in four categories:

1. Small unilamellar vesicles
2. Large unilamellar vesicles
3. Multilayered vesicles
4. Multiple vesicular vesicles

Among these, "large unilamellar vesicles" are considered to be the most suitable liposomes for the food and cosmetic industries due to their high yield and stability as well as their simple production methods (Fig. 3.16) (Sharma and Sharma, 1997).

Reverse phase evaporation, thin film drying, and membrane extrusion were used in liposome production by traditional methods. Recently, green synthesis methods have been developed to obtain liposomes using less organic solvent such as microfluidics, reverse phase evaporation, and supercritical reverse phase evaporation (Tsai and Rizvi, 2016).

There are some studies in the food industry to create liposomes using some vitamins, minerals, enzymes, proteins, antimicrobial substances, or food additives (Mozafari et al., 2008). In addition, vitamins E and C have been encapsulated and made more resistant to heat treatments (Marsanasco et al., 2011). Recent research has also focused on the inclusion of encapsulated forms of bioactives in real food matrices. In a study conducted using yogurt, it was aimed to mask the smell of fish oil by forming nanoliposomes of fish oil, and successful results were obtained (Ghorbanzade et al., 2017).

3.3.9 Electrostatic Precipitation

Electrostatic precipitation (ESP) is based on the principle of positively or negatively charging the particles to remove them from the gas stream using electrical energy. Then, the charged particles are attracted by the oppositely charged collector plates.

3.3.10 Electrohydrodynamic Processes

The basis of electrohydrodynamic process is the implementation of an electrostatic force either to spin a polymer solution into ultrafine fibers (electrospinning) or to spray small droplets (electrospraying). However, in order to obtain fibers or particles at certain sizes, electrostatic process conditions (applied voltage, flow rate, and distance between spinneret (including needle) and collector, solution properties (viscosity, concentration, electrical conductivity, and surface tension) and environment parameters (factors such as humidity, temperature, and airflow) need to be controlled (Jacobsen et al., 2018; Reneker et al., 2000).

In the electrospinning method, the prepared polymer solution is transformed into micron/nano-meter sized beads by applying a high potential field. During the formation of the encapsulation by the electrospinning method, gravity force, viscosity, and surface tension force prevent drag during the process. With this method, through a cone at the tip of the spinneret, a spiral-shaped jet is formed and agglutination is eliminated by drying the capsules under electrostatic forces applied during its flight from the cone to the collector and before it reaches the collector (Fig. 3.17). The most important advantage of the electrospinning method is that it provides the encapsulation of heat-sensitive compounds under mild conditions. For this reason, the solvents used in encapsulation applications by the electrospinning method are generally water and acetic acid; whereas preferred wall materials for the capsule are gelatin, zein, and whey (Seethu et al., 2020).

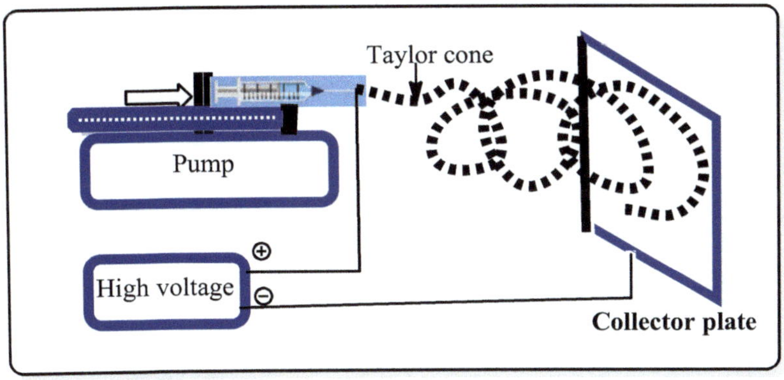

Figure 3.17 Scheme of the encapsulation process with the electrospinning method.

In order to increase oxidative stability, extend shelf life, and provide enhanced antioxidant activities, many bioactive compounds such as orange essential oil, cinnamaldehyde, flaxseed, cheese fish oil, dextran, six-juice protein concentrate, chitosan, cinnamaldehyde, linseed oil, curcumin, gallic acid, and fish oil were encapsulated by electrospraying (Cerqueira et al., 2016; Dhandayuthapani et al., 2012; Hadad and Goli, 2019; Jayan et al., 2019; Neo et al., 2013; Pérez-Masiá et al., 2015; Tavassoli-Kafrani et al., 2018). Resveratrol, used as a biopolymer in the food industry, was encapsulated in a smaller size at 100 nm by the electrospinning method. In addition, vitamin B could be encapsulated in a stable structure with the electrospinning method (Seethu et al., 2020). In general, micro-/nanostructured fiber and particle morphologies produced by the electrospinning method can offer adjustable release kinetics for encapsulated bioactive materials for various applications in the biomedical and food production sectors.

3.3.11 Hydrogel-Based Microgel Particles

Hydrogel-based microgels can be manufactured using natural or synthetically prepared polymers or combinations thereof. The main features of hydrogels are their formation of a highly hydrated insoluble structure by crosslinking polymer chains (Fig. 3.18). While water constitutes 98–99% of the polymer structure, it may contain 1% or 2% of a bio-based active polymer. Hydrogels can be dehydrated by removing the water in their structure, and then they can be swollen in water and regain their hydrated form. Soft contact lenses made from hydrogels are an example of common materials in use. Recently, hydrogel-based microgel structures have become a highly preferred material in the encapsulation of cells because hydrogels can be designed, they bear superior mechanical properties, they are bio-degradable, and they exhibit exceptional biological performance (Song and Li, 2018).

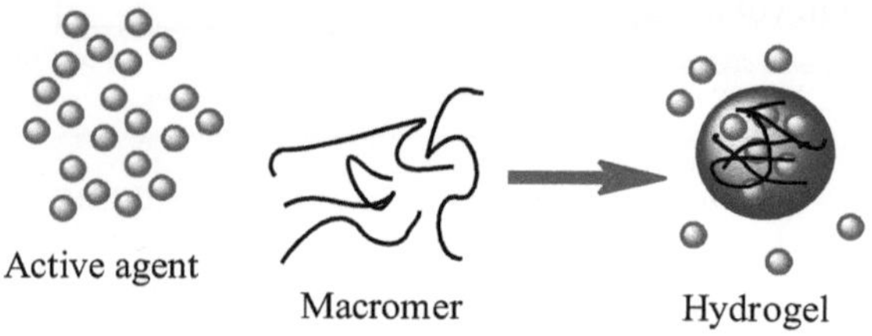

Figure 3.18 Hydrogel formation reaction.

Within the hydrogel structures, physical bonds (van der Waals bonds, dipole–dipole interactions, ionic bonds, hydrogen bonds, etc.) and covalent bonds (chemical crosslinking, dimerization with thiol groups, formation of disulfide bridges, etc.) can be formed. The most commonly used alginate-like polysaccharides for physical bonding are crosslinks formed by iodo-tropic gelation with Ca^{2+} (Augst et al., 2006). While the main advantage of hydrogel structures formed by the physical method is that they can be formed within the cell structure, the biggest disadvantage is that they are chemically and mechanically unstable due to their reversible bonding structures. For this reason, more studies have been conducted with hydrogel capsules formed by covalent crosslinks in a permanent and irreversible structure (Vermonden et al., 2012).

By using some biological molecules (drugs and peptides, proteins and oligonucleotides, etc.), artificial cells were synthesized using hydrogel-formed structures together with the cells. Biopolymeric microgels made from proteins and/or polysaccharides can be used with great advantages for encapsulation and controlled release applications in the pharmaceutical and food industries. In addition, these synthesized hydrogel structures can be protected from reactions related to the immune system (Nafea et al., 2011).

Hydrogel-based microgels are also produced for the purpose of stabilizing lipophilic molecules in order to enable their use in the food industry. With the work done, it has been determined that hydrogels and emulsion hydrogels can be easily used in the production of new functional foods because of their ease of design and control and their ability to be manufactured with food-grade components (Filippidi et al., 2014).

3.4 Physicochemical Methods

Physicochemical processes, as the name suggests, are hybrid processes that involve at least a physical method in combination with a chemical one, or vice versa. For instance, a physical process, say phase separation, is used in combination with a chemical process like condensation in order to achieve the microencapsulation of core material. Among the most commonly used physicochemical methods are complex coacervation and sol–gel methods (Alva et al., 2017; Al Shannaq and Farid, 2015).

3.4.1 Complex Coacervation/Phase Separation

The formation of tiny spherical droplets of assorted organic molecules (coacervates) is called coacervation, which can be achieved either by adding a desolvation agent for phase separation (simple coacervation) or by the complexation of two polymers carrying opposite charges (complex coacervation). Complex coacervation is carried out by the combination of physical method, phase separation, and the chemical method. The basis of the synthesis is that the oil phase is first prepared by dispersing it into a polymer solution (cationic solution) to form an emulsion, and then a polyanion (usually a substance like Gum Arabic) is added to the reaction medium while being mixed at medium speed. The pH of the medium is reduced by adding acid, whereby the phase separation is achieved. Thus, a polymer-rich phase is formed. The shell material (polyanion) deposits onto the core material upon the reduction in the pH of the medium and hence two polymers form a complex. The cationic polymer is not meant to be a part of the resulting capsule, but to trigger phase separation, which is achieved thanks to the immiscibility of the two polymers with each other. The immiscibility may be due to a phenomenon called polymer–polymer incompatibility, or due to opposite charges (Mishra, 2015).

Complex coacervation is an encapsulation technique in which two combined polymers are used as encapsulation agents. In this technique, a protein polymer structure is encapsulated with a polysaccharide derivative. The polymers used are oppositely charged polyelectrolytes (polycations and polyanions) (Al Shannaq and Farid, 2015). Electrostatic interactions occur between them. In addition, weak interactions, such as van der Waals forces and hydrophobic interactions, hydrogen bonds, disulfide bridges, and dipole–dipole interactions, thanks to the functional groups in their structures, also contribute to the formation of coacervates (Ach et al., 2015). Charged groups in resins that are present in the structure of polysaccharides, steric hindrance, and confessional changes resulting from changes in environmental conditions (pH, ionic strength, temperature, etc.) and structural differences also affect encapsulation by complex coacervation (Xiao et al., 2014).

The encapsulation process with complex coacervation begins with the dissolution of the polymer in an aqueous environment. Next, a lipophilic biomaterial is added to the aqueous solution and an emulsion is formed. The coacervate complexes formed are very effective in stabilizing emulsions. The formed complexes increase the layer thickness of the interfacial layers as they become adsorbed at the interfaces. Coacervate complexes formed turn into a viscose gel-like material by the differentiation of environmental conditions (such as pH, pI, and temperature). This optimization affects the efficiency of the encapsulation with complex coacervation. The resulting gel-like structure is cured and dried in the form of microcapsules using different methods, such as spray drying, fluidized bed drying, or freeze drying (Sánchez et al., 2016). Figure 3.19 shows a basic flow diagram of complex coacervation.

In encapsulations by this method, pectins, chitosan, Gum Arabic, and other gums (Bayés-García et al., 2010) were used as polysaccharide materials (i.e., shell material), whereas whey proteins, caseins, and soy proteins were used for the protein part (i.e., capsule core). The obtained capsules have been used in the synthesis of functional food products. In addition, the complex coacervation method was used in the encapsulation of essential oils, fish oils, palm oil, and lipophilic extracts (Gomez-Estaca et al., 2016;

Jun-xia et al., 2011; Moschakis et al., 2010; Rutz et al., 2017). In the encapsulation of antimicrobial components, such as some nisin-like compounds, and heat treatment sensitive components have also been successfully performed with this method (Calderón-Oliver et al., 2017). In addition to oils, this method has been applied in the encapsulation of some hydrophilic compounds such as vitamin C (Comunian et al., 2013). With complex coacervation, microcapsules having a size in the range of 10 to 250 μm can be obtained.

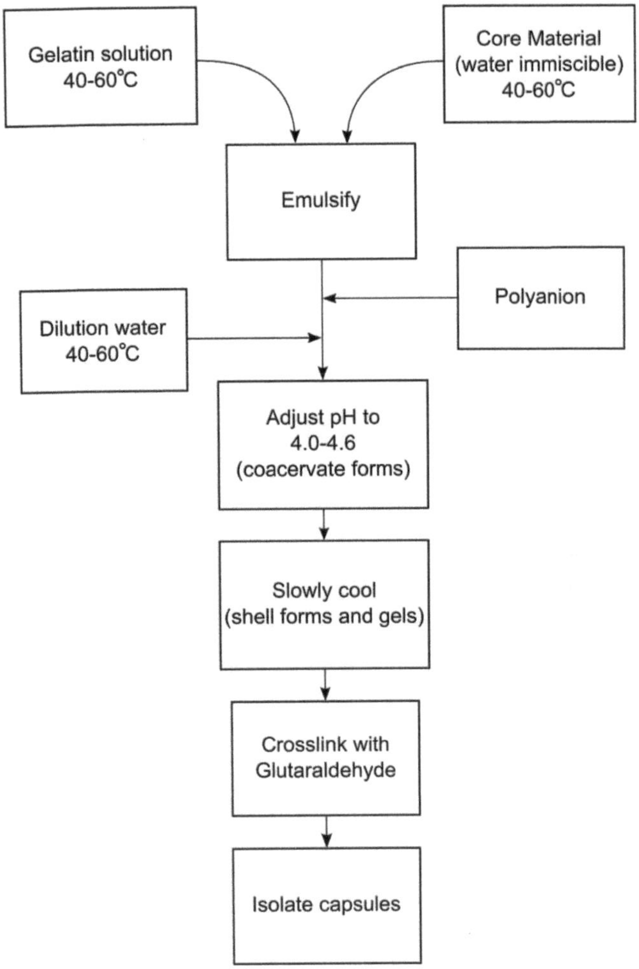

Figure 3.19 Flow diagram of the basic complex coacervation encapsulation process (Smith, 2009).

3.4.2 Sol–Gel Method

The sol–gel method requires milder conditions than other methods during nanocapsule preparation. The main processes of this method are as follows (Fig. 3.20): First, the metal alkoxide solvent, catalyst, and complexing agent are mixed homogeneously with the precursor material. A stable and transparent colloidal solution is obtained after hydrolysis and condensation reaction. In order to promote hydrolysis, the pH should be kept as low as 2–3 in the sol solution. Dropwise addition of the sol solution to an oil-in-water emulsion being stirred to prevent any continuous gelling will allow discrete silica gel walls around droplets to form, resulting in the encapsulation of the core material. Micro- or nano-sized encapsulated PCMs are obtained after drying, sintering, and vulcanization processes. The sol–gel method is a suitable method for encapsulating alkanes, palmitic acid, and indium with silicon dioxide shell materials (Erdoğan, 2017; Jamekhorshid et al., 2014; Alva et al., 2017; Macwan et al., 2011).

Nanocapsulated PCMs of SiO_2/palmitic acid (Tahan Latibari et al., 2013) and titania/stearic acid (Tahan Latibari et al., 2015) were produced by the sol–gel method. Hong et al. (2011) synthesized nanocapsules made of silica shell and indium using the sol–gel method.

3.5 Conclusion

Micro-/nano-encapsulation has been realized, with the formation of microcapsules having size between millimeter and micrometer, and nano-CAPs in the nano size. Basically, the process of enclosing the active ingredient in the capsule is defined as encapsulation. Encapsulation can be carried out by physical, chemical, and physicochemical methods. The capsules obtained are widely used in many different industries such as textile, detergent, pharmacology, automotive, etc. Capsules obtained by different methods can successfully protect their inner material from environmental negative effects. Then, the active material is released by effects such as melting, diffusion, tearing, or dissolution of the shell material. Micro-/nano-encapsulation encompasses both science and art. The encapsulation of active ingredients is successfully carried out with new methods that are developed or improved day by day and are offered to use in different industries.

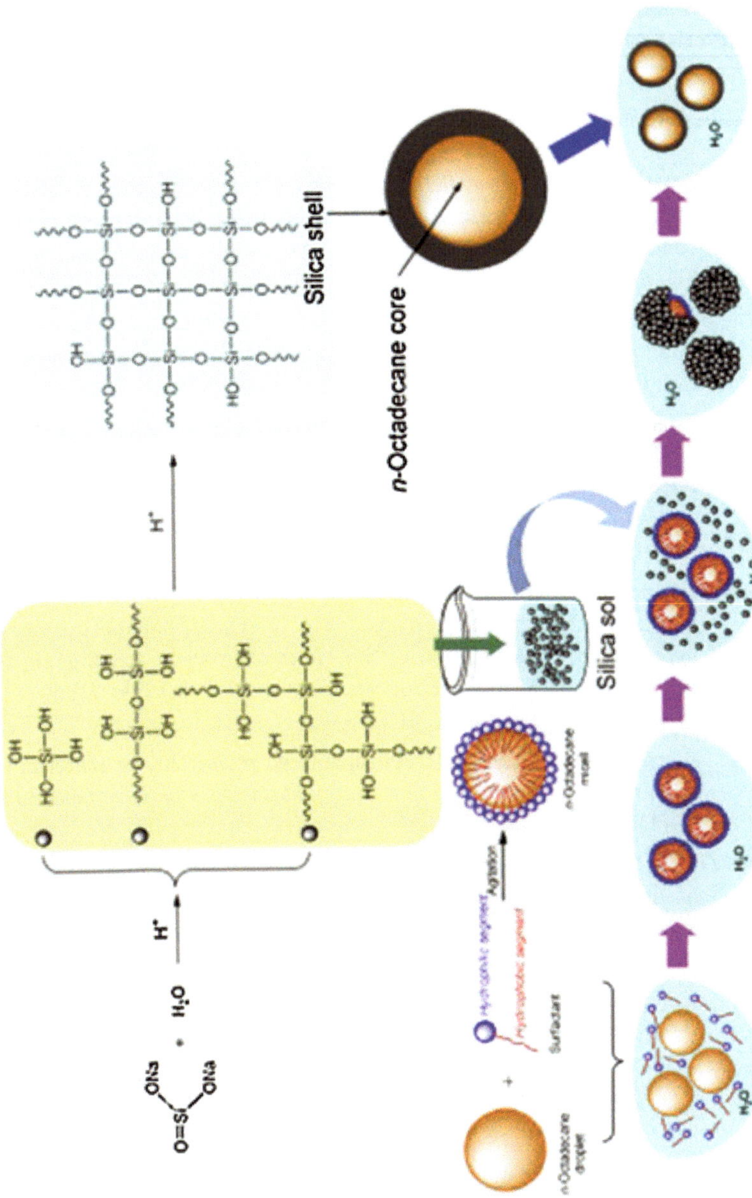

Figure 3.20 Sol–gel method (Alva et al., 2017).

In the scope of this chapter, it is aimed to compile the most common and applied methods used in micro-/nano-encapsulation. It has also been written to present technological approaches to scientists and industry working on micro-/nano-encapsulation.

References

Ach D, Briançon S, Dugas V, Pelletier J, Broze G, and Chevalier Y. Influence of main whey protein components on the mechanism of complex coacervation with Acacia gum. *Colloids and Surfaces A: Physicochemical and Engineering Aspects* 2015; **481**: 367–74.

Aghbashlo M, Mobli H, Madadlou A, and Rafiee S. Influence of wall material and inlet drying air temperature on the microencapsulation of fish oil by spray drying. *Food and Bioprocess Technology* 2013; **6**(6): 1561–9.

Al Shannaq R and Farid MM. Microencapsulation of phase change materials (PCMs) for thermal energy storage systems. In: Cabeza LF (ed), *Advances in Thermal Energy Storage Systems*. Woodhead Publishing series in energy. 66. Cambridge: Elsevier/Woodhead Publishing; 2015. p. 247–84.

Alexandridou S and Kiparissides C. Production of oil-containing polyterephthalamide microcapsules by interfacial polymerization. An experimental investigation of the effect of process variables on the microcapsule size distribution. *Microencapsulation* 1994; **11**: 603–14.

Alkan C, Sari A, Karaipekli A, and Uzun O. Preparation, characterization, and thermal properties of microencapsulated phase change material for thermal energy storage. *Solar Energy Materials and Solar Cells* 2009; **93**(1): 143–7.

Alva G, Lin Y, Liu L, and Fang G. Synthesis, characterization and applications of microencapsulated phase change materials in thermal energy storage: A review. *Energy and Buildings* 2017; **144**: 276–94.

Andrade B, Song Z, Li J, Zimmerman SC, Cheng J, Moore JS, et al. New frontiers for encapsulation in the chemical industry. *ACS Applied Materials & Interfaces* 2015; **7**(12): 6359–68.

Appelqvist I, Golding M, Vreeker R, and Zuidam NJ. Emulsions as delivery systems in foods. *Encapsulation and Controlled Release Technologies in Food Systems* 2007; **3**: 41–81.

Augst AD, Kong HJ, and Mooney DJ. Alginate hydrogels as biomaterials. *Macromolecular Bioscience* 2006; **6**(8): 623–33.

Aytac Z, Kusku SI, Durgun E, and Uyar T. Encapsulation of gallic acid/cyclodextrin inclusion complex in electrospun polylactic acid nanofibers: Release behavior and antioxidant activity of gallic acid. *Materials Science & Engineering C, Materials for Biological Applications* 2016; **63**: 231–9.

Bayés-García L, Ventolà L, Cordobilla R, Benages R, Calvet T, and Cuevas-Diarte MA. Phase change materials (PCM) microcapsules with different shell compositions. *Solar Energy Materials and Solar Cells* 2010; **94**(7): 1235–40.

Benelli L and Oliveira WP. System dynamics and product quality during fluidized bed agglomeration of phytochemical compositions. *Powder Technology* 2016; **300**: 2–13.

Bleiel SB, Kent RM, and Brodkorb A. Encapsulation efficiency and capacity of bioactive delivery systems. In: Roos Y. and Livney Y. (eds), *Engineering Foods for Bioactives Stability and Delivery. Food Engineering Series.* Springer, New York, NY 2017.

Boh B, Knez E, and Staresinic M. Microencapsulation of higher hydrocarbon phase change materials by in situ polymerization. *Journal of Microencapsulation* 2005; **22**(7): 715–35.

Bonaui C, Dumoulin E, Raoult-Wack A-L, Berk Z, Bimbenet JJ, Courtois F, et al. Food drying and dewatering. *Drying Technology* 1996; **14**: 2135–70.

Cal K, Sollohub K. Spray drying technique. I: Hardware and process parameters. *Journal of Pharmaceutical Sciences* 2010; **99**(2): 575–86.

Calderón-Oliver M, Pedroza-Islas R, Escalona-Buendía HB, Pedraza-Chaverri J, and Ponce-Alquicira E. Comparative study of the microencapsulation by complex coacervation of nisin in combination with an avocado antioxidant extract. *Food Hydrocolloids* 2017; **62**: 49–57.

Cao F and Yang B. Supercooling suppression of microencapsulated phase change materials by optimizing shell composition and structure. *Applied Energy* 2014; **113**: 1512–8.

Cerqueira MA, Fabra MJ, Castro-Mayorga JL, Bourbon AI, Pastrana LM, Vicente AA, et al. Use of electrospinning to develop antimicrobial biodegradable multilayer systems: Encapsulation of cinnamaldehyde and their physicochemical characterization. *Food and Bioprocess Technology* 2016; **9**(11): 1874–84.

Chen L, Wang T, Zhao Y, and Zhang X-R. Characterization of thermal and hydrodynamic properties for microencapsulated phase change slurry (MPCS). *Energy Conversion and Management* 2014; **79**: 317–33.

Chen Z-H, Yu F, Zeng X-R, and Zhang Z-G. Preparation, characterization and thermal properties of nanocapsules containing phase change material n-dodecanol by miniemulsion polymerization with polymerizable emulsifier. *Applied Energy* 2012; **91**(1): 7–12.

Comunian TA, Thomazini M, Alves AJ, de Matos Junior FE, de Carvalho Balieiro JC, and Favaro-Trindade CS. Microencapsulation of ascorbic acid by complex coacervation: Protection and controlled release. *Food Research International* 2013; **52**(1): 373–9.

Consoli L, Grimaldi R, Sartori T, Menegalli FC, and Hubinger MD. Gallic acid microparticles produced by spray chilling technique: Production and characterization. *Food Science and Technology* 2016; **65**: 79–87.

Constantin M., Bortolotti F., Cortesi R., Ascenzi P, and Menegatti E. Cellulose acetate butyrate -pH/ ... containing aminated poly (vinyl alcohol) microspheres for oral administration of DNA. *European Journal of Pharmaceutics and Biopharmaceutics* 2007; **66**: 11–20.

Cordeiro P, Temtem M, and Winters C. Spray congealing: Applications in the pharmaceutical industry. *Chemistry Today* 2013; **31**: 69–72.

Coronel-Aguilera CP and San Martín-González MF. Encapsulation of spray dried β-carotene emulsion by fluidized bed coating technology. *Food Science and Technology* 2015; **62**(1): 187–93.

da Silva Malheiros P, Daroit DJ, and Brandelli A. Food applications of liposome-encapsulated antimicrobial peptides. *Trends in Food Science and Technology* 2010; **21**(6): 284–92.

Delgado M, Lázaro A, Mazo J, and Zalba B. Review on phase change material emulsions and microencapsulated phase change material slurries. *Renewable and Sustainable Energy Reviews* 2012; **16**(1): 253–73.

Desai KGH and Park HJ. Recent developments in microencapsulation of food ingredients. *Drying Technology* 2005; **23**: 1361–94.

DeZarn TG. Food ingredients encapsulation: An overview. In: Risch SJ and Reineccius GA (eds), *Encapsulation and Controlled Release of Food Ingredients. ACS Symposium Series* 1995; Vol. 590, pp. 74–86.

Dhandayuthapani B, Anila M, Ravindran Girija A, Yutaka N, Venugopal K, Yasuhiko Y, et al. Hybrid fluorescent curcumin loaded zein electrospun nanofibrous scaffold for biomedical applications. *Biomedical Materials* 2012; **7**(4): 45001.

Drusch S and Diekmann S. Microencapsulation by spray drying. In *Handbook of Encapsulation and Controlled Release*. CRC Press 2015; p. 35–46. https://doi.org/10.1201/b19038-6

Duan B. Microencapsulation via in situ polymerization. In: Mishra M (ed), *Handbook of Encapsulation and Controlled Release. CRC Press* 2015; pp. 307–314.

El Bahri Z and Taverdet JL. Elaboration and characterisation of microparticles loaded by pesticide model. *Powder Technology* 2007; **172**(1): 30–40.

Erdoğan T. Isı Transfer Uygulamaları İçin Nanokapsüllenmiş Faz Değiştiren Madde İçeren Süspansiyonların Hazırlanması ve Karakterizasyonu. Çankırı, 2017. 78 p.

Ezhilarasi PN, Indrani D, Jena BS, and Anandharamakrishnan C. Freeze drying technique for microencapsulation of Garcinia fruit extract and its effect on bread quality. *Journal of Food Engineering* 2013; **117**(4): 513–20.

Fang G, Li H, Yang F, Liu X, and Wu S. Preparation and characterization of nano-encapsulated n-tetradecane as phase change material for thermal energy storage. *Chemical Engineering Journal* 2009; **153**(1–3): 217–21.

Fang X, Zhang Z, and Chen Z. Study on preparation of montmorillonite-based composite phase change materials and their applications in thermal storage building materials. *Energy Conversion and Management* 2008; **49**(4): 718–23.

Fang Y, Liu X, Liang X, Liu H, Gao X, and Zhang Z. Ultrasonic synthesis and characterization of polystyrene/n-dotriacontane composite nanoencapsulated phase change material for thermal energy storage. *Applied Energy* 2014; **132**: 551–6.

Fang Y, Yu H, Wan W, Gao X, and Zhang Z. Preparation and thermal performance of polystyrene/n-tetradecane composite nanoencapsulated cold energy storage phase change materials. *Energy Conversion and Management* 2013; **76**: 430–6.

Fang, Z. and Bhandari, B. Encapsulation of polyphenols: A review. *Trends in Food Science & Technology* 2010; **21**: 510–23.

Favaro-Trindade CS, Okuro PK, Eustáquio de Matos Jr, and Fernando. Encapsulation via spray chilling/cooling/congealing. In: Mishra M (ed), *Handbook of Encapsulation and Controlled Release*. CRC Press; 2015. pp. 71–87.

Filippidi E, Patel AR, Bouwens ECM, Voudouris P, and Velikov KP. All-natural oil-filled microcapsules from water-insoluble proteins. *Advanced Functional Materials* 2014; **24**(38): 5962–8.

Fujita D. Challenges to large molecular encapsulation. *Pure and Applied Chemistry* 2014; **86**(1): 3–11.

Garti N and McClements DJ. *Encapsulation Technologies and Delivery Systems for Food Ingredients and Nutraceuticals*. Oxford, Cambridge, Philadelphia, New Delhi: WP Woodhead Publishing, 2012. 612 p.

Ghorbanzade T, Jafari SM, Akhavan S, and Hadavi R. Nano-encapsulation of fish oil in nano-liposomes and its application in fortification of yogurt. *Food Chemistry* 2017; **216**: 146–52.

Ghosh SK. Functional coatings and microencapsulation: A general perspective. In: Ghosh SK (ed), *Functional Coatings by Polymer Microencapsulation*. WILEY-VCH Verlag GmbH & Co. KGaA 2006.

Gomez-Estaca J, Comunian TA, Montero P, Ferro-Furtado R, and Favaro-Trindade CS. Encapsulation of an astaxanthin-containing lipid extract from shrimp waste by complex coacervation using a novel gelatin–cashew gum complex. *Food Hydrocolloids* 2016; **61**: 155–62.

Gökmen S, Palamutoğlu R, and Sarıçoban C. Gıda endüstrisinde enkapsülasyon uygulamaları [Encapsulation Application in Food Industry]. *Electronic Journal of Food Technologies* 2012; **7**(1): 36–50.

Hadad S and Goli SAH. Improving oxidative stability of flaxseed oil by encapsulation in electrospun flaxseed mucilage nanofiber. *Food and Bioprocess Technology* 2019; **12**(5): 829–38.

Hall HS and Pondell RE. The Wurster process. In: *Controlled Release Technologies: Methods, Theory, and Applications*, 1980. p. 133–54.

Hallaj-Nezhadi S and Hassan M. Nanoliposome-based antibacterial drug delivery. *Drug Delivery* 2015; **22**(5): 581–9.

Havaić T, Đumbir A-M, Gretić M, Matijašić G, and Žižek K. Droplet impact phenomena in fluidized bed coating process with a Wurster insert. *International Journal of Chemical Engineering* 2018; **2018**: 1–11.

Hong Y, Wu W, Hu J, Zhang M, Voevodin AA, Chow L, et al. Controlling supercooling of encapsulated phase change nanoparticles for enhanced heat transfer. *Chemical Physics Letters* 2011; **504**(4–6): 180–4.

Jacobsen C, García-Moreno PJ, Mendes AC, Mateiu RV, and Chronakis IS. Use of electrohydrodynamic processing for encapsulation of sensitive bioactive compounds and applications in food. *Annual Review of Food Science and Technology* 2018; **9**(1): 525–49.

Jafari SM, Assadpoor E, Bhandari B, and He Y. Nano-particle encapsulation of fish oil by spray drying. *Food Research International* 2008; **41**(2): 172–83.

Jamekhorshid A, Sadrameli SM, and Farid M. A review of microencapsulation methods of phase change materials (PCMs) as a thermal energy storage (TES) medium. *Renewable and Sustainable Energy Reviews* 2014; **31**: 531–42.

Jayan H, Maria Leena M, Sivakama Sundari SK, Moses JA, and Anandharamakrishnan C. Improvement of bioavailability for resveratrol through encapsulation in zein using electrospraying technique. *Journal of Functional Foods* 2019; **57**: 417–24.

Jun-xia X, Hai-yan Y, and Jian Y. Microencapsulation of sweet orange oil by complex coacervation with soybean protein isolate/gum Arabic. *Food Chemistry* 2011; **125**(4): 1267–72.

Kandansamy K and Somasundaram PD. Microencapsulation of colors by spray drying: A review. *International Journal of Food Engineering* 2012; **8**(2).

Kawase T and Kurata H. Ball-, bowl-, and belt-shaped conjugated systems and their complexing abilities: Exploration of the concave-convex π–π interaction. *Chemical Reviews* 2006; **106**(12): 5250–73.

Kfoury M, Lounès-Hadj Sahraoui A, Bourdon N, Laruelle F, Fontaine J, Auezova L, et al. Solubility, photostability and antifungal activity of phenylpropanoids encapsulated in cyclodextrins. *Food Chemistry* 2016; **196**: 518–25.

Khandbahale SV. Microencapsulation: A novel approach in drug delivery: A Review. *Asian Journal of Research in Pharmaceutical Sciences* 2020; **10**(1): 39.

Koç M, Sakin M, and Kaymak-Ertekin F. Mikroenkapsülasyon ve gıda teknolojisinde kullanımı. Denizli: Pamukkale Üniversitesi Mühendislik Fakültesi, 2010. 11 sayfa.

Koehler. In-situ Polimerizasyon Yöntemi ile Mikrokapsülleme, https://www.koehlerinnovative.com/kis-en/microencapsulation/microencapsulation.php [accessed December 23, 2019].

Kurkov SV and Loftsson T. Cyclodextrins. *International Journal of Pharmaceutics* 2013; **453**(1): 167–80.

Lakkis JM. *Encapsulation and Controlled Release Technologies in Food Systems*, John Wiley & Sons, ISBN 9781118733523, 2016. pp. 116–7.

Liu C, Rao Z, Zhao J, Huo Y, and Li Y. Review on nanoencapsulated phase change materials: Preparation, characterization and heat transfer enhancement. *Nano Energy* 2015; **13**: 814–26.

Maa YF and Hsu C. Liquid–liquid emulsification by rotor-sator homogeneization. *Journal of Controlled Release* 1996; **38**: 219–28.

Macwan DP, Dave PN, and Chaturvedi S. A review on nano-TiO_2 sol–gel type syntheses and its applications. *Journal of Material Science* 2011; **46**(11): 3669–86.

Mahdavi SA, Jafari SM, Ghorbani M, and Assadpoor E. Spray-drying microencapsulation of anthocyanins by natural biopolymers: A review. *Drying Technology* 2014; **32**(5): 509–18.

Marquis M, Davy J, Cathala B, Fang A, and Renard D. Microfluidics assisted generation of innovative polysaccharide hydrogel microparticles. *Carbohydrate Polymer* 2015; **116**: 189–99.

Marsanasco M, Márquez AL, Wagner JR, del V. Alonso S, and Chiaramoni NS. Liposomes as vehicles for vitamins E and C: An alternative to fortify orange juice and offer vitamin C protection after heat treatment. *Food Research International* 2011; **44**(9): 3039–46.

Matos-Jr FE, Di Sabatino M, Passerini N, Favaro-Trindade CS, and Albertini B. Development and characterization of solid lipid microparticles loaded with ascorbic acid and produced by spray congealing. *Food Research International* 2015; **67**: 52–9.

Matos-Jr FE de, Comunian TA, Thomazini M, and Favaro-Trindade CS. Effect of feed preparation on the properties and stability of ascorbic acid microparticles produced by spray chilling. *LWT* 2017; **75**: 251–60.

Mishra MK. Overview of encapsulation and controlled release. In: Mishra M (ed), *Handbook of Encapsulation and Controlled Release*. CRC Press; 2015. pp. 3–19.

Moschakis T, Murray BS, and Biliaderis CG. Modifications in stability and structure of whey protein-coated o/w emulsions by interacting chitosan and gum Arabic mixed dispersions. *Food Hydrocolloids* 2010; **24**(1): 8–17.

Mozafari MR, Khosravi-Darani K, Borazan GG, Cui J, Pardakhty A, and Yurdugul S. Encapsulation of food ingredients using nanoliposome technology. *International Journal of Food Properties* 2008; **11**(4): 833–44.

Nafea EH, Marson A, Poole-Warren LA, and Martens PJ. Immunoisolating semi-permeable membranes for cell encapsulation: Focus on hydrogels. *Journal of Controlled Release* 2011; **154**(2): 110–22.

Neo YP, Ray S, Jin J, Gizdavic-Nikolaidis M, Nieuwoudt MK, Liu D, et al. Encapsulation of food grade antioxidant in natural biopolymer by electrospinning technique: A physicochemical study based on zein–gallic acid system. *Food Chemistry* 2013; **136**(2): 1013–21.

Neta P and Evans DH. Rates of conformational changes in the anion radicals of bianthrones. *Journal of the American Chemical Society* 1981; **103**(24): 7041–5.

Oriani VB, Alvim ID, Consoli L, Molina G, Pastore GM, and Hubinger MD. Solid lipid microparticles produced by spray chilling technique to deliver ginger oleoresin: Structure and compound retention. *Food Research International* 2016; 80: 41–9.

Park S, Lee Y, Kim YS, Lee HM, Kim JH, Cheong IW, et al. Magnetic nanoparticle-embedded PCM nanocapsules based on paraffin core and polyurea shell. *Colloids and Surfaces A: Physicochemical and Engineering Aspects* 2014; **450**: 46–51.

Paucar OC, Tulini FL, Thomazini M, Balieiro JCC, Pallone EMJA, and Favaro-Trindade CS. Production by spray chilling and characterization of solid lipid microparticles loaded with vitamin D_3. *Food and Bioproducts Processing* 2016; **100**: 344–50.

Pelissari JR, Souza VB, Pigoso AA, Tulini FL, Thomazini M, Rodrigues CEC, et al. Production of solid lipid microparticles loaded with lycopene by spray chilling: Structural characteristics of particles and lycopene stability. *Food and Bioproducts Processing* 2016; **98**: 86–94.

Pérez-Masiá R, Lagaron JM, and Lopez-Rubio A. Morphology and stability of edible lycopene-containing micro- and nanocapsules produced through electrospraying and spray drying. *Food and Bioprocess Technology* 2015; **8**(2): 459–70.

Pikal MJ, Rambhatla S, and Ramot R. The impact of freezing stage in lyophilization: Effect of the ice nucleation temperature on process design and product quality. *American Pharmaceutical Review* 2002; **5**(4): 48–53.

Piñón-Segundo E, Llera-Rojas VG, Leyva-Gómez G, Urbán-Morlán Z, Mendoza-Muñoz N, and Quintanar-Guerrero D. The emulsification-diffusion method to obtain polymeric nanoparticles. In: *Nanoscale Fabrication, Optimization, Scale-Up and Biological Aspects of Pharmaceutical Nanotechnology*. Elsevier; 2018. pp. 51–83.

Puel F, Briançon S, and Fessi H. Industrial technologies and scale-up. In: Benita S (ed), *Microencapsulation: Methods and Industrial Applications*, Second Edition. CRC Press; 2005. pp. 149–82.

Qiu Z, Li P, Wang Z, Zhao H, and Zhao X. PCM and PCM slurries and their application in solar systems. In: Zhao X and Ma X (ed), *Advanced Energy Efficiency Technologies for Solar Heating, Cooling and Power Generation*. Cham: Springer International Publishing; 2019. pp. 101–41.

Rebek Jr. J. Host–guest chemistry of calixarene capsules. *Chemical Communications* 2000 (8): 637–43.

Reineccius GA. The spray drying of food flavors. *Drying Technology* 2004; **22**: 1289–324.

Reneker DH, Yarin AL, Fong H, and Koombhongse S. Bending instability of electrically charged liquid jets of polymer solutions in electrospinning. *Journal of Applied Physics* 2000; **87**(9): 4531–47.

Rokka S and Rantamäki P. Protecting probiotic bacteria by microencapsulation: Challenges for industrial applications. *European Food Research and Technology* 2010; **231**(1): 1–12.

Rutz JK, Borges CD, Zambiazi RC, Crizel-Cardozo MM, Kuck LS, and Noreña CPZ. Microencapsulation of palm oil by complex coacervation for application in food systems. *Food Chemistry* 2017; **220**: 59–66.

Sánchez FM, García F, Calvo P, Bernalte MJ, and González-Gómez D. Optimization of broccoli microencapsulation process by complex coacervation using response surface methodology. *Innovative Food Science & Emerging Technologies* 2016; **34**: 243–9.

Sari A, Alkan C, Kahraman Döğüşcü D, and Biçer A. Micro/nano-encapsulated n-heptadecane with polystyrene shell for latent heat thermal energy storage. *Solar Energy Materials and Solar Cells* 2014; **126**: 42–50.

Sartori T, Consoli L, Hubinger MD, and Menegalli FC. Ascorbic acid microencapsulation by spray chilling: Production and characterization. *LWT - Food Science and Technology* 2015; **63**(1): 353–60.

Seethu BG, Pushpadass HA, Emerald FME, Nath BS, Naik NL, and Subramanian KS. Electrohydrodynamic encapsulation of resveratrol using food-grade nanofibres: Process optimization, characterization and fortification. *Food and Bioprocess Technology* 2020; **13**(2): 341–54.

Shahidi F and Han XQ. Encapsulation of food ingredients. *Critical Reviews in Food Science and Nutrition* 1993; **33**(6): 501–47.

Sharma A and Sharma US. Liposomes in drug delivery: Progress and limitations. *International Journal of Pharmaceutics* 1997; **154**(2): 123–40.

Singh H, Thompson A, Liu W, and Corredig M. Liposomes as food ingredients and nutraceutical delivery systems. In: Garti N and McClements DJ (eds), *Encapsulation Technologies and Delivery Systems for Food Ingredients and Nutraceuticals*. WoodHead Publishing Ltd., Oxford, U.K; 2012. pp. 287–318.

Smith MC. *Microencapsulated Phase Change Materials for Thermal Energy Storage: Development, Evaluation, and Application*. Auckland, 2009.

Song X and Li J. *Functional Hydrogels as Biomaterials* 2018; **12**: 141–59.

Srinivasaraonaik B, Singh LP, Tyagi I, Rawat A, and Sinha S. Microencapsulation of a eutectic PCM using in situ polymerization technique for thermal energy storage. *International Journal of Energy Research* 2020; **44**(5): 3854–64.

Szejtli J. Introduction and general overview of cyclodextrin chemistry. *Chemical Reviews* 1998; **98**(5): 1743–54.

Tahan Latibari S, Mehrali M, Mehrali M, Afifi ABM, Mahlia TMI, Akhiani AR, et al. Facile synthesis and thermal performances of stearic acid/titania core/shell nanocapsules by sol–gel method. *Energy* 2015; **85**: 635–44.

Tahan Latibari S, Mehrali M, Mehrali M, Indra Mahlia TM, and Cornelis Metselaar HS. Synthesis, characterization and thermal properties of nanoencapsulated phase change materials via sol–gel method. *Energy* 2013; **61**: 664–72.

Tavassoli-Kafrani E, Goli SAH, and Fathi M. Encapsulation of orange essential oil using cross-linked electrospun gelatin nanofibers. *Food and Bioprocess Technology* 2018; **11**(2): 427–34.

Thies C. A survey of microencapsulation processes. In: Benita S (ed), *Microencapsulation Methods and Industrial Applications.* Marcel Dekker, Inc. 1996: pp. 1–19.

Tsai W-C and Rizvi SSH. Liposomal microencapsulation using the conventional methods and novel supercritical fluid processes. *Trends in Food Science & Technology* 2016; **55**: 61–71.

Tumirah K, Hussein MZ, Zulkarnain Z, and Rafeadah R. Nano-encapsulated organic phase change material based on copolymer nanocomposites for thermal energy storage. *Energy* 2014; **66**: 881–90.

Varhan E and Koç M. Encapsulation of food materials with spray cooling method. *Food Health* 2018: 202–12.

Vermonden T, Censi R, and Hennink WE. Hydrogels for protein delivery. *Chemical Reviews* 2012; **112**(5): 2853–88.

Warmuth R and Yoon J. Recent highlights in hemicarcerand chemistry. *Accounts of Chemical Research* 2001; **34**(2): 95–105.

Wegmüller R, Zimmermann MB, Bühr VG, Windhab EJ, and Hurrell RF. Development, stability, and sensory testing of microcapsules containing iron, iodine, and vitamin a for use in food fortification. *Journal of Food Science* 2006; **71**(2): S181–S187.

Werner SRL, Jones JR, Paterson AHJ, Archer RH, and Pearce DL. Air-suspension coating in the food industry: Part II — micro-level process approach. *Powder Technology* 2007; **171**(1): 34–45.

Xiao Z, Liu W, Zhu G, Zhou R, and Niu Y. A review of the preparation and application of flavour and essential oils microcapsules based on complex coacervation technology. *Journal of the Science of Food and Agriculture* 2014; **94**(8): 1482–94.

Yadav SK, Suresh AK, and Khilar KC. Microencapsulation in polyurea shell by interfacial polycondensation. *AIChE Journal* 1990; **36**(3): 431–8.

Yeşilyurt MK, Nadaroğlu H, and Çomaklı Ö. Phase change materials and selection thereof for heat transfer applications. *International Journal of Innovative Research and Reviews* 2019; **3**(2): 16–22.

Yoshizawa M, Klosterman JK, and Fujita M. Functional molecular flasks: New properties and reactions within discrete, self-assembled hosts. *Angewandte Chemie International Edition* 2009; **48**(19): 3418–38.

Young SL, Salda X, and Rosenberg M. Microencapsulating properties of whey proteins. 1. Microencapsulation of anhydrous milk fat. *Journal of Dairy Science* 1993; **76**(10): 2868–77.

Zhang H, Li W, Huang R, Wang N, Wang J, and Zhang X. Microstructure regulation of microencapsulated bio-based n-dodecanol as phase change materials via in situ polymerization. *New Journal of Chemistry* 2017; **41**(23): 14696–707.

Zhao CY and Zhang GH. Review on microencapsulated phase change materials (MEPCMs). *Renewable and Sustainable Energy Reviews* 2011; **15**(8): 3813–32.

Zhao X-H and Tang C-H. Spray-drying microencapsulation of CoQ 10 in olive oil for enhanced water dispersion, stability and bioaccessibility: Influence of type of emulsifiers and/or wall materials. *Food Hydrocolloids* 2016; **61**: 20–30.

Zondag AH, Smidt RP de, Veldhuis JBJ, and Boer R de. *Advanced Heat Transfer Fluids*, 2016; **76**.

Zuidam NJ and Shimoni E. Overview of microencapsulates for use in food products or processes and methods to make them. In: Zuidam NJ and Nedovic V (eds), *Encapsulation Technologies for Active Food Ingredients and Food Processing*. Springer, New York; 2010. pp. 3–29.

Chapter 4

Nanoplatforms for Anticancer Therapy Based on Reactive Oxygen Species

Shameer Pillarisetti,[a] Veena Vijayan,[a] Saji Uthaman,[b] and In-Kyu Park[a]

[a]*Department of Biomedical Sciences, Chonnam National University Medical School, BioMedical Sciences Graduate Program (BMSGP), Chonnam National University, 322 Seoyang-ro, Hwasun 58128, Republic of Korea*

[b]*Department of Chemical Biology and Engineering, IOWA State University, AMES, IOWA- 50011, USA*

pik96@jnu.ac.kr

4.1 Introduction

Reactive oxygen species (ROS) are highly reactive molecules that originate in oxygen sources such as superoxide (O_2^-), peroxide (O_2^{2-}), singlet oxygen ($^1O_2$), and the hydroxyl radical ($\cdot OH$). ROS represent a class of highly reactive molecules that regulate signaling pathways. Earlier, ROS were considered mere by-products of cellular respiration, including mainly hydroxyl ($\cdot OH$) and superoxide (O_2^-) free radicals, and nonradical molecules such as hydrogen peroxide

Science and Applications of Nanoparticles
Edited by Waqar Ahmed and Ehsan Nourafkan

ISBN 978-981-4267-34-2 (Hardcover), 978-1-003-28029-3 (eBook)
www.jennystanford.com

(H_2O_2) [1]. In the case of mammalian cells, ROS generation occurs primarily in the mitochondria when the oxygen concentration is low along the electron transport chain during aerobic respiration or via metal-catalyzed oxidation and oxidoreductase enzymes during the cell cycle [2]. Low levels of cellular ROS play a vital role in maintaining cell proliferation and metastasis. ROS levels are controlled by maintaining the balance between ROS formation and elimination to ensure normal cellular function. However, the imbalance in ROS production and accumulation of detoxicants in cells can cause oxidative stress, resulting in oxidative destruction of cellular components (e.g., DNA, lipids, and proteins) and necrosis or apoptosis [3].

Recent studies [4–8] have reported that most cancer cells carry higher ROS levels than normal cells [9]. The increased levels of endogenous antioxidants alter the redox status of cancer cells, affecting their phenotypic behavior and response to therapeutic interventions [10]. However, most cancer cells exhibit intrinsically flexible redox status that facilitates adaptation to oxidative stress and increase the resistance of malignant cells to exogenous stress conditions such as chemotherapy and radiotherapy [11]. Thus, controlling the redox mechanisms in cancer is a better approach to eliminating these cells [12]. Moreover, cancers in the advanced stages can adapt well to oxidative stress by upregulating antioxidant expression, thus enabling their survival under harsh drug treatments [13]. However, the use of organic and inorganic sources to generate exogenous ROS has contributed to the development of drugs for the treatment of cancer [13].

This chapter is focused on the role of ROS in cancer, based on recent anticancer therapies and fundamental research findings. Importantly, recent advances suggest that ROS can be used to develop anti-neoplastic drugs by targeting redox sensitivity, which is an appropriate therapeutic strategy against cancer, either alone or in combination with other therapies used to kill tumor cells. We have highlighted different anticancer therapeutic applications based on ROS (Fig. 4.1), along with combination therapies yielding good therapeutic efficacy with minimal side effects.

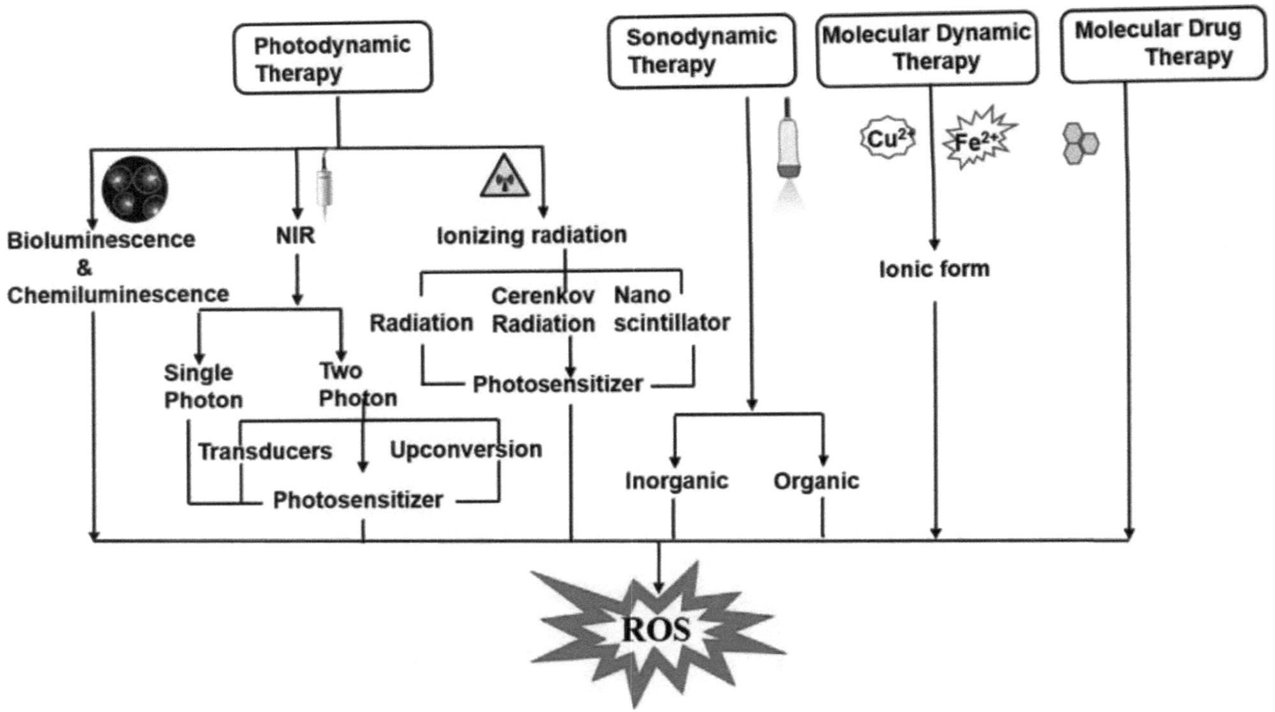

Figure 4.1 Overview of different modes of ROS therapy. Photodynamic therapy: therapeutic energy doses of light are employed at a specific wavelength to generate ROS by exciting photosensitizers. Sonodynamic therapy: activated sonosensitizers, the released energy can be transferred to oxygen to produce a large amount of ROS. Molecular dynamics: Fenton agents for enhancing ROS (hydroxyl radicals). Molecular drugs: chemical tools for targeted amplification of ROS.

4.2 Significance of ROS in Anticancer Therapy

The major sources of ROS in cells include mitochondria, NADPH oxidase, and myeloperoxidase (Fig. 4.2) [14]. In the mitochondria, ROS synthesis occurs in the presence of high levels of oxygen, during the production of adenosine triphosphate (ATP) and water. Superoxide anion is the first ROS produced upon the activation of NADH ubiquinone oxidoreductase (complex I) and co-enzyme Q, bc1 complex, and ubiquinone/cytochrome c reductase (complex III) in the matrix and intermembrane space of mitochondria. Metals such as copper and manganese catalyze the reactions in the intermembrane space that drive the superoxide anions to form stable H_2O_2 [15]. Oxidative stress leads to the generation of free radicals such as perhydroxyl and hydroxyl radicals, superoxide anions, and nitric oxides, along with additional nonradical but reactive species such as singlet oxygen, hydrogen peroxide, peroxynitrite, and hypochlorous acid [16]. The extent of ROS overproduction by mitochondria in malignant cells is functionally different compared with that of normal cells. Oxidative metabolism in mitochondria is a key factor influencing ROS production in eukaryotic cells. The generation of high levels of ROS in cancer cells may affect mitochondrial metabolism [17] and disrupt the signaling pathways [18] that are regulated by transcription factors STAT3 and NF-κB, growth factors, hypoxia-inducible factor-1α, kinases, cytokines, and other factors, including enzymes [19]. Some cancer cells show higher levels of ROS-producing NADPH oxidases and recruit antioxidant enzymes for ROS removal. Furthermore, studies [20–23] have shown that excessive ROS can lead to cancer cell death via apoptosis and/or autophagy. Cancer cells are more likely to be susceptible to ROS overproduction compared with normal cells, owing to biological effects on physicochemical changes. Therefore, increasing oxidative stress by generating exogenous ROS results in the targeted death of cancer cells without any deleterious effects on healthy cells [24].

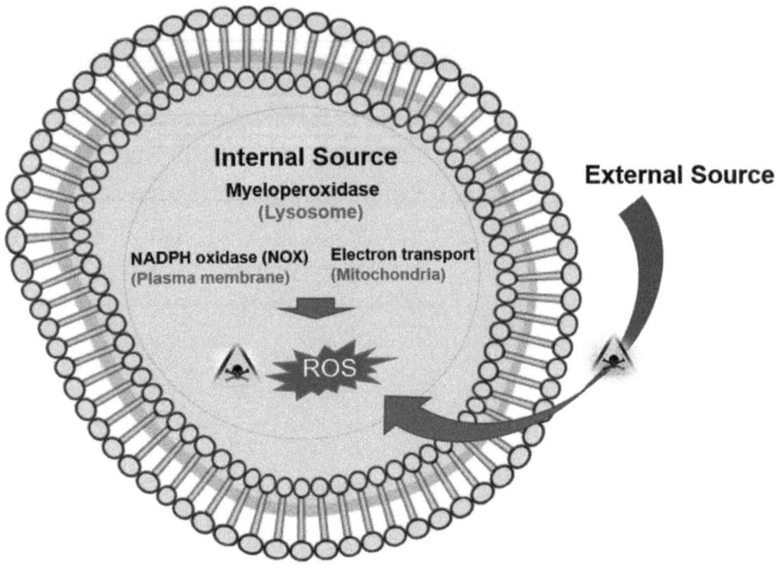

Figure 4.2 The schematic diagram represents the major exogenous and endogenous sources of ROS generation leading to cell death. Adapted by the permission of Ref. [14]. Copyright 2017, Redox Biology.

4.3 Strategies for ROS-Based Anticancer Therapy

Extracellular strategies to generate ROS, such as sonodynamic therapy (SDT) [25], photodynamic therapy (PDT) [26], and molecular dynamic therapy, have potentially lethal effects on cancer cells. Importantly, these methods trigger some of the intracellular mechanisms during ROS generation. Physical and chemical methods to promote ROS overproduction in biological systems are promising and can be utilized to inhibit cancer cell growth [27]. In terms of conventional technologies, ROS-based anticancer therapies are currently gaining considerable attention as a promising anticancer therapeutic strategy because of their ability to generate substantial amounts of ROS [28]. A selected list of ROS-based therapies and their sources is presented in Table 4.1.

Table 4.1 Sources of ROS therapy

Source	Stimuli	Reference
Photodynamic dyes	PDT	[26]
Sonosensitizers	SDT	[25]
Bioluminescent and chemiluminescent dyes	Self-luminescence	[27]
Metallic probes	Molecular dynamics (Fenton reaction)	[28]
Organic and inorganic probes	Ionization	[29]

4.4 Stimulation of ROS Production

Recently, the synthesis of ROS induced by stimuli other than light has attracted considerable attention as an anticancer therapy. Even though stimulation using light may still be considered the predominant strategy, advanced engineering techniques can further improve the safety of ROS-based therapies for cancer. In this section, we have discussed different approaches to induce ROS production [30].

4.4.1 Near-Infrared Irradiation

During the utilization of an energy source such as near-infrared (NIR) laser, the photosensitization mechanism involves the transfer of electrons (instead of energy). The light emitted by the sensitizer induces a non-absorbing substrate, which serves as a medium to generate ROS via a light-absorbing probe and results in PDT-induced cancer cell death [31]. Living tissues are far more complex owing to the moderate turbidity that may affect the penetration of light by changing the rates of transmission, scattering, reflection, and absorption (Fig. 4.2). The depth of light penetration in living tissues is limited by several factors, such as the intensity, wavelength, polarization, coherence, and physiological condition of the tissue. In the case of hair, the fibrotic structure, pigmentation, composition, and hydration affect the natural coherence of light. In addition, tissue fluorophores, as well as melanin and hemoglobin, absorb light in the visible range. Therefore, to overcome such limitations, the photosensitizer (PS) itself must have an excitation wavelength longer than the visible range (Fig. 4.3). However, NIR light of a longer wavelength (e.g., 2500 nm) has been known to be of limited practical use in exciting PSs in PDT studies, owing to a narrow energy gap and comparatively non-radiative fast conversion (e.g., thermal increase). The currently used PSs have excitation wavelengths ranging from 600 nm to 850 nm [32].

4.4.2 Ultrasound

Ultrasound is an alternative energy source with applications in spatiotemporally controlled noninvasive techniques, with effectiveness comparable to that of applied formulations in terms of biological parameters [33]. Yeh et al. [34] described an H_2O_2-encapsulated polymersome with ROS-based anticancer therapy exhibiting echogenic reflectivity that is activated by microbubble-based ultrasound diagnostics. The polymersome comprises a polymer (PLGA) entrapping H_2O_2 in the hydrophilic core and a shell packed with Fe_3O_4 nanoparticles. Exposure to ultrasound disrupts the PLGA polymersome, leading to the release of H_2O_2, which further reacted with the neighboring Fe_3O_4 nanoparticles embedded within the membrane of the polymersome. The production of OH^- resulting in a Fenton response was identified as an effective strategy with a lethal effect on cancer cells. ROS production in this study via an ultrasound-activated mechanism with a non-photon, non-thermal, PS-free, therapeutic activity resulted in complete cessation of tumor growth in a mouse model [34]. Moreover, hydrophilized titanium dioxide ($HTiO_2$) nanoparticles stimulated with a direct ultrasound energy source for high-efficiency ROS production have been reported [35]. The $HTiO_2$ nanoparticles exhibited successful reduction in tumor growth after ultrasound therapy and resulted in intense vascular damage and increased levels of proinflammatory cytokines [36]. Another study reported that the combination of ultrasound and light-activated PSs induced a higher level of ROS production than ultrasound- or light-activated PSs alone.

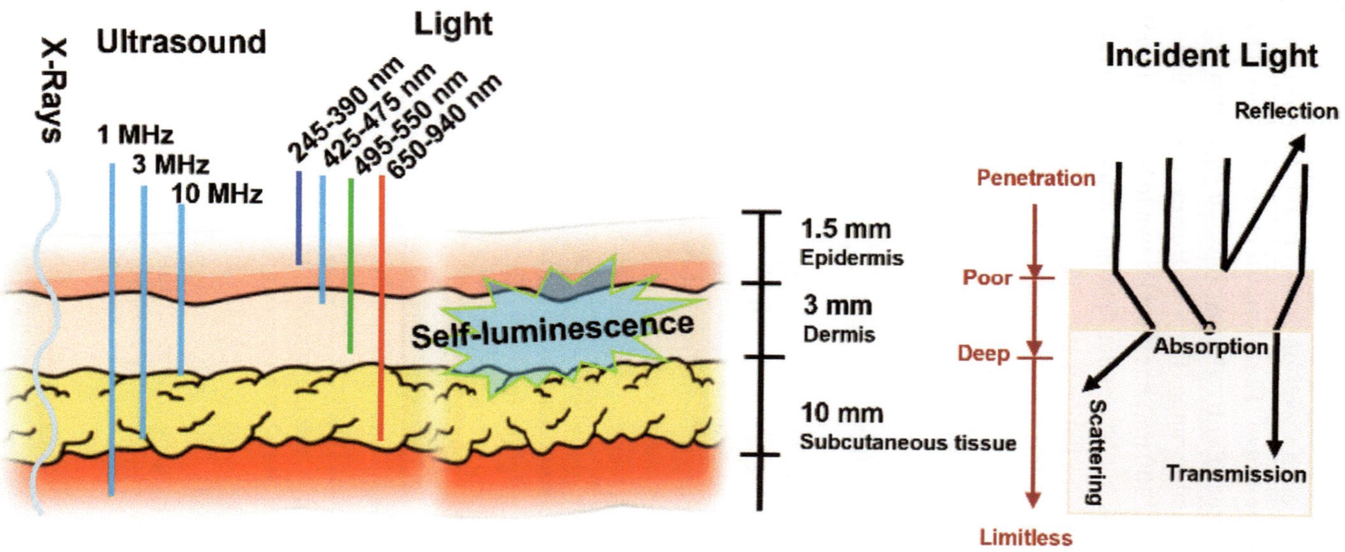

Figure 4.3 Comparison of penetration depths of different excitation sources. The depth of light penetration in living tissues depends on its absorption, scattering, transmission, and reflection rates and varying wavelength. Adapted by the permission of Ref. [29]. Copyright 2016, Chemical Society Reviews.

4.4.3 Self-Luminescence

Bioluminescence resonance energy transfer (BRET) and fluorescence resonance energy transfer (FRET) employ biochemical stimuli to trigger bioluminescent proteins that act as donors to stimulate molecules (e.g., proteins or fluorophores) in close proximity [37]. This phenomenon was first observed in marine organisms such as sea pansy (*Renilla reniformis*) and jellyfish (*Aequorea victoria*). Compared with FRET, BRET has additional benefits in that the activation and emission in BRET can be accomplished remotely through biochemical reactions. The unique features of BRET can facilitate crucial applications in biosensing and imaging *in vivo*. Artificial BRET platforms have promising inclusions as donor counterparts, wherein the enzyme *Renilla* luciferase (Rluc) and its substrate coelenterazine are the most common artificial BRET platforms. Owing to the unique optical properties, especially a large Stokes shift, quantum dots (QDs) can be used as potential candidates to execute the BRET mechanism, in which low intensity bioluminescence within close proximity can be transformed into a wavelength of higher fluorescence intensity. Therefore, considering the scope for application in deep-tissue bioimaging and sensing, QDs can substantially overcome the limitation of shallow tissue penetration of light in the case of conventional light-dependent platforms [18]. Lai et al. reported a study involving a BRET-based PDT system in which CdSe/ZnS QDs surface-modified with Rluc demonstrated the BRET phenomenon upon the addition of coelenterazine to generate ROS. BRET activated the PS (Foscan) in close proximity to the QDs, inducing fluorescence emission at 655 nm. The production of ROS was quantified at 40.8%, which was toxic to A549 cells *in vitro* and drastically reduced tumor growth *in vivo*.

4.4.4 Radiation

In contrast to traditional light, X-rays are electromagnetic waves with a frequency range of $3 \times 10^{16} - 3 \times 10^{19}$ Hz and energy of 0.1–100 keV. X-rays disrupt chemical bonds and ionize atoms because of their high energy. Based on their energy ranges, X-rays are classified into "hard" and "soft" X-rays with energies above 5–10 keV and below 5 keV, respectively. Despite the substantial risk of exposure to radiation and significant cancer-causing potential, X-rays are still considered the foremost approach to medical imaging owing to their substantial advantages in disease identification [29]. X-ray photons are a potential excitation source, enabling deep-tissue penetration [29]. The incoming X-rays are engaged with scintillators, which are materials with a wide bandgap and the capacity to transform X-rays into UV/visible light. These events are primarily attributed to the interaction between the scintillators and the photons with high energies and are entirely based on the photoelectric effect and the Compton scattering effects. Radiation therapy is analogous to Cerenkov radiation, which is indeed an electromagnetic radiation containing high-energy positrons (β+), which occur when a charged particle (β- or α-particle) gains sufficient potential to travel faster than the speed of light through a dielectric medium [38]. Polarization of the molecules in a medium by charged particles can induce excitation to a higher energy state. Cerenkov radiation emits Cerenkov luminescence (CL) due to the relaxation of polarized molecules that tend to relax back to the ground state [38]. The photons emitted during the relaxation of the polarized molecules constitute the light within a continuous spectrum (*ca.* 200–1000 nm) over a wide energy range (*ca.* 6.1–1.23 eV). Cerenkov radiation can penetrate effectively into deep tissues and has applications in phototherapeutic studies. Based on this technology, scintillator materials are used to convert the X-rays into UV/visible light as there are no PSs that can directly absorb X-ray energy [29]. The scintillation mechanism comprises three steps: (1) adaptation of inward radiation into a huge chunk of electron–hole pairs; (2) transmission of the energy of the electron–hole pairs to the luminescent ions; and (3) emission of the luminescent ions that radiatively return from an excited state to the ground state [29]. Herein, we have highlighted the recent progress in PDT for deep-seated tumors using scintillating nanoparticles (ScNPs) as energy mediators. These strategies are based on the excitation of the energy source and application of radionuclides for effective PDT.

4.4.5 Molecular Dynamics

Mitochondria are the primary organelles that produce singlet oxygen ($^1O_2$) by the removal of one electron from the oxygen (O_2)

molecule. In the cell, O_2 is the main source initiating oxidative cascades, and it acts as a precursor for the production of highly oxidizing or reducing species. ROS are constantly produced during cellular metabolism and are generated as by-products in a series of metabolic biochemical reactions. For example, the production of H_2O_2 from O_2 is catalyzed by superoxide dismutases (e.g., MnSOD, Cu/ZnSOD). Further, H_2O_2, a robust natural oxidant, may be partially reduced to hydroxide (HO^-) or fully reduced to water. Even during the overproduction of ROS, several elimination mechanisms in the cell are counteractive to maintain the oxidative balance. A profound imbalance in the generation and elimination of ROS is one of the underlying mechanisms for the elimination of cancer cells [13]. Indeed, caspase-mediated cell death during the later cell stages can be partially attributed to the increased production of ROS, which surpasses the ROS-eliminating ability of cells. Consequently, enabling the on-demand generation of ROS via engineered biochemical reactions may potentially trigger cancer cell death. Several approaches have been developed based on these strategies, including the Fenton reaction. Some of the cations such as Cu^{2+}, Mn^{2+}, V^{2+}, and Cr^{4+} exhibit Fenton-like reactions (Fig. 4.4) [28, 39]. Due to their robust reactivity with biomolecules, hydroxyl radicals (OH^-) induce more damage in biological systems than any other ROS [28]. Hydroxyl radicals ($OH^·$) are capable of attacking and oxidizing a maximum number of organic molecules with high rate constants, generally in the range of 10^6–10^9 M^{-1} s^{-1}. For example, in the case of cellular environment, OH^- radicals do not diffuse from the generation site, and in lieu, the biomacromolecules are oxidized rapidly in the surrounding areas. The high OH^- concentrations drive lipid peroxidation via the oxidation of polyunsaturated fatty acids in lipids and DNA damage and the oxidation of amino acids in the proteins of cancerous and healthy cells. Based on the high concentrations of OH^- radicals and H_2O_2 in cancer cells, it is essential to develop a therapeutic agent that responds to and generates high levels of OH^- via Fenton reaction in a tumor environment. Owing to their high reactivity, large surface area, and small size, nanoparticles can act as catalysts in the heterogeneous Fenton reaction to generate hydroxyl radicals [28]. Fe_3O_4 and α-Fe_2O_3 react in the presence of H_2O_2, leading to the production of OH^- at a low pH [28]. Certain other metal oxide nanoparticles, including gold, silver, copper, manganese, and palladium, are also capable of decomposing H_2O_2 to OH^- free radicals via Fenton-like reactions. The hydroxyl radicals exhibit strong reactivity toward biological molecules and are being investigated for specific biomedical applications, including the destruction of cancer cells [28].

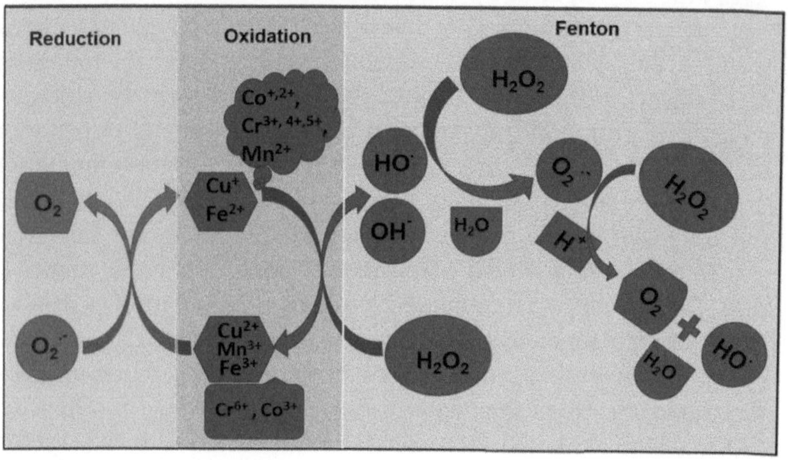

Figure 4.4 Molecular dynamics of the Fenton process. The metal ions are oxidized by hydrogen peroxide, thereby forming a hydroxyl radical and a hydroxide ion. Oxidized metal ions are then reverted to the initial form of metal ions in a cyclic process, another hydrogen peroxide molecule, forming a hydroperoxyl radical and proton.

4.5 ROS-Based Monotherapies

4.5.1 Photodynamic Therapy

PDT drives the exogenous production of ROS to destroy cancer cells by stimulating photosensitizing agents via light excitation [40]. This phenomenon was first described in the early 1890s and later rediscovered in 1975. PDT has been used widely as a primary approach in cancer treatment and tumor ablation over the past four decades. PDT involves three primary mechanisms: (1) the wavelength of the light source that provides the energy to excite PS based on their specific wavelength range; (2) the excited PS in the presence of oxygen-containing substrates (e.g., molecular oxygen) undergoes photodynamic reaction; (3) ROS are generated upon the donation of electrons from excited PSs. The first FDA-approved PDT PS was procarbazine, which was released in the year 1970. Since then, more than 16 drugs were developed and investigated in clinical trials, and some of them are currently available commercially [40]. Different types of PS have been developed for anticancer PDT and categorized into different generations. Nevertheless, the first-generation PSs satisfied several clinical criteria: (1) excitation by short wavelengths (visible range) with a low penetration depth, limiting the clinical applications of PDT to the treatment of superficial tumors [41]; (2) association with complex chemical synthesis and difficult purification methods; (3) poor water solubility of photofrin, which limits its biological use; and (4) poor bioavailability owing to their hydrophobicity, suggesting the need for high doses of first-generation PSs for improved PDT. These PSs in systemic circulation tend to accumulate in the skin, inducing skin photosensitization over a longer period [42]. Hence, the first-generation PSs had many shortcomings, which resulted in the synthesis of second-generation PSs with high $^1O_2$ production and NIR excitation. Stereotypically, these are macrocyclic moieties that result from porphyrin substituents or via changes in the porphyrin core or newly synthesized non-porphyrinoid PS compounds, including phthalocyanines, bacteriochlorins, porphycenes, chlorins, phenothiazines (methylene blue and toluidine blue), hypericin, xanthenes (Rose Bengal), metal derivatives of PSs (such as metalloporphyrins, $AlPcS_4$, Si(IV)-naphthalocyanine, and tin ethyl etiopurpurin), and cyanines (merocyanine-540) [42]. However, the limitations of second-generation PSs include non-specific localization in targeted cells/tissues, which led to the design of third-generation PSs. Normally, a tumor-targeting moiety is conjugated on the surface of the PS to develop a tumor-targeting PS. However, conventional PDT faces numerous challenges such as poor light penetration, PS localization, light-dependent activation, and O_2 reliance. Current advances in nanotechnology have enabled the development of an adaptable platform for PDT development [31].

4.5.2 Sonodynamic Therapy

Ultrasound techniques have been utilized extensively in medical applications for therapy and imaging [43]. The use of ultrasound-based medical diagnosis dates back to the 19th century, long before ultrasound was implemented in the 1950s. Among contemporary clinical practices, the advantages of a non-radiation and non-ionization technique such as ultrasound have facilitated the imaging of deeper tissues, especially under the skin, which has been effectively exploited in neonatal imaging. Ultrasound modulates the applied ultrasonic power and thereby serves as an external source of energy to inflict target damage. Consequently, ultrasound is an excellent source of alternative energy to improve the spatiotemporal efficacy of the administered formulations in a biological environment via noninvasive techniques [33]. The generation of ROS via ultrasound in an aqueous environment results in the formation of a unique phenomenon referred to as acoustic cavitation, which then activates the sensitizers to generate ROS. Sonoluminescence, a process in which light is produced via ultrasound irradiation of a solution to generate ROS, is based on acoustic cavitation [44, 45]. The cavitation mechanism comprises nucleation, growth, and implosive disintegration of gas-filled bubbles under appropriate ultrasound conditions. The process is further classified into stable and inertial cavitation. In stable cavitation, the bubbles oscillate to generate a surge in the surrounding liquid, to generate a mixture with the nearby media. In contrast, inertial cavitation induces the growth of gas bubbles to a near resonant size, resulting in the sudden expansion of the gas bubbles before a violent collapse [25]. The implosion of gas bubbles raises the temperatures up to 10000 K and elevates pressures of 81 MPa in the adjoining microenvironment, which represents a sonochemical reactor [46]. It has been suggested that sonodynamic therapy (SDT) can induce the cavitation phenomenon, based on two key factors describing the SDT mechanism of ROS generation. The primary key factor relies on "sonoluminescence" (SL), in which luminescence is produced by the irradiation of a solution with ultrasound. This phenomenon was demonstrated by Pickworth et al. [47] and others [48]. Although the mechanism underlying light generation is still disputed, it has been proposed that the bubble enlargement and collapse are likely related to bremsstrahlung radiation, blackbody radiation, recombination radiation, or complex radiation [48]. The second key factor relies on a concept known as pyrolysis. The process involves the irradiation of a fluid using ultrasound, which results in the enlargement of some of the bubbles and eventual rupture, resulting in a sudden rise in temporary heat production (10000 K) and an increase in the surrounding temperature. As a result of this temporary heat production and increase in temperature, the sound sensitizer is excited, resulting in the production of ROS [48]. Additional experiments showed that the sound energy from the sound sensitizer produced a significant volume of ROS that was sufficient to kill the cells.

Further, the addition of an ROS scavenger resulted in a remarkable decrease in the mortality of cells. The low cellular concentrations of ROS play a major role in modulating the intercellular signals, whereas high levels of ROS result in cytotoxicity.

4.5.3 Molecular Dynamic Therapy

The Fenton process involves the generation of highly reactive OH^- radical via a catalytic reaction between H_2O_2 and metals (e.g., iron, copper). The Haber–Weiss mechanism is similar to a Fenton reaction, but somewhat slower as it generates OH^- from H_2O_2 and O_2. OH^- interacts sequentially via controlled diffusion with biomolecules to produce a sequence of supplementary free radicals, e.g., peroxyl radicals ($(R)_3COO^-$), carbon-centered ($(R)_3C^-$, and thiyl (RS^-)), which may result in extensive cell damage. Fenton reaction-based ROS production has been explored as a potential anticancer therapy. Shi and coworkers [49] described a method to synthesize amorphous iron nanoparticles (AFe-NPs) via a hubble-bubble approach. In an acidic tumor microenvironment, these nanoparticles were swiftly ionized, and released Fe^{2+} ions on demand, which further initiated the Fenton reaction at these sites resulting in anticancer activity. Along with iron, nanoparticles or transition metals such as cobalt, manganese, cerium, copper, and nonmetallic carbons act as catalysts in a Fenton-like reaction. External sources of energy, including photothermal, sonodynamic, or electro-assisted Fenton reactions have also been extensively studied [50]. These therapies may deliver adaptable sources to retain Fenton-like reaction activity and stimulate ROS for anticancer therapy. Although the development of this strategy is in the preliminary stages, the application of the

Fenton reaction in anticancer therapy, based on the divergence between normal tissues and cancer microenvironment, increases the possibilities for cancer diagnosis via engineered biochemical reactions.

4.5.4 Molecular Drug Therapy

Doxorubicin (Dox) is an anthracycline drug that initiates redox cycling and generates ROS via several mechanisms [51]. Mechanistically, Dox may combine with trace metals (for example, Fe^{3+}) to form a free radical Fe^{2+}-doxorubicin complex, which reduces O_2 to $^1O_2$. Furthermore, Dox can be mitigated via a single electron transfer mechanism, *viz*, mitochondrial reductase to form anthracycline semiquinone free radicals by reducing the oxygen in O_2^- and H_2O_2 under aerobic conditions. Similarly, cisplatin-based drugs catalyze ROS via nicotinamide adenine dinucleotide phosphate hydrogen (NADPH). However, the dosage-dependent increase in ROS levels following the administration of these molecular drugs or the resulting anticancer effects have yet to be corroborated. These molecular-targeted drugs are known to enable intra-strand DNA crosslinking, which results in cellular toxicity and potential anticancer activity. During chemotherapy, ROS production is accelerated owing to the diversity of cellular signaling pathways [51]. The ROS alter the protein structure and enzyme function, which may induce apoptosis or necrotic cell death.

However, the correlated outcomes and mechanisms may vary with drug dose and type, as well as cell type. Drugs that induce ROS synthesis in cells can be used to delineate the cellular mechanism underlying their potential anticancer activities. Evaluation of the molecular mechanisms of drug-induced ROS generation is useful in the development of anticancer drugs [52].

4.6 ROS-Based Combination Therapy

4.6.1 Chemotherapy

Anticancer drugs, including cisplatin, doxorubicin, and mitomycin C, are used widely in cancer therapy. The major limitations of this approach include drug resistance and dose-limiting toxicities, which have led to the use of therapeutic combinations to minimize these drawbacks. Chemotherapy is one of the methods to treat cancer, and the aforementioned anticancer drugs are considered the best chemotherapeutic agents that can be used as single agents to eliminate various cancer cells [53]. The mechanisms underlying cancer cell death induced by chemotherapeutics depend on the binding of these drugs to DNA and the suppression of subsequent pathways of cell replication. However, drug resistance, both natural and assimilated, strongly limits the broad clinical applications of this therapeutic approach. Regardless of the extensive use of these anticancer drugs, enhancing the sensitivity of cancers cells to chemotherapeutic drugs is a major challenge in cancer research. ROS-based therapies can selectively improve the uptake of chemotherapeutic drugs by cancer cells, thereby reducing their toxic side effects in healthy cells and tissues [54].

4.6.2 Immunotherapy

Immunotherapy induces a strong immune response resulting in antitumor activity. ROS can potentially trigger an immune response against tumors by stimulating different components of the immune system [55]. ROS therapy can be combined with several other immunostimulatory boosters to activate tumor antigen presentation, by improving neutrophil infiltration, suppressing the expression of T-regs, and promoting T-cell activation. The administration of PSs with immunostimulants intravenously, intratumorally, or even topically could reduce malignancy [55]. Recent studies [56–58] have established the role of diverse immunostimulatory mediators and their outcomes in immune response when combined with ROS therapy in several malignancies. Anticancer studies have been conducted in several animal cancer models, including squamous cell carcinoma, colon cancer, breast cancer, lung cancer, and melanoma. Various approaches adopted in these studies included boosting of immune memory T-cells by draining the lymph nodes and increasing the infiltration of neutrophils, which results in the recruitment of activated macrophages to the tumor site, induction of immature DC maturation, and elevation of IL-12 (inflammatory cytokine) levels, to reduce tumor growth in *in vivo* studies [55]. Certain combination strategies with immunotherapeutics showed improved anticancer activity via effective rejection of tumor re-challenge or tumor reduction based on the abscopal effect. Photofrin was used in many studies as the PS agent; however, studies showed that tumor reduction using combination therapy was independent of the choice of PS [55]. However, several studies have reported different mechanisms of antitumor effect for immunogenic enhancement. It is also possible that some of these combination strategies further improve the efficacy of anticancer treatment. Hence, the immunostimulant treatment should be tailored to the clinical conditions along with extensive precautions because of the risk of systemic side effects such as severe autoimmune disorders.

4.6.3 Photothermal Therapy

Phototherapy is a minimally invasive approach, in which therapeutic energy doses of light are employed at a specific wavelength to treat diseases such as infection and cancer [59]. Currently, the most common methods of phototherapy are photothermal therapy (PTT) and photodynamic therapy (PDT) for the treatment of cancer. PTT acts by activating PSs via electromagnetic radiation such as RF, microwaves, NIR, or visible light to release heat in response to thermal ablation by selectively targeting cancer cells and abnormal tissues. The PSs, which are photoactivated molecules capable of generating ROS, undergo a series of photochemical reactions. Consequently, the increased oxidative stress in target cells triggers protein and DNA damage and induces intracellular lipid peroxidation, eventually resulting in cell death [60]. Recently, nanomaterials have been investigated to develop and employ nanomaterials in a variety of applications [61]. Many effective photothermal and photodynamic nanomaterials have been explored in the diagnosis and treatment of cancer (Table 4.2) [62].

Table 4.2 PDT and PTT dual-responsive organic dyes for cancer therapy: indo cyanine green (ICG) 5,10,15,20-tetrakis(3-hydroxyphenyl)chlorin (*mTHPC)*; boron dipyrromethene derivatives (BODIPY)

Dye	Excitation Wavelength	Mechanism	References
ICG	740–800 nm	PTT and PDT	[63]
IR780	780–800 nm	Process of photoinduced reactions with molecular oxygen to generate reactive oxygen species	[64, 65]
BODIPY Derivatives	Wide range		[66]

4.6.4 Combination Therapy

Conventional methods commonly used for the treatment of cancer involve a combination of invasive strategies with other therapies. In addition, substantial destruction of cancer cells in *in vitro* and *in vivo* studies has demonstrated the efficacy of combination therapies, including ROS-based therapy along with pharmacological therapies, compared with monotherapy against certain types of cancer [55]. The consequences are primarily reinforced by multiple mechanisms of action of ROS, which are generally not exhibited by monotherapy. The use of ROS in combination with chemotherapeutic drugs results in not only targeted cancer cell death by cytotoxic agents but also the destruction of the tumor microvasculature, and in some cases, the activation of a host immune response against cancer cells (Table 4.3). Further, booster doses are beneficial when combined with ROS therapy. Although ROS stimulate host immune activity, they can also induce the expression of immune suppressor cells [67]. Booster therapies enhanced the ROS-induced host immune response against cancer cells and overcame ROS-induced immune suppression.

Table 4.3 Sources of ROS

Nanoformulations	Combination Therapy	Cancer Model	References
Chondroitin sulfate-chlorin e6-lipoic acid nanoplatform loaded with docetaxel	SDT, chemotherapy, and immune response studies	B16F10 tumor-bearing mice	[68]
Core-shell nanoparticles carry oxaliplatin in the core and the photosensitizer pyropheophorbide-lipid conjugate (pyrolipid) in the shell	PDT, chemotherapy, and checkpoint PDL-1 inhibitor	CT26 tumor-bearing mice	[69]

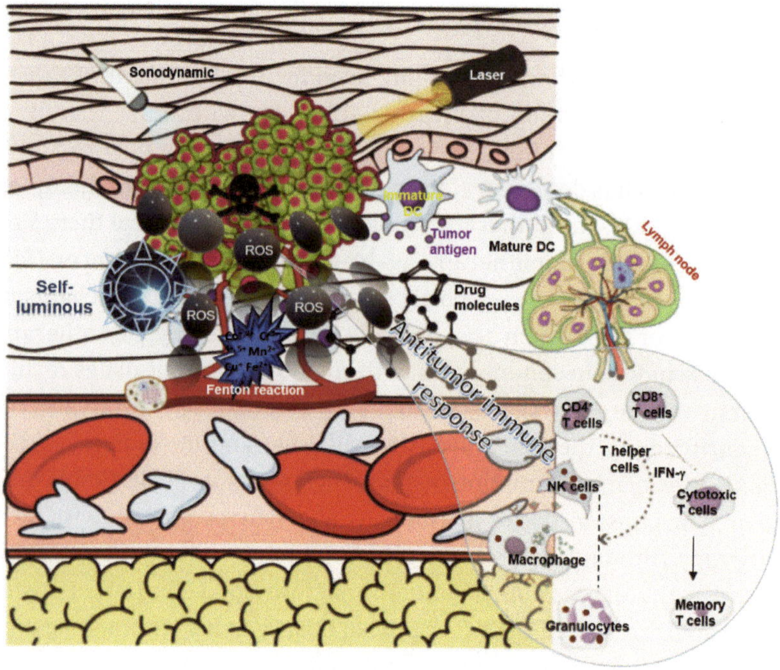

Figure 4.5 A schematic representation of combination therapy. ROS therapy can be combined with several other immunostimulatory boosters to activate tumor antigen presentation, by improving neutrophil infiltration, and promoting T-cell activation, which results in the recruitment of activated macrophages to the tumor site, induction of immature DC maturation, and elevation of inflammatory cytokine levels, to reduce tumor growth *in vivo*.

4.7 Clinical Implications of ROS Therapy

The clinical effectiveness of PDT depends on the mode of administration of light, exposure time, and dose to improve the therapeutic efficacy of PDT and ensure clinical outcomes at an appropriate depth of diffusion. Different sources of light have been used in PDT, such as NIR (700–1000 nm), red (600–700 nm), and ultraviolet (330–400 nm) wavelengths [68]. Light sources of different wavelength have been selected based on the PS. However, studies on PS are not specific to the therapy of cancer and other diseases.

Development of PSs that are effective against tumors represents a breakthrough in cancer research. Currently, no study has specifically investigated cancer cell death using a discrete wavelength of light. Most PSs reported in research articles are administered orally or intravenously, and several hours or days (notably in the case of PDT) later, the tumor is exposed to a specific wavelength of non-thermal light, which activates the PSs, with a lethal effect. A clinical study recommends a light penetration depth of at least 3 mm from the surface of the skin. Longer wavelengths (red and infrared) penetrate deeper than light with a shorter wavelength. Therefore, longer-wavelength light targets cells buried deeply within the tumor tissue [69]. Photobleaching is one of the issues affecting successful PDT, and a clinical study showed that a low-wavelength light source was not ideal for PDT [70].

4.8 Conclusion and Future Perspectives

We have discussed the characteristics and advantages of minimally invasive therapy using ROS and reinforced the applications of ROS and combination therapy in cancer treatment. Our observations substantiate the considerable increase in the number of studies investigating both metastatic tumors and minimally invasive approaches against the backdrop of anticancer therapies. Attention was mainly focused on the application of minimal invasive therapy in tumor treatment, which has greatly improved the therapeutic outcomes, with strong tumor-targeting properties, improved biocompatibility, and few side effects. In addition, a multidisciplinary approach, comprising synergistic drug combinations, is more advantageous than a single treatment. Therefore, synergistic therapies have garnered widespread attention in tumor therapy and have achieved considerable success. However, scientific challenges such as PS localization, light-dependent activation, and O_2 reliance, which often restrict tumor treatment or prognosis, should be addressed.

Table 4.4 FDA-approved drugs for ROS therapy

Generic Name	Clinical Trial Status	Treatment Modality	Cancer Type	References
Photofrin®	Approved	PDT, Ex: 630 nm	Esophageal cancer, lung adenocarcinoma, endobronchial cancer, cholangiocarcinoma	[71]
Foscan®	Approved	PDT, Ex: 652 nm	Advanced head and neck cancer	[72]
Lutex®	Approved	PDT, Ex: 730–770 nm	Prostate cancer with pre- and post-treatment with motexafin lutetium	[73]
Xcytrin	Phase II, not approved	X-Ray (PDT) 1.8 Gy	Non-small cell lung cancer with brain metastases	[74]
Tookad®	Approved	PDT, Ex: 762 nm	Prostate cancer	[75]
Verteporfin	Approved	PDT, Ex: 690 nm	Advanced pancreatic cancer	[76]
Metvix®/Metvixia®	Approved	PDT, Ex: 570–670 nm	Non-hyperkeratotic actinic keratosis and basal cell carcinoma	
Radachlorin®	Approved	PDT, Ex: 600 nm	Bladder cancer, skin cancer	[77]

Acknowledgments

This work was financially supported by the Bio & Medical Technology Development Program (No. NRF-2017M3A9E2056372) through the National Research Foundation of Korea (NRF) funded by the Korean government, MSIP. This work was also supported by the National Research Foundation of Korea (NRF) grant funded by the Korea government (MSIT) (No. 2020R1A2C2005620, 2018R1A5A2024181 and NRF-2019M3E5D1A02068082).

Compliance with Ethical Standards

Conflict of interest: All authors declare that they have no conflicts of interest.

Ethical Statement

No animal experiments were performed as part of the research described in this article.

References

1. Dickinson BC and Chang CJ. Chemistry and biology of reactive oxygen species in signaling or stress responses. *Nature Chemical Biology* 2011, **7**(8): 504.
2. Ray PD, Huang B-W, and Tsuji Y. Reactive oxygen species (ROS) homeostasis and redox regulation in cellular signaling. *Cellular Signalling* 2012, **24**(5): 981–990.
3. Cairns RA, Harris IS, and Mak TW. Regulation of cancer cell metabolism. *Nature Reviews Cancer* 2011, **11**(2): 85–95.
4. Yang Y, Karakhanova S, Werner J, and Bazhin A. Reactive oxygen species in cancer biology and anticancer therapy. *Current Medicinal Chemistry* 2013, **20**(30): 3677–3692.
5. Perillo B, Di Donato M, Pezone A, Di Zazzo E, Giovannelli P, Galasso G, Castoria G, and Migliaccio A. ROS in cancer therapy: The bright side of the moon. *Experimental & Molecular Medicine* 2020, 1–12.
6. Reczek CR and Chandel NS. The two faces of reactive oxygen species in cancer. *Annual Review of Cancer Biology* 2017, **1**: 79–98.

7. Aggarwal V, Tuli HS, Varol A, Thakral F, Yerer MB, Sak K, Varol M, Jain A, Khan M, and Sethi G. Role of reactive oxygen species in cancer progression: Molecular mechanisms and recent advancements. *Biomolecules* 2019, **9**(11): 735.
8. Liou G-Y and Storz P. Reactive oxygen species in cancer. *Free Radical Research* 2010, **44**(5): 479–496.
9. TP S and Nathan CF. Production of large amounts of hydrogen peroxide by human tumor cells. *Cancer Research* 1991, **51**: 794–798.
10. Sabharwal SS and Schumacker PT. Mitochondrial ROS in cancer: Initiators, amplifiers or an Achilles' heel? *Nature Reviews Cancer* 2014, **14**(11): 709–721.
11. Pervaiz S and Clément M-V. Tumor intracellular redox status and drug resistance-serendipity or a causal relationship? *Current Pharmaceutical Design* 2004, **10**(16): 1969–1977.
12. Trachootham D, Zhou Y, Zhang H, Demizu Y, Chen Z, Pelicano H, Chiao PJ, Achanta G, Arlinghaus RB, and Liu J. Selective killing of oncogenically transformed cells through a ROS-mediated mechanism by β-phenylethyl isothiocyanate. *Cancer Cell* 2006, **10**(3): 241–252.
13. Trachootham D, Alexandre J, and Huang P. Targeting cancer cells by ROS-mediated mechanisms: a radical therapeutic approach? *Nature Reviews Drug Discovery* 2009, **8**(7): 579–591.
14. Morry J, Ngamcherdtrakul W, and Yantasee W. Oxidative stress in cancer and fibrosis: Opportunity for therapeutic intervention with antioxidant compounds, enzymes, and nanoparticles. *Redox Biology* 2017, **11**: 240–253.
15. Prieto-Domínguez N, Ordóñez R, Fernández A, Méndez-Blanco C, Baulies A, Garcia-Ruiz C, Fernández-Checa JC, Mauriz JL, and González-Gallego J. Melatonin-induced increase in sensitivity of human hepatocellular carcinoma cells to sorafenib is associated with reactive oxygen species production and mitophagy. *Journal of Pineal Research* 2016, **61**(3): 396–407.
16. Park YI, Lee KT, Suh YD, and Hyeon T. Upconverting nanoparticles: A versatile platform for wide-field two-photon microscopy and multi-modal in vivo imaging. *Chemical Society Reviews* 2015, **44**(6): 1302–1317.
17. Chen G, Qiu H, Prasad PN, and Chen X. Upconversion nanoparticles: design, nanochemistry, and applications in theranostics. *Chemical Reviews* 2014, **114**(10): 5161–5214.

18. So M-K, Xu C, Loening AM, Gambhir SS, and Rao J. Self-illuminating quantum dot conjugates for in vivo imaging. *Nature Biotechnology* 2006, **24**(3): 339–343.

19. Ding Y, Wu F, Zhang Y, Liu X, de Jong EM, Gregorkiewicz T, Hong X, Liu Y, Aalders MC, and Buma WJ. Interplay between static and dynamic energy transfer in biofunctional upconversion nanoplatforms. *The Journal of Physical Chemistry Letters* 2015, **6**(13): 2518–2523.

20. Huang G and Pan S-T. ROS-mediated therapeutic strategy in chemo-/radiotherapy of head and neck cancer. *Oxidative Medicine and Cellular Longevity* 2020, **2020**: 5047987.

21. Wang J and Yi J. Cancer cell killing via ROS: To increase or decrease, that is the question. *Cancer Biology & Therapy* 2008, **7**(12): 1875–1884.

22. NavaneethaKrishnan S, Rosales JL, and Lee K-Y. ROS-mediated cancer cell killing through dietary phytochemicals. *Oxidative Medicine and Cellular Longevity* 2019, **2019**: 9051542.

23. Jia P, Dai C, Cao P, Sun D, Ouyang R, and Miao Y. The role of reactive oxygen species in tumor treatment. *RSC Advances* 2020, **10**(13): 7740–7750.

24. Massaro G, Hernando J, Ruiz-Molina D, Roscini C, and Latterini L. Thermally switchable molecular upconversion emission. *Chemistry of Materials* 2016, **28**(3): 738–745.

25. Rosenthal I, Sostaric JZ, and Riesz P. Sonodynamic therapy: A review of the synergistic effects of drugs and ultrasound. *Ultrasonics Sonochemistry* 2004, **11**(6): 349–363.

26. Hong EJ, Choi DG, and Shim MS. Targeted and effective photodynamic therapy for cancer using functionalized nanomaterials. *Acta Pharmaceutica Sinica B* 2016, **6**(4): 297–307.

27. Magalhaes CM, Esteves da Silva JC, and Pinto da Silva L. Chemiluminescence and bioluminescence as an excitation source in the photodynamic therapy of cancer: A critical review. *ChemPhysChem* 2016, **17**(15): 2286–2294.

28. Ranji-Burachaloo H, Gurr PA, Dunstan DE, and Qiao GG. Cancer treatment through nanoparticle-facilitated Fenton reaction. *ACS Nano* 2018, **12**(12): 11819–11837.

29. Kamkaew A, Chen F, Zhan Y, Majewski RL, and Cai W. Scintillating nanoparticles as energy mediators for enhanced photodynamic therapy. *ACS Nano* 2016, **10**(4): 3918–3935.

30. Zhou Z, Song J, Nie L, and Chen X. Reactive oxygen species generating systems meeting challenges of photodynamic cancer therapy. *Chemical Society Reviews* 2016, **45**(23): 6597–6626.

31. Cheng L, Wang C, Feng L, Yang K, and Liu Z. Functional nanomaterials for phototherapies of cancer. *Chemical Reviews* 2014, **114**(21): 10869–10939.

32. Luo S, Zhang E, Su Y, Cheng T, and Shi C. A review of NIR dyes in cancer targeting and imaging. *Biomaterials* 2011, **32**(29): 7127–7138.

33. Couture O, Foley J, Kassell NF, Larrat B, and Aubry J-F. Review of ultrasound mediated drug delivery for cancer treatment: Updates from pre-clinical studies. *Translational Cancer Research* 2014, **3**(5): 494–511.

34. Li W-P, Su C-H, Chang Y-C, Lin Y-J, and Yeh C-S. Ultrasound-induced reactive oxygen species mediated therapy and imaging using a fenton reaction activable polymersome. *ACS Nano* 2016, **10**(2): 2017–2027.

35. You DG, Deepagan V, Um W, Jeon S, Son S, Chang H, Yoon HI, Cho YW, Swierczewska M, and Lee S. ROS-generating TiO_2 nanoparticles for non-invasive sonodynamic therapy of cancer. *Scientific Reports* 2016, **6**(1): 1–12.

36. McCaughan B, Rouanet C, Fowley C, Nomikou N, McHale AP, McCarron PA, and Callan JF. Enhanced ROS production and cell death through combined photo-and sono-activation of conventional photosensitising drugs. *Bioorganic & Medicinal Chemistry Letters* 2011, **21**(19): 5750–5752.

37. Ward WW and Cormier MJ. Energy transfer via protein-protein interaction in Renilla bioluminescence. *Photochemistry and Photobiology* 1978, **27**(4): 389–396.

38. Tanha K, Pashazadeh AM, and Pogue BW. Review of biomedical Čerenkov luminescence imaging applications. *Biomedical Optics Express* 2015, **6**(8): 3053–3065.

39. Bokare AD and Choi W. Review of iron-free Fenton-like systems for activating H_2O_2 in advanced oxidation processes. *Journal of Hazardous Materials* 2014, **275**: 121–135.

40. Agostinis P, Berg K, Cengel KA, Foster TH, Girotti AW, Gollnick SO, Hahn SM, Hamblin MR, Juzeniene A, and Kessel D. Photodynamic therapy of cancer: An update. *CA: A Cancer Journal for Clinicians* 2011, **61**(4): 250–281.

41. Kamkaew A, Lim SH, Lee HB, Kiew LV, Chung LY, and Burgess K. BODIPY dyes in photodynamic therapy. *Chemical Society Reviews* 2013, **42**(1): 77–88.

42. Ormond AB and Freeman HS. Dye sensitizers for photodynamic therapy. *Materials* 2013, **6**(3): 817–840.

43. Wells PN. Ultrasound imaging. *Physics in Medicine & Biology* 2006, **51**(13): R83.

44. Niedre M, Patterson MS, and Wilson BC. Direct near-infrared luminescence detection of singlet oxygen generated by photodynamic therapy in cells in vitro and tissues in vivo. *Photochemistry and Photobiology* 2002, **75**(4): 382–391.

45. Kremkau FW, Kaufmann JS, Walker MM, Burch PG, and Spurr CL. Ultrasonic enhancement of nitrogen mustard cytotoxicity in mouse leukemia. *Cancer* 1976, **37**(4): 1643–1647.

46. Mišík V and Riesz P. Free radical intermediates in sonodynamic therapy. *Annals of the New York Academy of Sciences* 2000, **899**(1): 335–348.

47. Costley D, Mc Ewan C, Fowley C, McHale AP, Atchison J, Nomikou N, and Callan JF. Treating cancer with sonodynamic therapy: A review. *International Journal of Hyperthermia* 2015, **31**(2): 107–117.

48. Rengeng L, Qianyu Z, Yuehong L, Zhongzhong P, and Libo L. Sonodynamic therapy: A treatment developing from photodynamic therapy. *Photodiagnosis and Photodynamic Therapy* 2017, **19**: 159–166.

49. Zhang C, Bu W, Ni D, Zhang S, Li Q, Yao Z, Zhang J, Yao H, Wang Z, and Shi J. Synthesis of iron nanometallic glasses and their application in cancer therapy by a localized Fenton reaction. *Angewandte Chemie* 2016, **128**(6): 2141–2146.

50. Chen X, Ma W, Li J, Wang Z, Chen C, Ji H, and Zhao J. Photocatalytic oxidation of organic pollutants catalyzed by an iron complex at biocompatible pH values: Using O_2 as main oxidant in a Fenton-like reaction. *The Journal of Physical Chemistry C* 2011, **115**(10): 4089–4095.

51. Deavall DG, Martin EA, Horner JM, and Roberts R. Drug-induced oxidative stress and toxicity. *Journal of Toxicology* 2012, **2012**: 645460.

52. Renschler MF. The emerging role of reactive oxygen species in cancer therapy. *European Journal of Cancer* 2004, **40**(13): 1934–1940.

53. Yang H, Villani RM, Wang H, Simpson MJ, Roberts MS, Tang M, and Liang X. The role of cellular reactive oxygen species in cancer chemotherapy. *Journal of Experimental & Clinical Cancer Research* 2018, **37**(1): 266.

54. Irwin ME, Rivera-Del Valle N, and Chandra J. Redox control of leukemia: From molecular mechanisms to therapeutic opportunities. *Antioxidants & Redox Signaling* 2013, **18**(11): 1349–1383.

55. Brodin NP, Guha C, and Tomé WA. Photodynamic therapy and its role in combined modality anticancer treatment. *Technology in Cancer Research & Treatment* 2015, **14**(4): 355–368.

56. Weinberg F, Ramnath N, and Nagrath D. Reactive oxygen species in the tumor microenvironment: an overview. *Cancers* 2019, **11**(8): 1191.

57. Hwang HS, Shin H, Han J, and Na K. Combination of photodynamic therapy (PDT) and anti-tumor immunity in cancer therapy. *Journal of Pharmaceutical Investigation* 2018, **48**(2): 143–151.

58. Poillet-Perez L, Despouy G, Delage-Mourroux R, and Boyer-Guittaut M. Interplay between ROS and autophagy in cancer cells, from tumor initiation to cancer therapy. *Redox Biology* 2015, **4**: 184–192.

59. Oh J, Yoon H, and Park J-H. Nanoparticle platforms for combined photothermal and photodynamic therapy. *Biomedical Engineering Letters* 2013, **3**(2): 67–73.

60. Wang C, Tao H, Cheng L, and Liu Z. Near-infrared light induced in vivo photodynamic therapy of cancer based on upconversion nanoparticles. *Biomaterials* 2011, **32**(26): 6145–6154.

61. Wen J, Xu Y, Li H, Lu A, and Sun S. Recent applications of carbon nanomaterials in fluorescence biosensing and bioimaging. *Chemical Communications* 2015, **51**(57): 11346–11358.

62. Luo M, Shen C, Feltis BN, Martin LL, Hughes AE, Wright PF, and Turney TW. Reducing ZnO nanoparticle cytotoxicity by surface modification. *Nanoscale* 2014, **6**(11): 5791–5798.

63. Senge MO and Brandt JC. Temoporfin (Foscan®, 5, 10, 15, 20-tetra (m-hydroxyphenyl) chlorin): A second-generation photosensitizer. *Photochemistry and Photobiology* 2011, **87**(6): 1240–1296.

64. Mai Y, Qu X, Ding S, Lv J, Li X, Gao P, Liu Y, and Yuan Z. Improved IR780 derivatives bearing morpholine group as tumor-targeted therapeutic agent for near-infrared fluorescence imaging and photodynamic therapy. *Dyes and Pigments* 2020, **177**: 107979.

65. Wang K, Zhang Y, Wang J, Yuan A, Sun M, Wu J, and Hu Y. Self-assembled IR780-loaded transferrin nanoparticles as an imaging, targeting and PDT/PTT agent for cancer therapy. *Scientific Reports* 2016, **6**(1): 1–11.

66. Sun W, Zhao X, Fan J, Du J, and Peng X. Boron dipyrromethene nano-photosensitizers for anticancer phototherapies. *Small* 2019, **15**(32): 1804927.

67. Zuluaga M-F and Lange N. Combination of photodynamic therapy with anti-cancer agents. *Current Medicinal Chemistry* 2008, **15**(17): 1655–1673.

68. Meuthen D, Rick IP, Thünken T, and Baldauf SA. Visual prey detection by near-infrared cues in a fish. *Naturwissenschaften* 2012, **99**(12): 1063–1066.

69. Keereweer S, Van Driel PB, Robinson DJ, and Lowik CW. Shifting focus in optical image-guided cancer therapy. *Molecular Imaging and Biology* 2014, **16**(1): 1–9.

70. Baskaran R, Lee J, and Yang S-G. Clinical development of photodynamic agents and therapeutic applications. *Biomaterials Research* 2018, **22**(1): 1–8.

71. Schweitzer VG and Somers ML. PHOTOFRIN-mediated photodynamic therapy for treatment of early stage (Tis-$T_2N_0M_0$) SqCCa of oral cavity and oropharynx. *Lasers in Surgery and Medicine: The Official Journal of the American Society for Laser Medicine and Surgery* 2010, **42**(1): 1–8.

72. Lorenz KJ and Maier H. Photodynamic therapy with meta-tetrahydroxyphenylchlorin (Foscan®) in the management of squamous cell carcinoma of the head and neck: Experience with 35 patients. *European Archives of Oto-Rhino-Laryngology* 2009, **266**(12): 1937.

73. Thomas SR and Khuntia D. Motexafin gadolinium: A promising radiation sensitizer in brain metastasis. *Expert Opinion on Drug Discovery* 2011, **6**(2): 195–203.

74. Bradley KA, Zhou T, McNall-Knapp RY, Jakacki RI, Levy AS, Vezina G, and Pollack IF. Motexafin-gadolinium and involved field radiation therapy for intrinsic pontine glioma of childhood: A children's oncology group phase 2 study. *International Journal of Radiation Oncology Biology Physics* 2013, **85**(1): e55–e60.

75. Azzouzi A-R, Vincendeau S, Barret E, Cicco A, Kleinclauss F, van der Poel HG, Stief CG, Rassweiler J, Salomon G, and Solsona E. Padeliporfin vascular-targeted photodynamic therapy versus active surveillance in men with low-risk prostate cancer (CLIN1001 PCM301): An open-label, phase 3, randomised controlled trial. *The Lancet Oncology* 2017, **18**(2): 181–191.

76. Huggett MT, Jermyn M, Gillams A, Illing R, Mosse S, Novelli M, Kent E, Bown S, Hasan T, and Pogue B. Phase I/II study of verteporfin photodynamic therapy in locally advanced pancreatic cancer. *British Journal of Cancer* 2014, **110**(7): 1698.

77. Kochneva EV, Filonenko EV, Vakulovskaya EG, Scherbakova EG, Seliverstov OV, Markichev NA, and Reshetnickov AV. Photosensitizer Radachlorin®: Skin cancer PDT phase II clinical trials. *Photodiagnosis and Photodynamic Therapy* 2010, **7**(4): 258–267.

Chapter 5

Nanotechnology in Textiles

Subhankar Maity,[a] Sukumar Roy,[b] and Ashis Narayan Banerjee[c]

[a]*Assistant Professor, Department of Textile Technology, Uttar Pradesh Textile Technology Institute, India*
[b]*Senior Scientific Officer, Man Made Textiles Research Association, Surat, India*
[c]*Senior Lecturer (Textiles Processing), Indian Institute of Handloom Technology, Guwahati, India*
maity.textile@gmail.com

The demand for textile materials is increasing exponentially beyond their application as apparel and garments. Gradually, textiles are occupying the market for various functional and high-performance applications due to the development of performance properties. The advantage of the application of nanomaterials onto the textile surface is the generation of an active surface with high functionality and performance without affecting the textile properties such as durability, flexibility, and comfort. Various efforts have been made worldwide to create the nanostructure surface of textiles of high precision for obtaining the super functionality. The advent of

Science and Applications of Nanoparticles
Edited by Waqar Ahmed and Ehsan Nourafkan

ISBN 978-981-4267-34-2 (Hardcover), 978-1-003-28029-3 (eBook)
www.jennystanford.com

nanoscience and nanotechnology gives us an immense opportunity to produce functional textiles with spectacular properties suitable for various applications, including water repellency, wrinkle resistance, antistatic, electrical conductivity, antibacterial effect, sensors, UV protection, self-cleaning, and many more. This chapter reviews all the advanced nanomaterial for textiles, their methods of application, performance, and end use. The toxic effect of the nanomaterials due to exposure and their release during the washing of textiles, risk factors, and environmental effects are also discussed.

5.1 Introduction

The word "nano" is derived from the Greek word "nanos," which means "dwarf," i.e., a small or insignificant entity in comparison. So, nanotechnology refers to the technology that relies on the properties of materials at a very, very tiny scale [1–4]. Nanotechnology deals with many complex concepts that cannot be visualized or comprehended easily. Nanoscience and nanotechnology cover a wide range of domains, including physics, chemistry, biology, medicine, electronics, engineering, etc. Nanotechnology is defined as the design, characterization, production, and application of materials, devices, and systems by controlling shape and size in nanoscale. The materials, devices, and systems produce fundamentally new properties and functions by the deliberate and controlled manipulation, precision placement, measurement modeling, and production of the materials at the nanoscale range. Some other scientists defined the nanotechnology as a group of emerging technologies in which the structure of the material is controlled at the nanometer scale to produce novel materials and devices that are useful and unique in the property. It is a technology of atomic precision.

Nanoscience and nanotechnology provide many opportunities for product development, promising a bright future of textile industries. Scientists, researchers, and manufacturers are attracted toward exploring various nanomaterials and applications of the same onto textile surfaces for performance enhancement and new business dimensions. In today's scenario, conventional textile products are not satisfying the consumers' needs in the competitive market, with specific demand and performance, thereby resulting in the low-profit margin. Textile business is, therefore, struggling in spite of high-speed production machinery and enormous capital investment. To compete in the market, the development of new products with high performance is required, and the nanotechnology route can be a fruitful way of finding the path for product development. Therefore, many researchers started and established a suitable methodology for the development of featured products with improved quality and performance by employing the concept of nanoscience and technology using nanomaterials. Though there are tremendous researches that have been published in the literature; a few have been adopted by industries, and the research is still in the infant stage that requires more attention. Research with nanotechnology has commenced in almost all fields of textile applications, including tissue engineering, self-cleaning, wettability, stain resistance, soil release, antibacterial, abrasion resistance, UV protection, etc.

Nanotechnology deals with the particles that have a size smaller than 100 nm. The property of a particle drastically changes when its size reduces in the nanometer scale. Because, as the diameter decreases, the specific surface area of the particle increases, which increases the more accessible surface area of the active agent particles. As a result, though the nanoparticles are not visible by the naked eye, their activity and performance can be felt and measured. This is the principle of enhancing the performance of conventional textiles by employing nanotechnology.

In this chapter, various nanomaterials that are used for textile finishing, their application methods, performance properties, durability properties, and limitations are discussed.

5.2 Nanofibers and Nanoengineered Textiles Manufactured by Electrospinning

Normally nanofibers are manufactured by the electrospinning technology, which is described in Fig. 5.1. A basic electrospinning system consists of a charged polymer solution (or melt) that is fed through a small opening or nozzle (usually a needle or pipette tip). The polymer solution or melt is highly charged; a high potential difference is created between the polymer and the collecting plate. Because of the potential difference, the polymer solution is extruded and drawn toward a grounded collecting plate, which is usually a metal screen, plate, or a rotating mandrel. The distance between the plate and nozzle is typically 5–30 cm, and as a result, during traveling from the nozzle, the polymer solution dries up due to solvent evaporation and nanofilaments are produced, which are accumulated onto the plate in a controlled manner as required. The manner of accumulation decides the formation of nanofilaments or nanoweb composed of nanofibers with diameters between 50 nm and 10 microns [5]. These nanofibers and nanotextiles are used for various functional and high-performance applications, as mentioned in the following paragraphs:

- Preparation of textile structural composites that are strong, tough, durable, and have good damping quality.
- Preparation of filaments for incandescent light sources with decreased susceptibility to mechanical damage because of yarn toughness and mechanical damping ability.
- Manufacturing of protective clothing that provides antiballistic, antistatic, UV protection, radio and microwave protection, etc.
- Development of chemically or electrically powered artificial muscles for prosthetics and robots, morphing air vehicles, and minimally invasive catheters with enhanced functionality for medical applications.

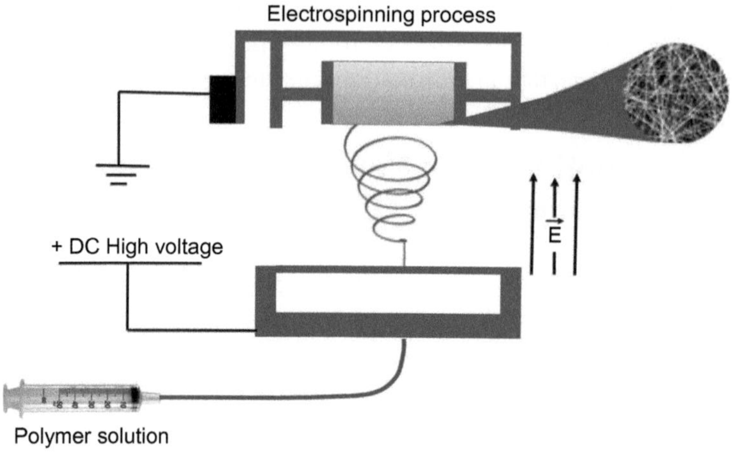

Figure 5.1 Electrospinning technology.

- Fabrication of supercapacitors, batteries, and fuel cells in the form of yarn structures that is weavable into textiles for storing or harvesting electrical energy.
- Developments of heat pipes that provide both structural reinforcement and better heat dissipation.
- Fabrication of electrical wiring and distributed sensors for electronic and smart textiles, etc.

Other than these nanofibers, conventional textile materials can be coated with various nanomaterials, nanoparticles, and nanofilms of active agents to develop enhanced functionality such as water and oil repellence, easy case, softening, antimicrobial, UV resistance, antistatic, wrinkle resistance, strength enhancements, etc. A few such applications of nanofibers and nanofinished textiles are shown in Fig. 5.2 [6].

Figure 5.2 Applications of nanotechnology in textiles.

5.3 Nanofinishing of Textiles

Nanofinishing is a new field of research in the area of textile finishing where novel functional textile products have been developed with an active surface that can serve one or a few specific properties with much better efficiency and durability. The textile surface can be activated by embedding various nanomaterials like man-sized ZnO, TiO_2, SiO_2, Al_2O_3, Ag, carbon nanotubes (CNT), etc. By embedding these nanomaterials on the textile surface, various functional effects can be achieved. Depending on nanomaterials and their degree of loading, such activated textiles can possess various properties like water and oil repellence, self-cleaning, softening, UV resistance, antimicrobial, etc.

5.3.1 Achieving Water and Oil Repellence Property through Nanotechnology Route

There are many shreds of evidence in nature on water repellence like a lotus leaf and bird feather. These naturally existing designs inspired various researchers and scientists to fabricate superhydrophobic textile materials, and the technology is called bioinspired water-repellent materials [7]. Water repellency or superhydrophobic property of textiles can be achieved by introducing nanowhiskers made of hydrocarbons on the surface of the textile fibers. The dimension of the nanowhiskers is three orders of magnitude smaller than the diameter of textile fibers. Due to the presence of nanowhiskers on the textile surface, a peach fuzz effect is created, which is called the lotus effect, where water droplets cannot wet the surface and roll out [8]. The surface tension of the superhydrophobic textile surface is enhanced and is responsible for this repellency effect [9, 10]. The nanowhiskers can allow air to pass through the textile fabric, which exhibits breathability property that is essential for garments. There are other approaches to developing such superhydrophobic textiles using nanotechnology. A 3D surface structure is created on the fabric surface by coating with nanoparticulate films comprising gel-forming additives [11]. Cotton textiles are coated with fluorocarbon derivative nanoparticles by an audio frequency plasma treatment [12]. In this process, superhydrophobicity is created by introducing roughness on the surface where the abrasion resistance and softness of the fabric do not deteriorate. Silica (SiO_2) nanoparticles can also be used to combine water-repellent agents to create hydrophobicity in textiles [13]. Silica (SiO_2) nanoparticles, when coated over cotton fabrics combined with a water-repellent agent by a sol–gel process, can achieve water contact angles above 130°. When perfluorooctylated quaternary ammonium silane coupling agent (PQASCA) is used with SiO_2 nanoparticles, surface roughness on the cotton increases, surface energy decreases, and as a result, water contact angle increases to 145° [14]. This cotton fabric shows oil repellency with a contact angle of 131°. Epoxy-containing poly (glycidyl methacrylate) (PGMA) and SiO_2 nanoparticles are coated on the textile surface to create primary reactive nanolayers, and then different functional groups such as carboxy, amino, hydroxyl, etc. are grafted for achieving the lyophilic effect. Silicone-nanofilaments-coated polyester fabric treated with plasma fluorination to generate superoleophobic textiles shows oil repellency effect [15]. The textile surface is also coated with amphiphilic Janus micro and nanoparticles that form crosslinks between fibers to immobilize them and exhibit water repellence [16]. Chitosan is used for coating textile fabrics to simulate the microstructure of bird feathers using a surface solution precipitation method followed by improvement with a silicone compound resulted in low surface energy [17]. Cotton textiles are coated with pristine and surface-modified carbon nanotubes (CNTs) to create a lotus effect as shown in Fig. 5.3 [18]. The cotton surface possesses a water contact angle of more than 150°. Combined hydrophobic and oleophobic properties can be imparted to cotton textiles by coating with silica nanoparticles to develop surface roughness, followed by hydrophobization with the attachment of poly (dimethylsiloxane) (PDMS). The coated fabric shows the static water contact angle of 155° [19]. To introduce oleophobicity, the silica particles are treated with a perfluoroalkyl chain on the fibers, which exhibit oleophobicity with a static contact angle of 140° for oil droplets.

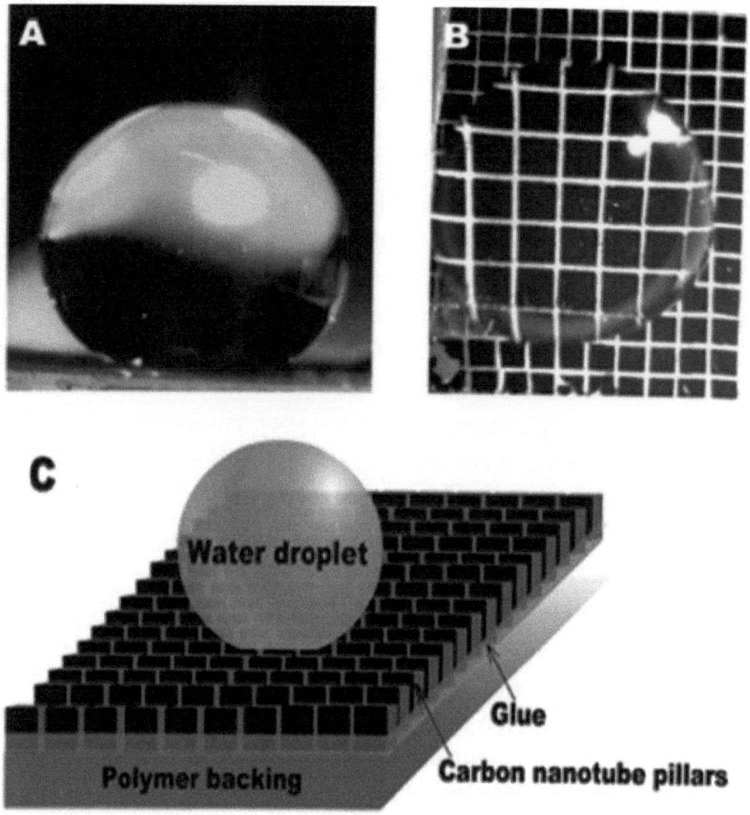

Figure 5.3 Superhydrophobic behavior of micropatterned carbon nanotube pillars of 250 μm in width and 100 μm in height. (A) A 10 μL water droplet sitting on the surface of vertically aligned carbon nanotubes. (B) Top view of water droplet sitting on the micropatterned carbon nanotube pillars. (C) Schematic diagram showing carbon nanotube pillars held at base by a polymer adhesive and a water droplet sitting on the top of pillars [18].

5.3.2 Antistatic Properties by Nanotechnology

Synthetic textile fibers such as polypropylene, polyester, nylon, acrylic, etc. exhibit high static charge due to hydrophobicity and that causes various nuisance during processing and use of the textiles. Antistatic nanofinish can be applied onto such textile materials with various nanoparticles like TiO_2 [19], ZnO [20], antimony (Sb) doped tin oxide (SnO_2), etc. These nanoparticles bring about sufficient electrical conductivity to the textiles, and as a result, quick static dissipation occurs. Another approach to the development of the antistatic property to synthetic textiles is generating hydrophilic groups on the textile surface. The application of silences on the textile surface by the nano-sol technique is an example of such a technique. The silanes can absorb moisture from the atmosphere and help to enhance electrical conductivity. In another approach, poly (tetrafluoroethylene) (PTFE) is coated over the textile surface as an antistatic membrane [21]. PTFE creates a frictionless surface and reduces the mechanical friction between the surface and thereby reduce the generation of static charge.

5.3.3 Enhancement of Wrinkle and Crease Recovery Properties of Textiles

Cotton textiles are prone to crease formation and shrinkage due to rigorous inherent intermolecular crosslinking upon drying. Nanocoatings that block all active cites responsible for the formation of crosslinks can prevent crease formation. Usually, crosslinking resins are used as the anti-wrinkle agents for this purpose, but the treatment decreases tensile strength, abrasion resistance, and dyeability of the fabrics. Nanoparticles can successfully produce wrinkle resistance finish to cotton and silk fabrics. TiO_2 nanoparticles with carboxylic acid as a catalyst are found to produce crosslinks between cellulose molecules and the acidic groups to produce the effect [22]. TiO_2 nanoparticles in chitosan used to treat *Bombyx mori* silk by the crosslinking reactions of citric acid and maleic anhydride [23]. The treated silk exhibited dry and wet delay-wrinkle recovery angles of 267 and 250° compared to the untreated fabric of 235° and 178°, respectively.

5.3.4 Enhancement of Mechanical Properties of Textiles by Nanotechnology

Various textile composites can be fabricated using various nanomaterials as either reinforcement or filler materials. Carbon nanotubes (CNTs) in various forms like single-walled (SWCNT) and multi-walled (MWCNT) are the most successful material used in composites for enhancing the strength, toughness, and modulus of textile materials [24]. CNTs can be coated over cotton textiles by a simple dipping method [7]. Some binders such as poly (butyl acrylate) can be useful for the surface grafting of CNTs over the textile surface by the processes of dipping, drying, and subsequent curing. The CNT-coated cotton fabrics have improved tensile strength in both weft and warp directions [7]. Nanostructured polycaprolactone nonwoven mats are manufactured by melt-spinning with improved mechanical properties. Melt-spinning of thermoplastic polymers can be done with the dope-mixing of various nanomaterial to enhance mechanical properties and texture. Wet–dry spinning or jet melt-spinning processes are developed to manufacture microfibers (1–100 μm in diameter) but for the manufacturing of nanofibers or filaments, the electrospinning technology is the most successful. CNTs can be added to the solution of other polymers for electrospinning to produce nanofibers with various levels of twist. Such CNT composite highly twisted fibers exhibited high strength, toughness, and energy-damping capabilities, which are favorable for application in electronic textiles such as sensors, actuators, energy storage, electrostatic discharge protection, heating, radio, and microwave absorption, etc.

5.3.5 Nanofinished Textiles for Protection from Ultraviolet (UV) Radiation

Conventional UV blockers are toxic and thermally unstable. The nano-UV blockers are inorganic, nontoxic, and chemically and thermally stable. TiO_2 and ZnO nanoparticles are such materials that are very efficient to absorb and scatter UV radiation, as shown in Fig. 5.4 [25–28]. These nanoparticles have a size range of 20–40 nm and can be able to scatter UV radiation of 200–400 nm wavelength. Textile fabrics are successfully coated with such nanomaterials by the sol–gel method, and as a result, the textile fabric becomes a UV shield, and the effect can sustain up to 50 launderings cycles [29–31].

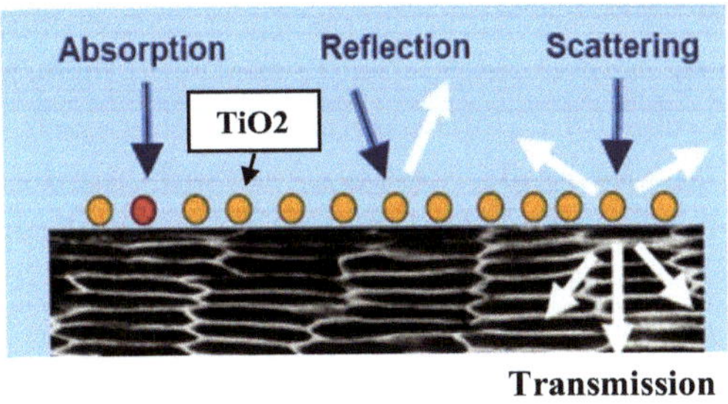

Figure 5.4 Mechanism of UV protection by TiO_2-nanoparticles-coated textiles.

5.3.6 Antibacterial Nanofinishing of Textiles

Natural textile fibers like cotton, wool, silk, etc. are liable to bacterial attacks for prolonged unwashed use in day-to-day life. Synthetic fibers are also helping the growth of the microbes in cloths. As preventive measures, antibacterial disinfection and finishing techniques have been developed by treating various natural and synthetic textile with silver nanoparticles, TiO_2 nanoparticles, zinc oxide (ZnO) nanoparticles, etc. In most cases, the treatment process is the preparation of their colloidal solutions (25–50 ppm) followed by a dip-dry-cure process. These nanosized metallic ions and compounds trigger a photocatalysis reaction in the presence of UV or visible light and atmospheric oxygen and generate O^{2-} radical ions that destroy or kill microorganisms which are in contact. The advantage of the use of these nanosized particles is that huge numbers of particles can be accumulated per unit area over the textile surface. As a benefit, the antibacterial efficiency can be maximized [27–35].

5.3.7 Self-Cleaning Textiles

Self-cleaning textiles and garments retain their original texture and feel after the chemical treatment. Textiles that can keep themselves clean can save a lot of water, detergent, and energy by avoiding frequent washing. Using the technologies mentioned above, self-cleaning fabrics can prevent dirt, oil, and also act as a disinfectant. Moreover, the self-cleaning effect keeps the textiles long lasting and fresh looking than ordinary fabrics. Anti-dirt, antibacterial, and self-cleaning clothes can be used in medical, sports, defense, and home textiles widely. A few commercial products are available in the global market with this self-cleaning quality. One example is Mincor® TX TT, which can be used for tailoring outdoor textiles like tents, umbrellas, sunshades, flags, and sails. NanoTex® is another product suitable for apparel like men's shirts, dress materials, etc. Nanosphere® is mainly used for the preparation of men's shirts with a self-cleaning effect. It is a product where water drops or particles of dirt lie only on the peaks of the nanoparticles and, therefore, have a lower contact area. Adhesion is significantly reduced, water runs off, dirt is repelled, or can simply be rinsed off. Textiles require less frequent washing and can be washed at lower temperatures. It has a high level of abrasion resistance. The function is maintained even after numerous washing or cleaning cycles. Aesthetic look, feel, and breathability are not affected. NanoSphere® is ideal for use in clothing for outdoor, sportswear, men's wear, women's wear, workwear, as well as for shoes and home furnishings (www.nano-sphere.ch).

5.4 Nanopretreatments of Textile: Nanoscouring, Nanobleaching, Nanosoftening

With the advent of nanoscience and nanotechnology, a new field of research and innovations in textile wet processing has emerged, called nanopretreatment and nanofinishing. Pretreatment processes of textiles like scouring, bleaching, and softening can be achieved by applications of various nanomaterials and processes. The scouring is an essential pretreatment process for all natural fibers like cotton, wool, silk, jute, flax, etc. where there is natural wax protecting the fiber surface, and removal of the same is difficult. In the case of human-made fibers, oils are manually applied on the surface of the fibers as spin finishes to improve the spinning process. Nanotechnology is a suitable alternative process for scouring textile fibers.

Nanosized TiO_2 particles are the best nanoparticles explored in material science due to their low bandgap energy. When TiO_2 nanoparticles are irradiated with light rays possessing energy higher than their bandgap, electrons and holes pairs are liberated from their surface. This phenomenon is called the photocatalytic effect. Attributing to this photocatalytic effect, TiO_2 has been explored for various applications of textiles such as water purification, UV protection, self-cleaning, antibacterial effect, etc. The same photocatalytic event of TiO_2 nanoparticles can be used for the decomposition of natural oil and wax impurities and coloring substances from grey cotton fabrics to improve whiteness and hydrophilicity. The process is termed nanophotoscouring. This hydrophilicity of cotton fabrics is attributed to the decomposition of the surface wax through photocatalytic action of the TiO_2 nanoparticles under sunlight [36]. Nanophotoscouring is also applicable for protein fibers like wool and silk. Wool has natural waxes consisting of esters of various long-chain fatty acids having long-chain alcohols and steroids. There are esters of 1- and 2-alkanols and 1,2-diols, triterpene alcohols, sterol esters, free acids, and sterols available in the wool wax. The photocatalytic decomposition of the esters occurs during sunlight irradiation, and esters are converted into aldehyde and alcohols first and further decomposed until removed from the wool surface. Various oxidizing radicals, such as OH*, RO*, and RO_2*, which are generated on TiO_2 surface during sunlight irradiation, are responsible for the photocatalytic decomposition [37]. ZnO nanoparticles are *in situ* synthesized over the raw wool surface, and treated wool fabrics are kept in sunlight for 7 days for obtaining scoured and hydrophilic wool fabrics [38]. The nano-ZnO particles generate positive holes and electrons due to the irradiation of UV rays present in sunlight. The electrons are responsible for reduction reactions of Zn^{2+} to Zn^{1+} and generation of the anions of oxygen (O^{2-}). Whereas the holes are responsible for the oxidation reaction of water molecules and the anions of oxygen (O^{2-}) and hydroxyl (–OH) groups. The generation of hydrophilic hydroxyl groups improves the hydrophilic properties of the wool fabrics, as shown in Fig. 5.5. Therefore, the photocatalytic activity generates abundant hydroxyl groups on the wool surface, resulting in improved hydrophilicity and a smaller water contact angle.

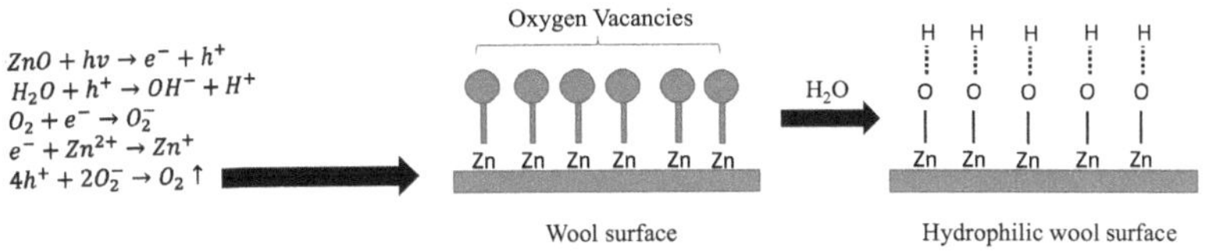

Figure 5.5 Generation of the hydrophilic surface of wool by ZnO nanoparticle-induced nanophotoscouring.

Although surface wax, oil, and impurities can be removed by the scouring process, coloring matters remain on the fiber surface, attributing to the yellow and brown appearance of the fibers. These coloring matters need to be destroyed to obtain whiter appearance without causing a detrimental effect on the fiber structure and properties. Scoured cotton fabrics need to be coated with nano-TiO_2 particles and kept in sunlight or irradiated with UV rays in the laboratory. As a result, oxidizing radicals produced by the photocatalytic effect of TiO_2 in the presence of UV rays attack the coloring matters, pectin, and hemicellulose possessing polysaccharide and break them down. The bleaching mechanism by discoloring natural pigments on natural fibers is attributed to the generation of oxidizing radicals such as OH*, RO*, and RO^{2*} by the TiO_2 nanoparticles. This mechanism of bleaching is called nanophotobleaching, as shown in Fig. 5.6.

Wool fabrics are used for photobleaching by treatment with nano-crystallized TiO_2 particles [36]. TiO_2 particles absorb UV-C rays and prevent the photo-oxidation of wool, which is responsible for the yellowing. In wool fibers, the natural pigments are eumelanin and pheomelanin, which give wool a creamy or pale yellow color. When UV rays of energy more than the bandgap of ZnO are incident on ZnO particles, they generate electrons in their conduction band and holes in their valence band. Holes have strong oxidation potentials and oxidize water molecules to produce hydroxyl radicals (*OH). In contrast, electrons have strong reduction potential and reduce oxygen molecules into the anions of superoxide radicals. These hydroxyl radicals and anions of superoxide radicals destroy various coloring substances present on the wool surface [38].

Softness is the most important quality of textile cloths from the consumers' point of view. The comfort of the clothing is mostly related to the softness of the fibric, and the purchase decision of the garment solely depends on this softness. Household softeners are available in the retail market, which are added during washing or after wash to get the gentle feel of the fabric for more comfortable wear. These treatments are temporary and require application during each washing cycle. There are semi-permanent and permanent softeners used in the industry to obtain cost-effective softening treatment, which is wash durable, resistant to high temperatures, non-yellowing, and non-detrimental for dye-shade. Like nanopretreatments

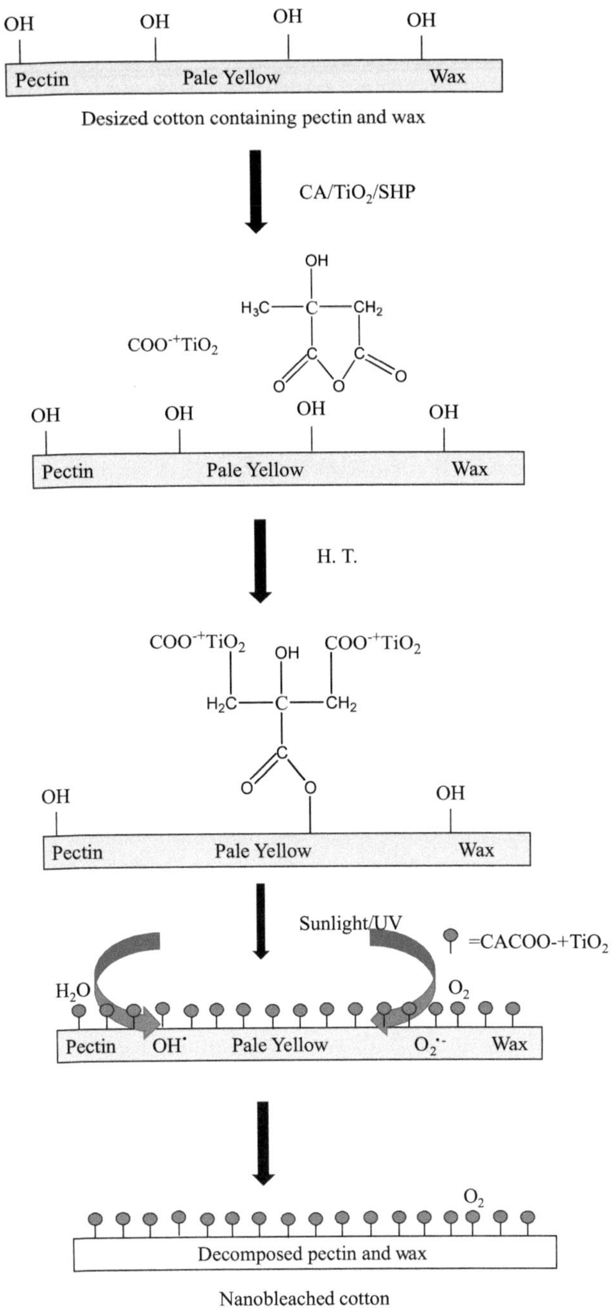

Figure 5.6 The schematic presentation of nanophotoscouring and nanophotobleaching of cotton fiber by TiO_2 nanoparticles.

and nanofinishing methods, nanosoftening methods are also available through the science of nanotechnology. The advantage of nanosoftening agents is their nanometer size, which enables them to easily penetrate the fiber structure and result in a plasticizing action by enhancing the molecular mobility and bringing down the glass transition temperature of the fiber. As a result, the mechanical response of the fiber becomes easier at relatively low temperature, and the material behaves softer. In this regard, nanoemulsions of silicone-based softeners have been successfully applied to textile materials for achieving high levels of softness. By another approach, the softness of textile materials can be achieved by reducing fiber-to-fiber friction. In this approach, nanoplatelets such as nanolayers of silicate (nanoclay) or nanographene platelets are applied on the textile surface, which forms their nanolayers and reduces friction [39–41].

5.5 Nanodyeing

Nanotechnology seeks to supply and apply the knowledge of the behavior of objects within the nanometer (nm) range to the assembly of complex structures to be used during a sort of practical application. The tiniest substances promise to rework industry and make an enormous market. In chemicals, cosmetics, pharmaceuticals, technology, and textiles, businesses are researching and manufacturing products to support nanotechnology, which uses bits of matter measured in billions of meters. Nanotechnology adds its labels to popular clothing brands with various products: Resists Spills, Resists Static, Coolest Comfort, and Repels and Releases Stains. Researchers all around the world are watching all kinds of metal additives and polymer additives, inorganic, organic materials to require them at the nanoscale to impart many interesting properties to textiles.

Nanodyeing is done for the following reasons:

- To reduce power consumption, heating, and process time in the process of textile dyeing [42].
- To reduce water consumption in the process of dying.
- To decolorize the effluent water with the discharging of wastewater with lower TDS.

- To refrain from using harmful chemicals or less emission of harmful material to the environment.
- Begin to earn a big profit in process time.

The revolutionary dyeing technology with nano dyes now enables cotton exhaust dye jets to use up to 99% of dyestuffs without salt and exhaust (eliminating solid waste), using 75% less water and 90% less oil while providing more shade reproducibility and better-quality fabric in all colors. Furthermore, rising the waste in the effluent allows for zero discharge.

On the other hand, plant colorants can also be used as an alternative to synthetic coloring to dye natural silk fibers [43]. Using a Ball Mill grinder, natural colors are processed into nanoscale powder. Silk fabrics are used as a substratum and dyed with various natural and synthetic mordants. The use of multiple solvents for extraction and mordant for dyeing resulted in different shades of color on silk fabrics. Colorfastness to washing tests reported colorfastness to wash from 3 to 4/5 and color change from 3/4 to 4/5. Colorfastness to the washing of dyed cotton yarn was assessed in a launder-o-meter as per a method prescribed in IS 105-C10 (2006).

The nanoscale dye powder would help the dyeing and printing industries such as batik manufacturers and the local fashion industry by using natural colors. Synthetic fibers such as polyester create a major component among synthetic fibers, commonly dyed with dispersed colors. The powder was developed mainly in the size of μm (10–6), but some were in the form of nanosized powder in the range of 200 to 800 nm (10–9), making it possible to dye polyester fabrics at a lower temperature.

It can be concluded as a whole that the nano-dye method needs only one "drop-in" pretreatment step for the greige products directly from knitting and a modification of the exhaust dye jet process, raising the performance and reducing the overall cost of dyeing. Due to their concentration in emerging economies, the dyeing industry needs best practices and disruptive approaches to prevent further harm to our quantum climate. The future of sustainable exhaust dyeing has arrived with the nano-dye system after years of research and development. The resulting fabric has a cleaner tone of color, a smoother hand, and a pleasant smell and color.

5.6 Process Involved in the Preparation of Dye Nanoparticles

5.6.1 Ultraviolet and Microwave Treatment

Ultraviolet and microwave treatment can impart better colorfastness properties to the modified fabrics. Some researchers have found that microwave treatment improves the extraction of color from their natural source, which offers deep shading to the fabrics. Microwaves can equitably and effectively enter a medium and consistently heat the substance. During microwave treatment, polar groups of the medium are excited, which results in heat creation. Past investigations have demonstrated that microwave treatment did not change the microfibrils of cellulose, and the physical properties, for example wrinkling and tensile strength, were highly improved [44].

5.6.2 Ultrasound Treatment

Conventional dyes are subjected to an innovative process to convert them to a nanoscale size by sonication via an ultrasonic homogenizer. Ultrasound improves a wide range of both chemical and physical processes; mostly because of the phenomenon known as cavitation inside the liquid medium, which is the development and breakdown of tiny bubbles that can create a hot spot, for example, restricted high temperature and pressure, shock waves, and extremely shear forces equipped for breaking bonds (http://www.egr.msu.edu/cmsc/biomaterials/research.html). Many researchers have been investigating this method in textile coloration as it is a noteworthy wet process, which requires high energy and water and releases a huge amount of effluent into the environment. Besides, any enhancements in coloration processes are generally ascribed to the cavitation phenomena, and, as an outcome, other mechanical and chemical impacts are listed below:

- Decrease any agglomerations with high molecular mass;
- Remove any dissolved/entrapped air from cavities;
- Accelerate the diffusion rate of dye;
- Powerful agitation of the liquid;

- Break down the diffusion layer interfaces;
- Produce free radicals.

5.6.3 Mechanical Milling

The dyes are ground using an energy Ball Mill with a speed of 50 cycles/min. The dye powder was sealed in a hardened steel vial (AISI 44°C stainless steel) using hardened steel balls of 6 mm diameter. Milling was performed employing a ball: the powder mass ratio was 4:1. The dye was milled at different intervals; after each milling interval, the particle size of the resulting dye powder was measured. The smallest particle size of around 40 nm was chosen to be used in the present study and was obtained from milling the dye powder for 30 days. A stock solution of concentration x% was prepared using the milled dye particles, where x gram dye powder was dispersed in $(100 - x)$ cm^3 of distilled water. The suspension was irradiated afterward with ultrasound waves (720 kHz) and stirred at 80°C for different periods (4, 6, and 8 h). The behavior of the fabrics dyed with the ultrasonic method was also studied, and the build-up of dyes was measured as color strength in terms of K/S values. From this, the ultrasound treatment of 8 h was chosen to proceed with since it gave the best K/S values. Higher K/S values are obtained when the dye particles are anchored in the fiber matrix. The K/S (Kubelka-Munk function) values using the Spectra Scan 5100+ computer color matching system 17 give the receptivity of textiles fabrics dyed with nanodyes according to Eq. (5.1).

$$\frac{K}{S} = \frac{(1-R)^2}{2R} \tag{5.1}$$

where K is the absorption coefficient, S is the scattering coefficient, and R is the reflectance of the dyed fabric at the wavelength of maximum absorption.

5.7 Testing and Analysis of Nanomaterials

The morphological features of nanomaterials and structures must be determined in various stages of production, like size distribution, porosity, pore size distribution, surface structure, and composition, which are critical to making sure the materials and structures are in nanoscale to achieve special properties. Such structural feature can be characterized by a range of techniques and instruments such as particle size analyzer, atomic force microscopy (AFM), scanning tunneling microscopy (STM), scanning electron microscopy (SEM), high-resolution transmission electron microscopy (TEM), X-ray photoelectron spectroscopy (XPS), auger electron spectroscopy (AES), near field scanning optical microscopy (NSOM or SNOM), X-ray and neutron diffraction, and other spectroscopic techniques [45]. The properties of nanomaterials and structures/composites also need to be measured to test whether specific and unique properties can be derived due to the nanoscale structures. Properties and characterization should be administered in various aspects about specific functions to be achieved, like in physical, mechanical, electrochemical, optical, and thermal properties. Finally, the functional performances of ultimate products must be characterized to check the roles and impact of nanomaterials and nanoscale structures. Only when new and/or significantly enhanced properties are introduced, due to the nanoscale sizes or structures, can nano-products be claimed, based on the definition of nanotechnology.

5.8 Economic and Environmental Aspects of Using Nanoparticles

The unique properties of nanomaterials have attracted not only scientists and research workers but also businesses. By using fewer resources in terms of energy, water, chemicals, and auxiliaries without sacrificing performance, nanodyeing may save raw materials and also upgrade the quality of life (http://www.nanophase.com/applications/, 2002). Moreover, this process emits less aqueous pollutants and harmful chemicals to the water bodies [46].

While nanotreated textiles have been found effective in many applications, their contamination in wastewaters may have certain environmental risks. Due to the large surface area of nanomaterials, they are very reactive to other materials when released from any stages of their life cycle. Therefore, we need to consider both the opportunity and risks of the nanomaterials used for an application. Though Ag nanoparticles are suspected toxic for aquatic life, there are controversial reports available in the literature as mitochondria and DNA damage, or weak toxicity to humans [47]. TiO_2 and ZnO nanoparticles are toxic in the water bodies due to their photocatalytic activity. Whereas carbon nanotubes are less toxic in the water bodies. Considering human exposure, these nanoparticles have poor stabilization properties and form aggregates or agglomerates to become bigger; thus the chance of entering the body is less. But the nanomaterials can be carriers of other harmful materials due to their high active surface area, which will then be dangerous. However, no concerning report has been published regarding this issue. Researchers have tried to carry out cytotoxicity to examine the toxicity of nanotreated textiles specially designed for medical and wound-care applications. But there are controversial reports published. Harmful effects of nanotreated textiles against the environment and human health have not been comprehensively proved yet, and the available data are very controversial depending on nanoparticles' dosage level, size, porosity, shape, stability, and end use. The lack of regulations for the restriction of nanomaterial-based treatments is recognized. Ecolabeling is required for the nanomaterials when they are incorporated into clothes, especially in direct contact with the human body. Efforts are required to explore environment-friendly preparation steps using green chemistry, use of *in situ* synthesis methods discharging low effluent and enhanced durability, and use of nano-based materials with less toxicity for the treatment and finishing of textile materials.

5.9 Conclusion

Overviewing the history of textile pretreatment, dyeing, and finishing processes during the few last decades, there is considerable concern about complexity, energy consumption, wastage of water and chemicals, and effluent load to the environment. Tremendous efforts have been made to minimize those for the development of sustainable processes. In the advent of nanotechnology, suitable nano-treatment processes show the path for sustainable scouring, bleaching, cleaning, dyeing, softening, UV-blocking, antimicrobial, water-repellent finishing, etc. for textiles. These nano-treatment processes guarantee minimal chemical wastage, energy saving, low toxicity, maximum efficiency, and performance. The nano-treatment processes avoid harsh chemical treatment conditions, and as a result, mechanical and comfort properties of the textiles do not deteriorate. The nanopretreatment, nanodyeing, and nanofinishing of textiles have already been proved to be successful alternative methods of existing processes and have a bright future ahead.

References

1. Cherenack, K. and van Pieterson, L. (2012). Smart textiles: Challenges and opportunities. *Journal of Applied Physics*, **112**(9), 091301.
2. Sawhney, A. P. S., Condon, B., Singh, K. V., Pang, S. S., Li, G., and Hui, D. (2008). Modern applications of nanotechnology in textiles. *Textile Research Journal*, **78**(8), 731–739.
3. Gould, P. (2003). Textiles gain intelligence. *Materials Today*, **6**(10), 38–43.
4. Paradiso, R., Loriga, G., Taccini, N., Gemignani, A., and Ghelarducci, B. (2005). WEALTHY-a wearable healthcare system: New frontier on e-textile. *Journal of Telecommunications and Information Technology*, 105–113.
5. Avila, A. G. and Hinestroza, J. P. (2008). Smart textiles: Tough cotton. *Nature Nanotechnology*, **3**(8), 458–459.
6. Yetisen, A. K., Qu, H., Manbachi, A., Butt, H., Dokmeci, M. R., Hinestroza, J. P., et al. (2016). Nanotechnology in textiles. *ACS Nano*, **10**(3), 3042–3068.
7. Liu, Y., Tang, J., Wang, R., Lu, H., Li, L., Kong, Y., and Xin, J. H. (2007). Artificial lotus leaf structures from assembling carbon nanotubes and their applications in hydrophobic textiles. *Journal of Materials chemistry*, **17**(11), 1071–1078.
8. Russell, E. (2002). Nanotechnologies and the shrinking world of textiles, *Textile Horizons*, **9**(10), 7–9.
9. Marmur, A. (2004). The lotus effect: Superhydrophobicity and metastability. *Langmuir*, **20**(9), 3517–3519.
10. Gao, L. and McCarthy, T. J. (2006). The "lotus effect" explained: Two reasons why two length scales of topography are important. *Langmuir*, **22**(7), 2966–2967.
11. El-Khatib, E. M. (2012). Antimicrobial and self-cleaning textiles using nanotechnology. *Research Journal of Textile and Apparel*, **16**(3), 156–174.

12. Zhang, J., France, P., Radomyselskiy, A., Datta, S., Zhao, J., and van Ooij, W. (2003). Hydrophobic cotton fabric coated by a thin nanoparticulate plasma film. *Journal of Applied Polymer Science*, **88**(6), 1473–1481.

13. Bae, G. Y., Min, B. G., Jeong, Y. G., Lee, S. C., Jang, J. H., and Koo, G. H. (2009). Superhydrophobicity of cotton fabrics treated with silica nanoparticles and water-repellent agent. *Journal of Colloid and Interface Science*, **337**(1), 170–175.

14. Yu, M., Gu, G., Meng, W. D., and Qing, F. L. (2007). Superhydrophobic cotton fabric coating based on a complex layer of silica nanoparticles and perfluorooctylated quaternary ammonium silane coupling agent. *Applied Surface Science*, **253**(7), 3669–3673.

15. Artus, G. R., Zimmermann, J., Reifler, F. A., Brewer, S. A., and Seeger, S. (2012). A superoleophobic textile repellent towards impacting drops of alkanes. *Applied Surface Science*, **258**(8), 3835–3840.

16. Synytska, A., Khanum, R., Ionov, L., Cherif, C., and Bellmann, C. (2011). Water-repellent textile via decorating fibers with amphiphilic janus particles. *ACS Applied Materials & Interfaces*, **3**(4), 1216–1220.

17. Liu, Y., Chen, X., and Xin, J. H. (2008). Hydrophobic duck feathers and their simulation on textile substrates for water repellent treatment. *Bioinspiration & Biomimetics*, **3**(4), 046007.

18. Sethi, S., Ge, L., Ci, L., Ajayan, P. M., and Dhinojwala, A. (2008). Gecko-inspired carbon nanotube-based self-cleaning adhesives. *Nano Letters*, **8**(3), 822–825.

19. Weiguo, D. (2002). Research on properties of nano polypropylene/ TiO_2 composite fiber. *Journal of Textile Research*, **23**(1), 22–23.

20. Zhou, Z., Chu, L., Tang, W., and Gu, L. (2003). Studies on the antistatic mechanism of tetrapod-shaped zinc oxide whisker. *Journal of Electrostatics*, **57**(3–4), 347–354.

21. Hoefnagels, H. F., Wu, D., De With, G., and Ming, W. (2007). Biomimetic superhydrophobic and highly oleophobic cotton textiles. *Langmuir*, **23**(26), 13158–13163.

22. Shishoo, R. (2002). Recent developments in materials for use in protective clothing. *International Journal of Clothing Science and Technology*, **3**(4), 201–215.

23. Wang, C. C. and Chen, C. C. (2005). Physical properties of crosslinked cellulose catalysed with nano titanium dioxide. *Journal of Applied Polymer Science*, **97**(6), 2450–2456.

24. Lu, Y. H., Lin, H., Chen, Y. Y., Wang, C., and Hua, Y. R. (2007). Structure and performance of Bombyx mori silk modified with nano-TiO_2 and chitosan. *Fibers and Polymers*, **8**(1), 1–6.

25. Kumar, S., Doshi, H., Srinivasarao, M., Park, J. O., and Schiraldi, D. A. (2002). Fibers from polypropylene/nano carbon fiber composites. *Polymer*, **43**(5), 1701–1703.

26. Xin, J. H., Daoud, W. A., and Kong, Y. Y. (2004). A new approach to UV-blocking treatment for cotton fabrics. *Textile Research Journal*, **74**(2), 97–100.

27. Yang, H., Zhu, S., and Pan, N. (2004). Studying the mechanisms of titanium dioxide as ultraviolet-blocking additive for films and fabrics by an improved scheme. *Journal of Applied Polymer Science*, **92**(5), 3201–3210.

28. Saito, M. (1993). Antibacterial, deodorising, and UV absorbing materials obtained with zinc oxide (ZnO) coated fabrics. *Journal of Coated Fabrics*, **23**(2), 150–164.

29. Xiong, M., Gu, G., You, B., and Wu, L. (2003). Preparation and characterisation of poly (styrene butylacrylate) latex/nano-ZnO nanocomposites. *Journal of Applied Polymer Science*, **90**(7), 1923–1931.

30. Daoud, W. A. and Xin, J. H. (2004). Low temperature sol-gel processed photocatalytic titania coating. *Journal of Sol-Gel Science and Technology*, **29**(1), 25–29.

31. Wang, R., Xin, J. H., Tao, X. M., and Daoud, W. A. (2004). ZnO nanorods grown on cotton fabrics at low temperature. *Chemical Physics Letters*, **398**(1–3), 250–255.

32. Kathirvelu, S., D'souza, L., and Dhurai, B. (2009). UV protection finishing of textiles using ZnO nanoparticles. *Indian Journal of Fibre &Textile Research*, **34**(3), 267–273.

33. Yeo, S. Y., Lee, H. J., and Jeong, S. H. (2003). Preparation of nanocomposite fibers for permanent antibacterial effect. *Journal of Materials Science*, **38**(10), 2143–2147.

34. Lee, H. J., Yeo, S. Y., and Jeong, S. H. (2003). Antibacterial effect of nanosized silver colloidal solution on textile fabrics. *Journal of Materials Science*, **38**(10), 2199–2204.

35. Lee, K. H., Kim, H. Y., Khil, M. S., Ra, Y. M., and Lee, D. R. (2003). Characterisation of nanostructured poly (ε-caprolactone) nonwoven mats via electrospinning. *Polymer*, **44**(4), 1287–1294.

36. Klasen, H. J. (2000). A historical review of the use of silver in the treatment of burns. II. Renewed interest for silver. *Burns*, **26**(2), 131–138.

37. Montazer, M. and Morshedi, S. (2012). Nano photo scouring and nano photo bleaching of raw cellulosic fabric using nano TiO_2. *International Journal of Biological and Macromolecules*, **50**(4), 1018–1025.

38. Gauthier, E., Boyer, C., Thivel, P.X., Delpech, F., and Rou, J. C. (2006). Experimental study of odorous ester photo catalysis. *Environmental Engineering and Management Journal*, **5**(5), 1001–1010.

39. Montazer, M., Maali Amiri, M., and Mohammad Ali Malek, R. (2013). In situ synthesis and characterisation of nano ZnO on wool: Influence of nano photo reactor on wool properties. *Photochemistry and Photobiology*, **89**(5), 1057–1063.

40. Sadeghian Maryan, A., Montazer, M., and Damerchely, R. (2015). Discoloration of denim garment with color free effluent using montmorillonite based nano clay and enzymes: Nano bio-treatment on denim garment. *Journal of Cleaner Production*, **91**, 208–215.

41. Sadeghian Maryan, A., Montazer, M., Harifi, T., and Mahmoudi Rad, M. (2013a). Aged-look vat dyed cotton with antibacterial/anti-fungal properties by treatment with nano clay and enzymes. *Carbohydrate Polymers*, **95**(1), 338–347.

42. Sadeghian Maryan, A., Montazer, M., and Rashidi, A. (2013b). Introducing old-look, soft handle, flame-retardant, and antibacterial properties to denim garments using nano clay. *Journal of Engineered Fibres and Fabrics*, **8**(4), 68–77.

43. Palamutcu, S. (2010). Electric energy consumption in the cotton textile processing stages. *Energy*, **35**(7), 2945–2952.

44. Ahmad, W., Yunus, W., Nor, M., Saim, N., Ab Kadir, M. I., and Ahmad, M. R. (2012). Nano natural dyes from *Melastoma malabathricum* L. *Advanced Materials Research*, **545**, 59–63.

45. Das, S. C., Paul, D., Hossan, S. M., Chowdhery, N., and Bain, S. (2013). Application of Nanotechnology in Textiles: A Review. *International Conference on Mechanical Engineering and Renewable Energy*, 1–3.

46. Oakes, J. and Dixon, S. (2004). Physical interactions of dyes in solution—Influence of dye structure on aggregation and binding to surfactants/polymers. *Coloration Technology*, **34**(1), 110–128.

47. Tarafder, N. (2018). Applications of nanotechnology for textile products: A review. *NanoR*, **1**(3), 15–22.

48. AshaRani, P. V., Mun, G. L. K., Hande, M. P., and Valiyaveettil, S. (2009). Cytotoxicity and genotoxicity of silver nanoparticles in human cells. *ACS Nano*, **3**(2), 279–290.

Chapter 6

An Update on Nanotechnology and Sustainable Agriculture

Allah Ditta[a,b] and Naseer Ullah[c]

[a]*Department of Environmental Sciences, Shaheed Benazir Bhutto University Sheringal, Dir (U), 18000, Khyber Pakhtunkhwa, Pakistan*

[b]*School of Biological Sciences, the University of Western Australia, 35 Stirling Highway, Perth, WA 6009, Australia*

[c]*Environmental Chemistry Laboratory, Department of Environmental Science and Engineering, School of Space and Environment, Beihang University, Beijing, P. R. China*

allah.ditta@sbbu.edu.pk, ad_abs@yahoo.com

Nanotechnology has revolutionized various sectors of the agro-industry over the last few decades and contributed a lot to sustainable agriculture. Progress has been made in the production of nanoparticles to be used as pesticides, fertilizers, quality stimulants, sensors, and herbicides, with several field applications. The processes of nano-encapsulation protect the elemental composition as well as influence the dispersion and interaction of nanoparticles

Science and Applications of Nanoparticles
Edited by Waqar Ahmed and Ehsan Nourafkan

ISBN 978-981-4267-34-2 (Hardcover), 978-1-003-28029-3 (eBook)
www.jennystanford.com

(NPs) with other materials. Despite various benefits obtained, the evaluation of harmful effects of NPs used in the agricultural sector is vital. The goal of this chapter is to provide an update on various applications of nanotechnology in sustainable agriculture, taking into account the negative effects on the environment. Moreover, an outlook with future research direction for research gaps identified has also been added.

6.1 Introduction

Farming has primarily led to human ecological progress, and more than 60% of the population of the world now rely on farming for their subsistence [1]. The area under agriculture is currently confronting a variety of regional and multilateral problems such as low productivity, climatic change, low organic matter, reduced nutrient-use efficiency, increased residues of pesticides, etc. These restrictions are further compounded by the rising demand for food at a pace of 4% that is needed for withstanding a worldwide populace of 9 billion by 2050 [2]. After its emergence, nanotechnology has made an enormous contribution in many scientific disciplines such as healthcare and pharmaceuticals, textiles, chemicals, information technology, and clean energy industries, owing to its ability to make more effective goods. Even then, the drive to employ nanotechnology for boosting the agriculture and food industry has emerged recently [3, 4]. The main contribution of nanotechnology in agriculture includes nanopesticides production, incorporating nanomaterials (NMs) for the amplification of green pesticides, NM-assisted micronutrient fertilizers and biofertilizers for gradual and regulated release, genetic materials transport, and recognition of pathogen with nanobiosensors [5–7].

Recently, emerging research trends in nanotechnology are primarily focusing on the green synthesis of NMs for various agricultural applications, e.g., gradual release of nutrients or chemical compounds through nano-encapsulation of fertilizers and pesticides, nano-powders for the preservation of soil moisture, rejuvenation of nano-seed, and remediation of pollutants on nanoscale present in soil and water ecosystems. Nevertheless, due to the enormous implementations of nanostructured materials in the near future, the agricultural industry is expected to undergo a dramatic transition. As humans understand that "Technology-Yes, but Safety-Must," they should also be mindful of the potential risks involved with these nanoparticles (NPs) before applying these new technologies.

These technologies have helped address potential agricultural demands, improved the quality and yield of crops, mitigated the emissions induced by contaminants, and indeed defended plants from climate change and ecological pressures [8]. While the use of nanomaterials in crop production is increasing, what negative effects such technologies would have on the environment are still not fully understood, especially in plants and living organisms [9]. Several forms of studies have shown that higher amounts of nanoparticles (NPs) can impose toxic impacts on various crops [10–13]. Therefore, it is necessary to educate the public regarding the benefits of this modern technology.

For the management of such complications, nanotechnology researchers must be conscious that their use without caution would lead to several problems for plants, animals, and eventually for humanity. Since stated by Mishra et al. [14], it is necessary to raise the understanding of nanoparticle safety and develop robust regulatory structures such that scientists should safely implement nanotechnology and prevent environmental catastrophes. The production of nanoparticles (NPs) has risen at an industrial scale such that around 10,000 ton of titanium dioxide (TiO_2) NPs have been generated globally during 2010 [15] and NPs of ZnO, CeO_2, AlO_x, FeO_x, and carbon nanotubes are approximately 500 to 1000 to have been generated yearly by the European countries alone [16]. In this regard, the present chapter summarizes the latest advancement in nanotechnology for sustainable agriculture. Moreover, the research gaps and future research needs have also been discussed.

6.2 Classification of Nanomaterials

The size of nanomaterials (NMs) ranges from 1 to 100 nm, distinguished by length, height, and width [17, 18]. Owing to the smaller size with large surface area and quantum properties, NMs exhibit diverse and robust chemical, optical, and mechanical characteristics [19]. Scientists have been able to find specific uses for these new properties. Based on dimensions at the macroscale,

NMs can be classified as follows: (1) 0D includes the NPs with all dimensions at the nanoscale; (2) 1D is the nanofibers and nanowires that have just one macroscale dimension; (3) 2D encompasses nanosheets and thin films, and ultimately; (4) 3D describes the materials in bulk. Other than categorizing employing dimensions, nanomaterials may be categorized through their chemical nature. This classification consists of four principal classes of nanomaterials, i.e., carbon, ceramic (oxides of metals), steel, and polymeric compounds [19]. Carbon-based nanomaterials consist of structures along with carbon nanotubes (CNTs), graphene, and fullerenes [20]. Ceramics are inorganic solids manufactured from metal–oxide compounds along with FeO_2, ZnO, and TiO_2 [21], and metals encompass nanomaterials based on Ni, Cu, Au, and, Ag. Nanomaterials composed of organic compounds consist of dendrimers that could be obtained from organic NPs, which are characteristically proportioned around the nucleus [20].

Owing to various structures and lengths, this group contains several well-known structures such as graphene, nanofibers, and nanotubes, and such compounds generally allude as nanomaterials [22, 23]. These are used in solar cells, structural materials, electronics, and semiconductors, primarily because of their fascinating characteristics, including certain semiconductor nature, thermal expansion, higher mechanical properties, and light absorption [24]. Carbon nanotubes are often categorized according to quantities of walls, which may be single-wall, double-wall, or multi-wall, and their structural and conductive properties are modified by the number of walls [25, 26].

6.3 Nanotechnology in Agriculture

Diverse implementations in agricultural engineering are identified by nanoparticles. Among the most developed countries in agricultural nanotechnology production, China is leading with 28 patents between 2011 and 2015 in nanotechnology for agricultural production [27]. Through developing nanofertilizers, nanoherbicides, nanofungicides, and nanosensors, nanotechnology can be used for sustainable agriculture [28–30]. By improving the availability of micronutrients (Zn, titanium dioxide, silica, and Cu) for appropriate growth, nanofertilizers promote successful crop development [31, 32], while dendrimers from polymeric nanoparticles function as nano-carriers [33]. The protection against biotic stresses is provided by nanopesticides since the encapsulation of chemicals ensures a managed and target-oriented release of pesticides, thereby reducing the cost of pesticides and greatly boosting the lifespan of the elemental composition of chemical substances [34].

With the latest advancements, nanotechnology has arisen among the essential aspects of the modern agricultural environment. Such innovation has become a strategic controlling factor toward reforming current farming practices. The potential of innovative plant enhancement and efficient distribution systems would eliminate the usage of bulk agrochemicals and can have a more affordable alternative. Figure 6.1 summarizes the major applications of nanotechnology in agriculture.

In the agricultural sector, the use of nanotechnology can be narrowly divided into two sectors: (1) precision farming and (2) enhancement in the production of crops. Both these topics have been discussed in the following sections:

6.3.1 Precision Farming

To optimize the yield of crops to utilize limited inputs, precision farming has already been apprehended. Global satellite positioning systems, supercomputers, and remote-sensing instruments take advantage of this extremely advanced strategy to detect crop problems that eventually result in reduced cost of productions. This technology relies on a detailed knowledge of the pasture ecosystem supported by the computers accompanied by data interpretation. This new technology allows agronomists in calculating the number of diverse variables responsible for the reduced crop production while creating computer models that direct the consumer to make a swift assessment in a way to apply input timely. It has been predicted that more precision farming would eliminate agricultural waste and would thus be more environmentally sustainable relative to traditional farming. This technology enables field conditions like fertility of the soil, growth of the crop, moisture level, weeds, temperature, the status of nutrients in the crop, attacks of insect, multiple plant diseases, etc. to be tracked in real time and can have a significant effect on the prospect of the market of agriculture [36, 37]. The collected statistics would mobilize critical agricultural practices such as time of plantation, harvest optimization, time of irrigation and control in combination with further agrochemical relevance, etc.

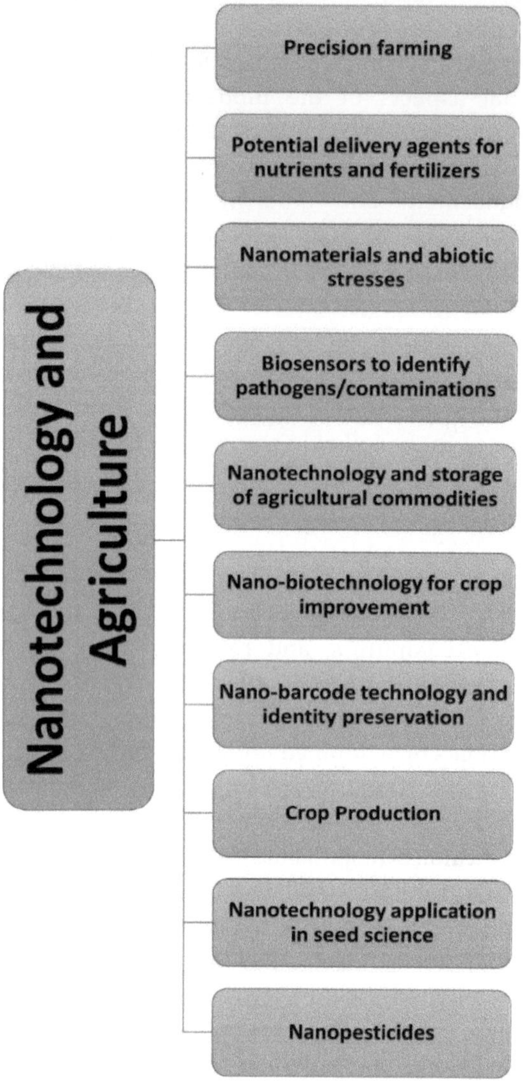

Figure 6.1 Summary of the major contribution of nanotechnology in agriculture [28, 35].

When food scarcity, lack of moisture, and pests are observed in any area, unique nutrients and water, as well as pesticide formulations, will be automatically modified for specific goals and location. Established in 2014, Cambridge startup CiBo Technologies has obtained approximately 30.3 million dollars investments to create digital forum models that seek to replicate the dynamic climatological aspects of each complex farming ecological unit wherever on the planet. Such a framework will produce in-essence pastures having inputs from the real world, allowing developers to provide assistance and guidance about how to optimize real fields. This allows consumers to enhance sustainability, minimize inventory, and improve the ratio of time to market alongside allowing the best economical use of the sector.

Certain companies such as the Michigan startup Farmobile and the Broomfield startup aWhere have shown their interest in the area of precision farming. One model of precision agriculture focused on nanotechnology requires the use of the sensors for Normalized Difference in Vegetative Index (NDVI). To produce red and near-infrared light, this unit uses diodes, which emit light based on the nanoscale, and NDVI values are measured accordingly. This is focused on the concept that stable plants' chlorophylls consume extra red lights in the period of photosynthesis with display greater quantities of near-infra-red light relative to diseased or unhealthy plants. NDVI values are determined depending on this disparity.

The improper use of industrial chemicals is considered to have a detrimental influence on the climate. The only approach to limit the use of agrochemicals is to monitor the employment of chemicals to particular plants depending on specifications. In this case, the use of agricultural robotics offers an intelligent approach to reduce the use of agrochemicals. Different researchers have developed this method for dismemberment into yield and wild plants in the area. This approach allows the application of treatments that are specific to plants and safe, and it is the ultimate aim of precision farming. For joint stem detection and crop-weed classification by employing convolutionary neural networks, a novel technique has been developed. For the management of specific conditions and plants, such as the addition of fertilizer to crops or the removal of weeds, this method would be effective. Also, for the deeper advancement of automated machines working in the agricultural sector, a comprehensive farming automaton database has now been developed [38].

6.3.2 Crop Production

Applications hinged on nanotechnology possess the ability to increase crop production via permitting flora to be properly handled. The involvement of nanomaterials in the cultivation of plants could be classified as (1) nanopesticides/nanoherbicides, (2) nanofertilizers, (3) synthesis of green nanoparticles for use in agriculture, (4) smart delivery systems based on nanoparticles for minor elements and regulators for growth of plant, (5) additives of feed, (6) pathogens and contaminant detection biosensors, etc.

6.3.2.1 Nanopesticides

The chemical/biological compositions employed to manage pests for higher performance and increased crop yield are known as pesticides. Chemical pesticides are most prevalent in this regard, as the public health and environmental threats posed by these have been well known by now. Chemical pesticides have disadvantages because (1) poor precision and elevated toxic effects and (2) endemic undecaying are the main disadvantages of chemical pesticides. The active component of the pesticide is vulnerable to oxidation by different environmental conditions. The biological and chemical effects of pesticides are influenced by an insentient situation such as salinity, temperature, light, and physicochemical properties of soil. Natural components, like microbes, crops, as well as certain pests, play an important part in the breakdown as well as accumulation of environmental contaminants. Microbes play a key role in the maintenance of agricultural concentration of reactants and the decomposing of xenobiotics in water and soil. Global pesticide regulation organizations such as the Organization for Economic Co-operation and Development (OECD), Environment Protection Agency (EPA), World Health Organization (WHO), and US Department of Agriculture, have well established the environmental impacts of pesticides and their toxicity data. A potential threat is known to be persistent quantities of pesticide in a concentration such as parts per billion (μg/L) and parts per trillion (ng/L) in the ecosystem, products of food, and tissues of the body. The option of nanopesticides is aimed at addressing the possible issues presented by the overuse of pesticides and the likelihood of reducing chemical input into the environment. Over the past several years, scientists around the world have been especially interested in developing pesticide formulations focused on nanotechnology for more complex behavior. The timescale for various developments in atrazine nanopesticides is shown is Fig. 6.2.

Every herbicide available today is primarily successful in managing or destroying the aboveground portion of the weeds. The effectiveness of herbicides can be increased through nanotechnology via directly marking the receptors in the ground and root components of weed. However, the glycolysis was inhibited with the application of nanoherbicides, which were supposed to create fresh locations for entry and translocation into various sections of the rhizome. Through conventional methods of application, the application dose of herbicide and other similar chemicals like imidazolinones or glyphosate will also be increased [39]. In this regard, significant quantities of nanomaterials have been used to formulate inorganic nanoparticles, polymeric nanoparticles, nanoherbicides, and nanoparticles based on waste from agricultural industries, etc. For the encapsulation of various chemical herbicides, nanoparticles composed of polymers in the form of nanocapsules and spheres have been extensively utilized. During herbicide atrazine encapsulation, Pereira et al. [40] produced and differentiated nanoparticles of poly epsilon-caprolactone and appraised the same herbicide and genotoxic action. The study directed toward *Brassica* sp., maize, and non-targeted species indicated that the nanoherbicide formulation demonstrated excellent effect on the target species. This has been subsequently proposed that the application of nanoparticles would significantly decrease the volatility of atrazine in the root zone. Also, a decreased amount of herbicide genotoxicity was recorded in *Allium cepa* as revealed through chromosome aberration assay [40]. Similarly, Grillo et al. [41] utilized nanocapsules of poly (γ-caprolactone) (PCL) to decrease ecological toxicity issues. The researchers applied three herbicides of triazine such as simazine, ametrine, and atrazine by using the process of encapsulation. All the triazines were employed with the stability of almost 270 days, and the encapsulation efficacy was found to be >84%. The nanocapsules employed released triazenes slowly and constantly as observed under *in vitro* conditions. Also, the relatively low genotoxic effect of the developed nanoherbicides was observed [41].

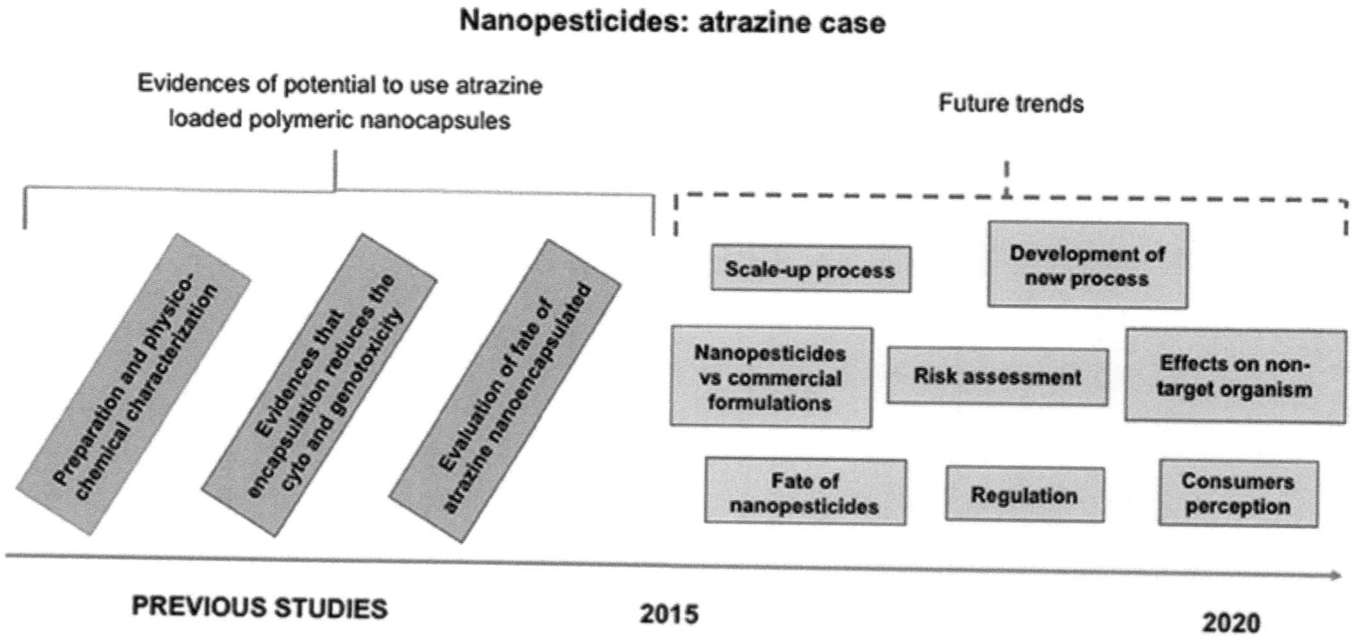

Figure 6.2 Timescale for developments in atrazine nanopesticides. Adapted from Ref. [145].

Dos Santos Silva et al. [42] employed paraquat-functionalized herbicide, generated and categorized using different spectrometry and microscopic strategies on nanomaterials of alginate-chitosan (Ag-Cs). The results from the study conducted by Falsini et al. [43] suggested that the toxicity assessment of nanopesticides is important for the production of environmentally safe substances designed for agricultural applications. Pour et al. [44] conducted an experiment to examine the consequences of nano-encapsulation with plant growth-promoting rhizobacteria while utilizing nanoparticles of silica and carbon nanotubes as well as its metabolic compounds on the improvement of UCB1 Pistachio micropropagation. The results showed that inclusion of both nanocapsules of *Pseudomonas fluorescens* and *Bacillus subtilis* greatly strengthened the prevalence and size of root. In addition, metabolites released from *Pseudomonas fluorescens* nanoformulation increased the length of shoot (3.34 cm) and root (6.26 cm). In another study, a new technique to prolong foliar fluorescent carbon dots (CD) retention and to enhance lettuce photosynthetic activity was used by implementing elastic polyvinyl alcohol (PVA) to encapsulate fluorescent CDs [45]. It was revealed that photosynthetic activity, quality of growth, total carbohydrate contents as well as soluble proteins in lettuce were dramatically improved by sprinkling 3900 mg/L nanocapsules of the CD on the leaves. It was suggested that this study provides a base for sprinkling leaves with light-harvesting materials as light fertilizer to increase metabolic processes, while encouraging crop production and physiological mechanisms for contextual nano-engineering.

The goal of modern nano-formulations is to reduce the harmful impacts of herbicides and insecticides on beneficial living organisms in our environment. These innovative materials attempt to prolong and enhance the life of such molecules using targeted delivery, and to have itself defense against external factors like degradation through UV radiation, as well as greater selectivity to protect other beneficial living organisms [46]. The most widely used encapsulating chemical compounds are polymeric NPs, which enhance herbicide spread and release rates [47]. To encapsulate methyl viologen, which is a non-selective prevalent herbicide employed currently, chitosan/tripolyphosphate NPs were used [48]. This nanoformulation, despite retaining the efficacy following encapsulation, had demonstrated less hazardous behavior than pure compound and showed significant ability to protect certain species of plants following the application of herbicides. On the contrary, the utilization of photocatalytic nanoparticles has focused on herbicide breakdown, which in the presence of sunlight leads to mineralization of the herbicide diquat [49] and depending on the herbicide-to-photocatalyst ratio, they could trim down 70% to 90% of the herbicide dependent on glyphosate [50]. Under the solar irradiation of the imazethapyr herbicide, titanium dioxide demonstrated a higher degradation rate that might decompose quickly at low temperature, demonstrating physisorption on the surface of photocatalyst [51]. The nanocomposite of WO_3-TiO_2 also displayed the degradation of imazapyr herbicide completely when exposed to UV after 120 min [52].

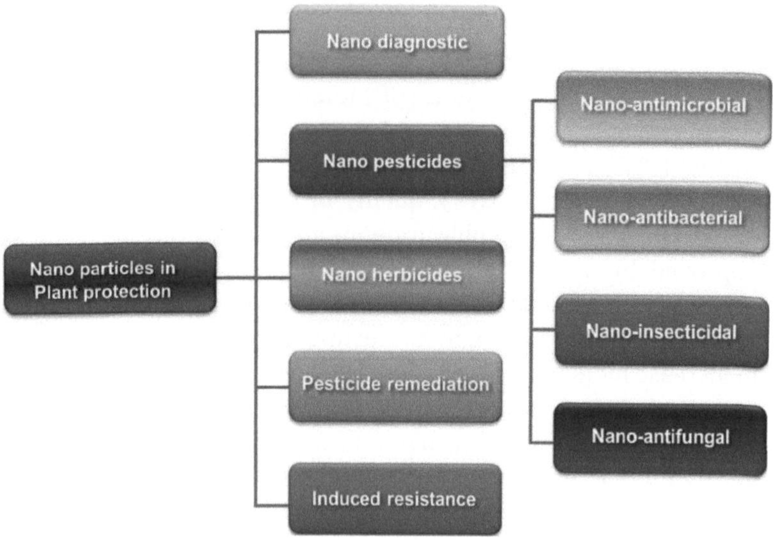

Figure 6.3 Nanoparticles and plant protection. Adapted from Ref. [140].

The toxic impact of pesticides relies on solubility, absorption, photodecomposition, chemical stability, and bioavailability in soil [53]. Nanotechnology's key objective is to reduce the effects of nano-carriers that enable pesticides to be slowly released on the target species [54]. Figure 6.3 gives an overview of the role of nanoparticles in plant protection. Cu-based nanopesticides ($Cu(OH)_2$) have demonstrated adverse impacts on spinach, causing deterioration in the profile of metabolic reactions, minimizing antioxidant molecules

(29–85%) like alpha-tocopherol, total phenolic contents, ascorbic acid, 4-hydroxybutyric acid, threonic acid, and ferulic acid [55]. From these findings, the exposure of nanoparticles to living species must be identified to maintain a healthy interaction between nanotechnology and the environment. The benefits provided by such nanostructured compounds added to plants are greater yield, through better nutrient availability. However, the deep implications of such polymers are not yet well known, since these nano-compounds are capable of providing nutrients and protection against pests. These are also capable of creating stress in other ecosystem organisms, resulting in ecological risk [56]. TiO_2 NPs, subsequently in 30 min, under UV irradiation, degraded pesticides such as chlorpyrifos, cypermethrin, and chlorothalonil [57]. Also, ZnO doped with Cu was examined for the breakdown of monocrotophos pesticides where Cu produced an interim band to stimulate an electron between the valence band and conduction band that leads to a decrease in the bandgap, boosted optical absorption, and demonstrated a tremendous degradation of monocrotophos pesticides [58].

Recently, the role of nanomaterials in plant pathology has been studied. Carbon nanomaterials, metals, and metal oxide nanoparticles, other polymeric nanoparticles in functionalized forms and nonmetals. Lamsal et al. [59] found that before anthracnose outbreaks, peppers were subjected to 100 μg/mL of Ag nanoparticles that lead to successful suppression of the disease. The possibility of postharvest treatments with nano-Ag was proposed by Jagana et al. [60]. Upon integration to fruits of banana at differing rates, the nanoparticles of Ag, which were prepared biologically from the extract of plants, demonstrated postharvest regulation of *Colletotrichum musae*. It was found that the infection rate was minimum (6.7%) with the maximum dose of 2000 μg/mL while control treatment had 75.6%. An experiment conducted by Giannousi et al. [61], developed composites of nano-copper oxide, Cu_2O, and Cu/Cu_2O for their ability to inhibit *Phytophthora infestans* in tomatoes and measure their effectiveness against fungicides based on copper. The findings showed that compounds made up of copper nanoparticles appeared equally or better than industrial products based on copper. The antibacterial and disinfectant function of pathogens of the plant with nanoparticles of Zn was similarly analyzed. A good number of *in vitro* tests have shown that several pathogenic fungi are inhibited by nano-Zn, including *S. sclerotiorum*, *F. oxysporum*, *A. alternata*, *M. plumbeus*, *B. cinerea*, *R. solani*, *P. expansum*, and *R. stolonifera* [62–64].

There are tremendous losses in crop production due to fungal damage every year. This dilemma can be solved by nanomaterials with particular antifungal activities. According to Sharma et al. [65], nanoparticles of nickel ferrite ($NiFe_2O_4$) and cobalt ($CoFe_2O_4$) demonstrated antimycotic activities against three pathogenic fungi attacking crops, namely *Fusarium oxysporum*, *Colletotrichum gloeosporioides*, and *Dematophora necatrix*, and their use was recommended for the management of plant diseases. Nanoparticles of copper synthesized by green technology have been used *in vitro* studies against *Fusarium solani*, *Neofusicoccum* sp., and *Fusarium oxysporum* as effective antifungal agents. Two different ions and nanoparticles of silver have previously been studied for their antifungal action against two fungi inducing pathogenic diseases in the plant, *Bipolaris sorokiniana* and *Magnaporthe grisea* [66]. Elmer and White [67] and Elmer et al. [68] have outlined the overall use of nanoparticles in disease control.

Viral infections have been among the main threats to farming systems, posing a huge threat to human food safety. Unfortunately, the situation has been made worse by the lack of any successful control policy. Due to their small size and higher effectiveness, nanoparticles could indeed serve a significant function in this field. According to Cai et al. [69], tobacco mosaic virus (TMV) infection could be managed under *in vitro* and *in vivo* environments, by the application of nanoparticles of silica and zinc oxide. TMV particles were found to be clustered and demonstrated deformation of the outermost surface when mixed for 2 h with these NPs. Once inoculated into tobacco plants, this complete mixture displayed lesser colonization of the virus in comparison to the control. Similarly, Cai et al. [70] in another study reported that continuous application to Fe_3O_4 NPs encourages the production of *Nicotiana benthamiana* and provokes virus-defensive response. Abdelkhalek and Al-Askar [71] reported that the extract of *Mentha spicata* had been an important reduction agent for the green synthesis of ZnO nanoparticles, which demonstrated substantial antiviral activity, suggested the defensive and curative activity against TMV, and its application for the mitigation of crop viral diseases.

6.3.2.2 Potential delivery agents for nutrients and fertilizers

By developing smart distribution systems that can routinely discharge pollutants to multiple identified sites in a carefully managed and distantly controlled way by eliminating many natural obstacles, the scarcity of nutrients in crops might be resolved. By improving production, soil organic matter, and decreasing fertilizations in excessive quantities, and alleviating deficiencies of many diverse nutrients, nanofertilizer formulations would greatly improve crop productivity [35, 72]. Fertilizer is an intrinsic or artificial material containing the elemental components needed for plant growth, productivity, and natural fertility to be improved by addressing micronutrient deficiencies [73]. ZnO NPs showed ideal results through the foliar application of NPs to alleviate zinc deficiency, particularly for rice. Enhanced growth and developmental parameters, as well as intensified activity of dehydrogenase enzyme, were noted with the application of ZnO [74]. Nanocomposites could also be used to provide macronutrients like nitrogen through nanoparticles encapsulation using hydroxyapatite modified with urea in montmorillonite's layers [75]. Such molecules ensured the gradual delivery of nitrogen and ultimately increased the productivity of rice (*Oryza sativa*). Bioengineered nanoparticles, like membrane vesicles, have also been employed to supply *Brassica oleracea* with Zn by fertilization through foliar, resulting in good-quality and protoplasmic circulation [76].

Miranda-Villagómez et al. [77] reported that in roots and shoots, nano-KH_2PO_4 facilitated greater physiological productivity of phosphorus that caused greater accumulation in such tissues. The concentration of phosphorus in shoots was positively associated with the photosynthesis process factor. Also, in rice plants, nano-KH_2PO_4 improved instant water usage quality. Tarafdar et al. [78] reported that tri-calcium phosphate ($Ca_3P_2O_8$) could be utilized as phosphorous precursor salts for the production of the crop. Likewise, ZnO nanoparticles are very efficient in the fungal synthesis of polysaccharides and phosphatase relative to ZnO bulk nanoparticles [79]. By minimizing nitrogen leakage and increasing absorption by porous nanoscale sections of plants, nanotechnological principles could likewise be employed to maximize the effectiveness of the usage of nitrogen in conventional fertilizers. Kumaraswamy et al. [80]

prepared a novel and sophisticated chitosan-silicon nanofertilizer (CS-Si NF) that used the nano-matrix of chitosan-tripolyphosphate (TPP) to encapsulate silicon (Si) for its gradual release and reported the immense promise of CS-Si NF toward potential deployment in a nanofertilizer of the coming decade for food production. Salako et al. [81] reported that zeolites could be used for the trapping, preservation, and gradual extraction of nitrogen in agriculture and for a regulated release of nitrogen in Tomato (*Lycopersycum esclentum*). Manikandan and Subramanian [82] reported the applications of zeolite-based nitrogen nanofertilizer for the slow release of nitrogen and its effects on growth, yield, and quality of maize. Iron has been an important micronutrient, which has a crucial function in plant metabolism, namely, in respiration and chlorophyll synthesis, and chlorosis can result from its shortage in crops. Therefore, the optimum supplies of iron in crops are required for proper development. In this respect, iron-based nanoparticles have revealed capable performance. Askary et al. [83] reported important changes in the growth parameters of plants, pigments of photosynthesis, and content of total protein after nanoparticles of iron oxide were integrated with *Catharanthus roseus* in various quantities like 0, 5, 10, 20, 30, and 40 mM for 70 days. Rui et al. [84] measured the efficacy of nanoparticles of iron oxide in *Arachis hypogaea* and statistically contrasted it with conventional Fe fertilizers. The results showed that nanoparticles of iron oxide augmented the values of SPAD, weight, and length of root, and biomass of plants. According to Das et al. [85], in beetroot, carrot, fenugreek, etc., nano-iron pyrite seed coating was used, which resulted in substantial growth promotion. Bastani et al. [86] reported that the nano-iron (Fe) complex has a beneficial impact on tobacco development with a small delivery of Fe. Likewise, molybdenum (Mo), another nutritional element, performs numerous essential functions, including the integration of ammonia in plants [87]. Consequently, Mo malnutrition will raise the ammonium concentration within crops, which further becomes harmful for human health [88]. Several nanoparticles of Mo had subsequently been utilized to increase its utilization performance. Synthesized Mo NPs have stimulated the interaction of nitrate reductase (NR) particularly in comparison to Mo in elemental form in *Spinacia oleracea* and significantly decreased the accumulation of

NO_3 [89]. In another experiment by Osman et al. [90], the spraying of molybdenum oxide nanoparticles at a concentration of 40 ppm enhanced the germination percentage of *Phaseolus vulgaris* by approximately 82–84%. Polymeric nanoparticles for the coating of biofertilizers are being used as previously mentioned, so these nanofertilizer formulations demonstrated excellent stability to decomposition and drainage. Throughout this context, the technique of emulsifier for oil and water are productively employed for the transport to the target sites of various microorganisms as well as encapsulation. This technique also decreases evaporation, increases cell viability, and improves the kinetics of controlled release.

6.3.2.3 Nanomaterials and abiotic stresses

Under natural field conditions, plants undergo various abiotic stresses, such as salinity [91–93], drought [94–97], temperature [98], and heavy metals [99–103]. Heavy metals are known to be metals with >5 g/m^3 density while 17 out of 53 heavy metals such as copper, molybdenum, iron, manganese, and zinc are essential for living organisms [104]. The remaining heavy metals such as chromium, arsenic, mercury, cadmium, and lead are harmful to the environment since these can affect various biochemical parameters of crop plants [105–108]. Metals known as plant micronutrients could likewise exert lethal impacts on vegetation at concentrations required than normal like zinc that prevents the uptake of Mg, Ca, Cu, Mn, and Fe in *Oryza sativa* (L.) under concentrations of 1.5–8.5 mM [109].

The elimination of hazardous heavy metals from polluted water has been a big problem for sustainable agriculture as these can enter the food chain through assimilation and ultimately pose a major risk to human health. Various studies have confirmed the contamination of agricultural soils with As, Cd, Pb, Ni, Cu, Pb, Hg, etc. [94, 110–113]. In this regard, many nanoparticles based on carbon, silica metal, and metal oxide have shown encouraging effects, and several literature papers have addressed these in depth [114–117]. In the future, more emphasis should be given to this area of research with a special focus on the mechanisms using modern state of the art technologies.

6.3.2.4 Biosensors to identify pathogens/contaminations

The biosensor is a computer that identifies any gene expression through a microprocessor into an electronic signal for statistical data acquisition. The key functional elements of such a system include living organisms or organic macromolecules functioning both as communication mechanism (bioengineering group), and efficient transmitter and signaling usability. Compared to other conventional technologies, the use of nanobiosensors enables fast sensing with improved sensitivity and precision, reducing time for simple and accurate measurement and allowing authentic signal monitoring using remotely controlled advanced sensors. Also, it is possible to use nanobiosensors to identify single or multiple analytes that may pose a potential hazard to humans, animals, and plants. Integration of nanocomposites into biosensors furthermore aims to apply emerging technology for signal transduction. Significant uses of nanobiosensors in the agricultural sector include the primary or secondary detection of contaminants, drug residues, foodborne microbial pathogens, neurotoxin compounds, and heavy metal ions, etc. The said innovations were even developed for the monitoring of plant growth, infectious diseases, plant stress, soil composition, nutritional content, or quality of food [118–120]. Walia and Acharya [121] reported nanosensors of fluorescent nature for the identification of pesticides of organochlorine dicofollica. Dar et al. [122] reported colorimetric nanosensors based on gold nanoparticles for pesticides of organophosphorus dimethoate. Amperometric-based nanosensors have also been used for other pesticides [123–125].

6.4 Nanotechnology Application in Seed Science

The yield of every plant depends on seed quality. In many other situations, before passing on to farmworkers, the germination percentage of seed is systematically ensured in laboratories. Yet, the predicted outcomes are rarely replicated in the farmlands. Therefore, the smart germination of seeds driven by modern technological methods takes an hour. It has been documented that carbon nanotubes (CNTs) reach the tough coating of sprouting tomato seeds and have shown growth enhancement. Additionally, the percentage of seed germination was improved with multi-walled carbon nanotubes (MWCNTs) sprinkled either on the seed surface or combined in the growth medium of corn, soybean, and barley [126]. Moreover, various nanocomposites have been utilized as a shield to protect seeds from seed-borne infections throughout the germination period [127].

6.5 Nanotechnology and Storage of Agricultural Commodities

To ensure a strong production process, the storage of agricultural products is equally important for the cultivation. However, agricultural weeds could lead to damages of products in pre- and postharvest. Nano-formulated chemicals have been found a good choice for the control of rice weevil, a foremost insect of grains present in stores [128]. Another big problem is the spoilage of crops, owing to the development of infectious agents. Silver NPs, however, are now incorporated in plastic to create food storage bins to mitigate the detrimental growth of bacteria [129, 130]. To boost mechanical or physical properties, nanomaterials may be coated or integrated on the polymers of plastic, which are among the widely employed components in the food-processing sector [131]. "Nanosensors" are also employed to monitor the oxidation phase in the food industry for packaging materials. Also, nanosensors can deliver physical manifestation and knowledge of appropriate dietary supplement knowledge in conjunction with reactive nano-layers or intelligent inks found on nanoparticles [130].

6.6 Nanobiotechnology for Crop Improvement

Recently, the main challenge for organic farmers to satisfy the escalating production of goods extracted from plants is crop enhancement. To accelerate the crop improvement program, a cutting-edge biotechnological approach such as next-generation sequencing (NGS) is being employed (Fig. 6.4). In Genome Aided

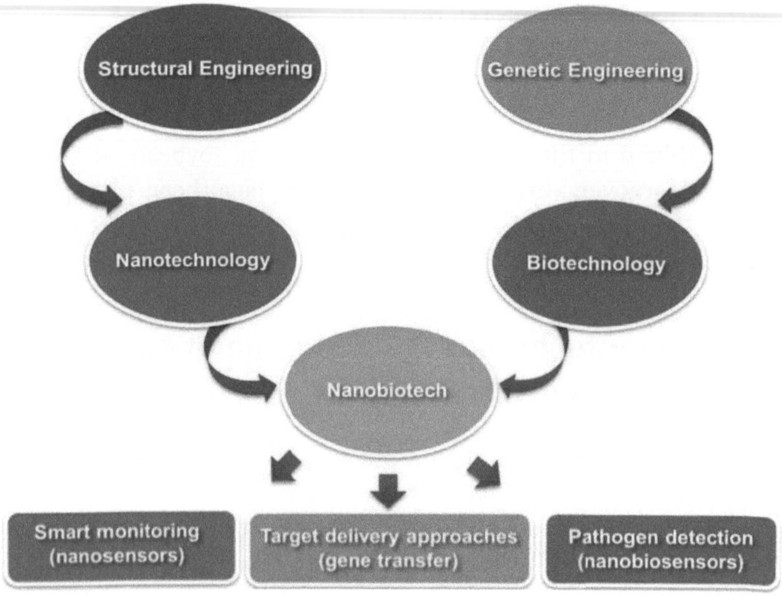

Figure 6.4 An overview of nanobiotechnology. Adapted from Ref. [140].

Breeding, an integrated method of recognizing as well as choosing specific genes, the NGS technique, is being used [132, 133]. By altering gene expression, parasites, infectious agents, and the growth of crops capable to tolerate stress could be accomplished through specific DNA and RNA [134]. Poor biological membrane infiltration as well as nucleus selection, respectively, though, entails complexities in developing plant-specific transportation [135]. Microinjection, transformations via agrobacterium, striking with microprojectile, and viral vectors for the distribution of gene are the approaches considered for plant genetic transformation [136]. Nanoparticles are currently being employed to build successful vehicles for gene transfer, and approaches to nanobiotechnology are projected to include innovative ways to fully substitute the genetic material between plants [137]. Nanoparticles of silicon dioxide are designed to provide tobacco (dicotyledonous) and maize (monocotyledonous) with DNA fragments/sequences without any adverse impacts [138]. Embedded carbon matrices of gold nanoparticles (5–25 nm) are employed to deliver DNA during the transition of plant cells that contain larger quantities of genetic content [139]. According to Shang et al. [140], the modern methods of gene transformations are further implemented with the unique distribution of technology, i.e., CRISPR/Cas9 single guide RNA (sgRNA) based on nanomaterials, which is fruitfully employed in editing plant genome [141]. Applications of nanobiotechnology would also make a major contribution to satisfying the potential need for plant-based goods by manipulating plants in a desirable way.

6.7 Nano-Barcode Technology and Identity Preservation

Identity preservation (IP) is characterized by the process that creates greater benefits via offering farmers knowledge concerning the programs and procedures required to cultivate particular seedlings or other horticultural products [142]. IP on the nanoscale can greatly increase the quality assurance of plant-based products [143]. Such innovations would increase the efficiency of agro-product maintenance and careful control of the production system. IP also enables data, documents, and supplier protocols to be accessed by stakeholders. Due to the potential of forming a wide number of variations, nano-barcode technology enables agricultural and livestock items to be tagged for identification. UV lamps and an optical microscope are used to monitor the barcodes of a glass-sized micrometer. These barcodes could be produced by the soaking light materials having metal ions [144]. Nano-barcodes of particles with basic computer-readable and encodable properties have been developed through modern scientific studies. Gold and silver nanoparticles are employed to create such varieties of barcodes.

6.8 Conclusion and Future Perspectives

Agriculture is the primary basis of good nutrition and is expected to have a significant effect on the introduction of new emerging technologies. It is important to keep up-to-date agricultural knowledge. To satisfy the rising food demands of the growing global population, agriculture needs to be modernized and innovative.

With all its recent developments, agricultural nanotechnology is expected to make a big contribution to the markets of agricultural food, although the detrimental issues of numerous nanoparticles are currently under review. Modern nanotechnological methods can resolve a wide range of issues related to traditional cultivation such as the reduced amount of micronutrient availability to the crop plants, inadequate management over the release of micronutrients, failure of high fertilizer rates, less effective use of plant nutrients, etc. To develop the agriculture industry, nanotechnology is an essential element.

Nanotechnology has emerged as a progressive, high-impact technology due to its immense potential. This field has been applied in various agricultural sectors ranging from the agro-food industry to the handling of multiple stresses and much more. The toxicity of nanomaterials is characterized by features such as external surface modification, appearance, accumulation, concentration, and dimension, and in turn, the potential varied biological consequences induced in plants. It is essential to know specifically the toxic potential that nanostructured structures could impose on species of plants, independent of their source, either as a commodity with a particular intent for agriculture or its potential application to the ecosystem through the mismanagement of waste containing nanomaterials. It is, therefore, important to establish stringent and effective regulations that prohibit their misuse, resulting in the protection of numerous crop species that encounter such nanomaterials with extremely negative impacts on their growth. More emphasis should be given to the processes through which crops could enhance the production of secondary metabolites or phytochemicals with the potential to cope with biotic and abiotic stresses such as drought, radiation, salinity, insect pests, temperature, floods, viruses, and microbes. There should be more awareness campaigns that could alleviate the concerns associated with the nanomaterials used in sustainable agriculture. Additional support through a trained workforce, as well as awareness of the topic, would encourage the domain to expand. The elicitors' application enhances the growth of phytochemicals and the epidemiological influence of various nanoparticles could be exploited for the alleviation of crop stresses, contributing to the generation of secondary metabolites, which could overtake the elicitors, being frequently applied. However,

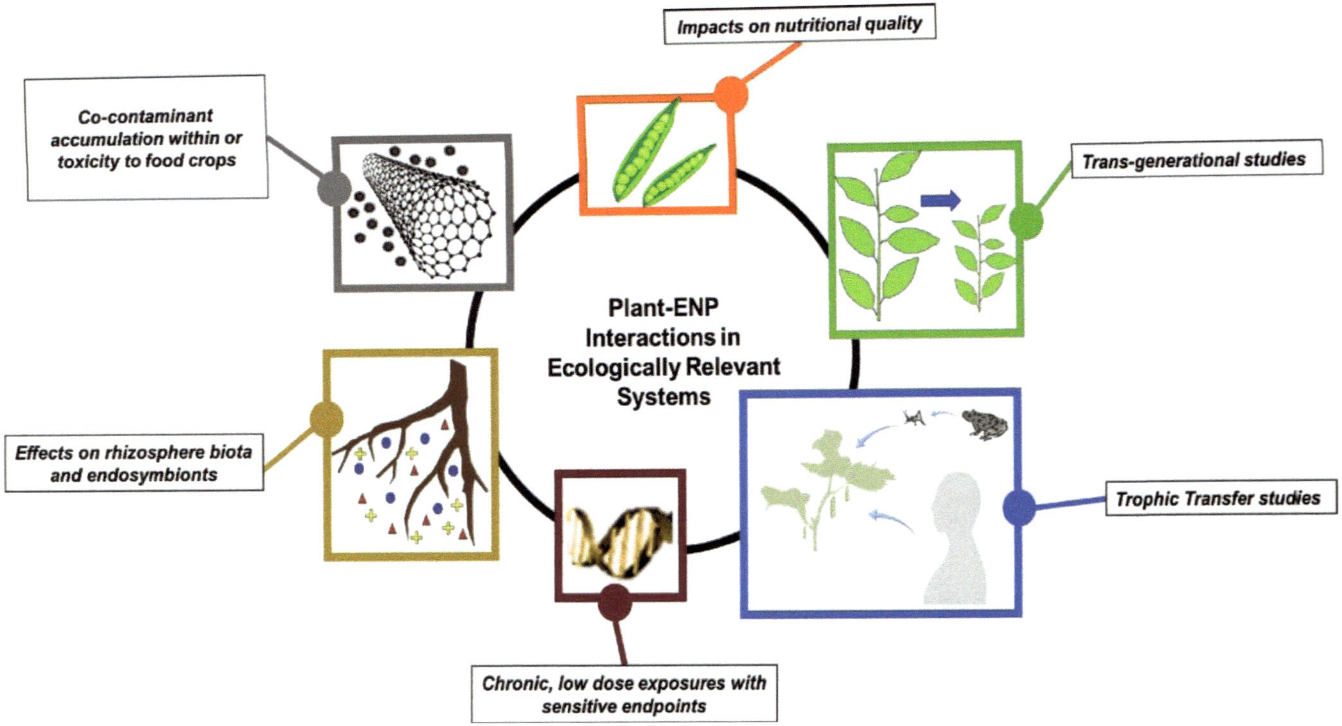

Figure 6.5 Key knowledge gaps and recommended research areas to characterize comprehensively the risks and benefits of nanomaterials' use in crop production. Adapted from Ref. [146].

widespread knowledge of possible dangers during the treatment of these nanomaterials is still low among the general population that might create considerable intimidation to human health and the environment. Therefore, it would be a long-term benefit for the people that planned studies, as well as successful implementation of policies or measures must be devised through a compilation of practicable scientific evidence (Fig. 6.5).

Recommendations

The basic recommendations would concentrate on the following while concerning the future strategic analysis:

1. Reinforcement of monitored nanoparticle synthesis through green synthesis.
2. Detailed knowledge of the vast number of nanoparticles, which are uncharacterized and unidentified and may perform a significant function in the production of plants.
3. How nanoparticles communicate with the plants and knowledge of the carrier system for nanoparticles in the host plant.
4. Formulation, classification, and implementation of biologically synthesized nanoparticles to be used in the economical production of crops.
5. A comparative appraisal of NPs' side impact on multiple variables of the environment and interpretation of the non-target consequences of nanoparticles.
6. Specialized seed coating strategies with micronutrient-enriched nanoparticles for the targeted processing of seeds.
7. Production of soil-, herb-, water-, and pesticide-monitoring tools based on the nanosensors.
8. Return on investment (ROI) estimate for pilot-scale nanotechnology-based agricultural activities and under field conditions.
9. Organizations with appropriate experience or infrastructure that regularly deal in the appraisal of nanomaterials for safety and health ought to be established. To track the research results properly, such organizations should operate through

an interconnected fashion with both biological and chemical research labs.

10. The Food Safety and Standards Authority shall rigidly enforce specific instructions for tracking and testing nanomaterials structures in compliance with FAO/WHO guidelines.
11. Until marketing approval, every nano-based food must be tested regarding potential dangers, so subsequent information must be documented with a significant amount of data.
12. Hazards from nanostructured materials dependent on marine species must be adequately measured and reported.
13. For the promotion of excellent nano-products as well as to mitigate the issues of farmland, qualified and knowledgeable human resources must be generated.

References

1. FAO. (2013). *Statistical Year Book*. World Food and Agriculture, FAO, Rome.
2. Chen, H. and Yada, R. (2011). Nanotechnologies in agriculture: New tools for sustainable development. *Trends Food Sci. Technol.* **22**, pp. 585–594.
3. Rutzke, C. J. (2003). Nanoscale science and engineering for agriculture and food systems: A report submitted to cooperative state research, education and extension service, the United States Department of Agriculture: National Planning Workshop, November 18–19, 2002, Washington, DC.
4. Kuzma, J. and Verhage, P. (2006). *Nanotechnology in Agriculture and Food Production: Anticipated Applications.* Woodrow Wilson International, Washington, DC, USA.
5. Salamanca-Buentello, F., Persad, D. L., Court, E. B., Martin, D. K., Daar, A. S., and Singer, P. A. 2005. Nanotechnology and the developing world. *PLoS Med.* **2**, pp. 383–386.
6. Subramanian, K. S. and Tarafdar, J. C. (2011). Prospects of nanotechnology in Indian farming. *Indian J. Agric. Sci.* **81**, pp. 887–893.
7. Iavicoli, I., Leso, V., Beezhold, D. H., and Shvedova, A. A. (2017). Nanotechnology in agriculture: Opportunities, toxicological implications, and occupational risks. *Toxicol. Appl. Pharm.* **329**, pp. 96–111.

8. Liu, R. and Lal, R. (2015). Potentials of engineered nanoparticles as fertilizers for increasing agronomic productions. *Sci. Total Environ.* **514**, pp. 131–139.

9. Tripathi, D. K., Shweta, S., Singh, S., Singh, S., Pandey, R., Singh, V. P., Sharma, N. C., Prasad, S. M., Dubey, N. K., and Chauhan, D. K. (2017). An overview on manufactured nanoparticles in plants: Uptake, translocation, accumulation and phytotoxicity. *Plant. Physiol. Biochem.*, **110**, pp. 2–12.

10. López-Moreno, M. L., Avilés, L. L., Pérez, N. G., Irizarry, B. Á., Perales, O., Cedeno-Mattei, Y., and Román, F. (2016). Effect of cobalt ferrite ($CoFe_2O_4$) nanoparticles on the growth and development of *Lycopersicon lycopersicum* (tomato plants). *Sci. Total Environ.*, **550**, pp. 45–52.

11. López-Moreno, M. L., De La Rosa, G., Hernández-Viezcas, J. A., Peralta-Videa, J. R., and Gardea-Torresdey, J. L. (2010). X-ray absorption spectroscopy (XAS) corroboration of the uptake and storage of CeO_2 nanoparticles and assessment of their differential toxicity in four edible plant species. *J. Agric. Food Chem.*, **58**, pp. 3689–3693.

12. Rastogi, A., Zivcak, M., Tripathi, D. K., Yadav, S., and Kalaji, H. M. (2019). Phytotoxic effect of silver nanoparticles in *Triticum aestivum*: Improper regulation of photosystem I activity as the reason for oxidative damage in the chloroplast. *Photosynthetica*, **57**, pp. 209–216.

13. Zhu, J., Zou, Z., Shen, Y., Li, J., Shi, S., Han, S., and Zhan, X. (2019). Increased ZnO nanoparticle toxicity to wheat upon co-exposure to phenanthrene. *Environ. Pollut.*, **247**, pp. 108–117.

14. Mishra, M., Dashora, K., Srivastava, A., Fasake, V. D., and Nag, R. H. (2019). Prospects, challenges and need for regulation of nanotechnology with special reference to India. *Ecotoxicol. Environ. Saf.*, **171**, pp. 677–682.

15. Bundschuh, M., Filser, J., Lüderwald, S., McKee, M., Metreveli, G., Schaumann, G., Schulz, R., and Wagner, S. (2018). Nanoparticles in the environment: Where do we come from, where do we go to? *Environ. Sci. Eur.*, **30**, pp. 1–17.

16. Piccinno, F., Gottschalk, F., Seeger, S., and Nowack, B. (2011). Industrial production quantities and uses of ten engineered nanomaterials in Europe and the world. *J. Nanoparticle Res.*, **14**, pp. 1109.

17. Thakkar, K. N., Mhatre, S. S., and Parikh, R. Y. (2010). Biological synthesis of metallic nanoparticles. *Nanomed. Nanotechnol. Biol. Med.*, **6**, pp. 257–262.

18. Jeevanandam, J., Barhoum, A., Chan, Y. S., Dufresne, A., and Danquah, M. K. (2018). Review on nanoparticles and nanostructured materials:

History, sources, toxicity and regulations. *Beilstein J. Nanotechnol.*, **9**, pp. 1050–1074.

19. Khan, I., Saeed, K., and Khan, I. (2019). Nanoparticles: Properties, applications and toxicities. *Arab. J. Chem.*, **12**, pp. 908–931.

20. Ealias, A. M. and Saravanakumar, M. P. (2017). A review on the classification, characterization, synthesis of nanoparticles and their application. *IOP Conf. Ser. Mater. Sci. Eng.*, **263**, pp. 1–15.

21. Thomas, S., Harshita, B. S. P., Mishra, P., and Talegaonkar, S. (2015). Ceramic nanoparticles: Fabrication methods and applications in drug delivery. *Curr. Pharm. Des.*, **21**, pp. 6165–6188.

22. Mathur, R. B., Singh, B. P., and Pande, S. (2017). *Carbon Nanomaterials Synthesis, Structure, Properties and Applications*, CRC Press: Bora Raton, FL, USA, pp. 284.

23. Zaytseva, O. and Neumann, G. (2016). Carbon nanomaterials: Production, impact on plant development, agricultural and environmental applications. *Chem. Biol. Technol. Agric.*, **3**, p. 17.

24. Dolez, P. I. (2015). Nanomaterials definitions, classifications, and applications. In: Dolez, P. I. (Ed.), *Nanoengineering*, Elsevier, Amsterdam, The Netherlands, pp. 3–40.

25. Rodríguez-Méndez, A., Guzmán, C., Elizalde, E., Eduardo, A., Escobar-Alarcón, L., Vega, M., Alvarado-Rivera, J., and Escalante, K. (2017). Effluent disinfection of real wastewater by Ag–TiO_2 nanoparticles photocatalysis. *J. Nanosci. Nanotechnol.*, **17**, pp. 711.

26. Sudha, P. N., Sangeetha, K., Vijayalakshmi, K., and Barhoum, A. (2018). Nanomaterials history, classification, unique properties, production and market. In: Barhoum, A. and Makhlouf, A. S. H. (Eds.), *Emerging Applications of Nanoparticles and Architecture Nanostructures*, Elsevier, Amsterdam, The Netherlands, pp. 341–384.

27. Yata, V. K., Tiwari, B. C., and Ahmad, I. (2018). Nanoscience in food and agriculture: Research, industries and patents. *Environ. Chem. Lett.*, **16**, pp. 79–84.

28. Ditta, A. (2012). How helpful is nanotechnology in agriculture? *Adv. Natural Sci. Nanosci. Nanotechnol.*, **3**, pp. 033002.

29. Ditta, A. (2019). Role of Nanoclay Polymers in Agriculture: Applications and perspectives. In: Sharma, S. K. (Ed.), *Nanohybrids in Environmental and Biomedical Applications*, Taylor and Francis (CRC Press), USA. pp. 323–334.

30. Ditta, A., Mehmood, S., Imtiaz, M., Rizwan, M. S., and Islam, I. (2020). Role of nanotechnology in soil fertility and nutrient management. In:

Husen, A. and Jawaid, M. (Eds.), *Nanomaterials for Agriculture and Forestry*, Woodhead Publishing Ltd. Elsevier UK. pp. 273–288.

31. López-Valdez, F., Miranda-Arámbula, M., Ríos-Cortés, A. M., Fernández-Luqueño, F., and De-la-Luz, V. (2018). Nanofertilizers and their controlled delivery of nutrients. In: *Agricultural Nanobiotechnology*, Springer International Publishing, Cham, Switzerland, pp. 35–48.
32. Vandevoort, A. R. and Arai, Y. (2018). Macroscopic observation of soil nitrification kinetics impacted by copper nanoparticles: Implications for micronutrient nanofertilizer. *Nanomaterials*, **8**, pp. 927.
33. Kah, M., Kookana, R. S., Gogos, A., and Bucheli, T. D. (2018). A critical evaluation of nanopesticides and nanofertilizers against their conventional analogues. *Nat. Nanotechnol.*, **13**, pp. 677–684.
34. Pascoli, M., Lopes-Oliveira, P. J., Fraceto, L. F., Seabra, A. B., and Oliveira, H. C. (2018). State of the art of polymeric nanoparticles as carrier systems with agricultural applications: A minireview. *Energy Ecol. Environ.*, **3**, pp. 137–148.
35. Ditta, A., Arshad, M., and Ibrahim, M. (2015). Nanoparticles in sustainable agricultural crop production: Applications and perspectives. In: Siddiqui, M. H., Al-Whaibi, M. H., and Mohammad, F. (Eds.), *Nanotechnology and Plant Sciences: Nanoparticles and Their Impact on Plants*, Springer, Switzerland. pp. 55–75.
36. Moskvin, G. A. and Spakovica, E. G. (2002). New method and low-cost intelligent instruments for the fraud detection and conformity control of agricultural products. In: ASAE Annual Meeting. ASAE Paper Number 023077. Nanotechnology for Environmental Engineering. 1:2.
37. Day, W. (2005). Engineering precision into variable biological systems. *Ann. Appl. Biol.* **146**, pp. 155–162.
38. Chebrolu, N., Lottes, P., Schaefer, A., Winterhalter, W., Burgard, W., and Stachniss, C. (2017). Agricultural robot dataset for plant classification, localization and mapping on sugar beet fields. *Int. J. Robot. Res.* **36**, pp. 1045–1052.
39. Pérez-de-Luque, A. and Rubiales, D. (2009). Nanotechnology for parasitic plant control. *Pest Manag. Sci.* **65**, pp. 540–545.
40. Pereira, A. E. S., Grillo, R., Mello, N. F. S., Rosa, A. H., and Fraceto, L. F. (2014). Application of poly(epsilon-caprolactone) nanoparticles containing atrazine herbicide as an alternative technique to control weeds and reduce damage to the environment. *J. Hazard. Mater.* **268**, pp. 207–215.

41. Grillo, R., dos Santos, N. Z. P., Maruyama, C. R., Rosa, A. H., de Lima, R., and Fraceto, L. F. (2012). Poly(γ-caprolactone) nanocapsules as carrier systems for herbicides: Physico-chemical characterization and genotoxicity evaluation. *J. Hazard. Mater.* **231–232**, pp. 1–9.

42. dos Santos Silva, M., Cocenza, D. S., Grillo, R., de Melo, N. F. S., Tonello, P. S., de Oliveira, L. C., Cassimiro, D. L., Rosa, A. H., and Fraceto, L. F. (2011). Paraquat-loaded alginate/chitosan nanoparticles: Preparation, characterization and soil sorption studies. *J. Hazard. Mater.* **190**, pp. 366–374.

43. Falsini, S., Tani, C., Schiff, S., Gonnelli, C., Clemente, I., Ristori, S., and Papini, A. (2020). A new method for the direct tracking of in vivo lignin nanocapsules in Eragrostis tef (Poaceae) tissues. *European Journal of Histochemistry: EJH*, pp. 64.

44. Pour, M. M., Saberi-Riseh, R., Mohammadinejad, R., and Hosseini, A. (2019). Nano-encapsulation of plant growth-promoting Rhizobacteria and their metabolites using alginate-silica nanoparticles and carbon nanotube improves UCB1 pistachio micropropagation. *J. Microbiol. Biotechnol*, **29**, pp. 1096–1103.

45. Xu, X., Mao, X., Zhuang, J., Lei, B., Li, Y., Li, W., Zhang, X., Hu, C., Fang, Y., and Liu, Y. (2020). PVA-coated fluorescent carbon dot nanocapsules as an optical amplifier for enhanced photosynthesis of lettuce. *ACS Sust. Chem. Eng.*, **8**, pp. 3938–3949.

46. Li, Z. Z., Chen, J. F., Liu, F., Liu, A. Q., Wang, Q., Sun, H. Y., and Wen, L. X. (2007). Study of UV-shielding properties of novel porous hollow silica nanoparticle carriers for avermectin. *Pest Manag. Sci.* **63**, pp. 241–246.

47. Abigail, M. E. A., Melvin, S. S., and Chidambaram, R. (2016). Application of rice husk nanosorbents containing 2,4-dichlorophenoxyacetic acid herbicide to control weeds and reduce leaching from soil. *J. Taiwan Inst. Chem. Eng.* **63**, pp. 318–326.

48. Grillo, R., Pereira, A. E., Nishisaka, C. S., de Lima, R., Oehlke, K., Greiner, R., and Fraceto, L. F. (2014). Chitosan/tripolyphosphate nanoparticles loaded with paraquat herbicide: An environmentally safer alternative for weed control. *J. Hazard. Mater.* **8**, pp. 163–171.

49. Shibin, O. M., Yesodharan, S., and Yesodharan, E. P. (2015). Sunlight induced photocatalytic degradation of herbicide diquat in water in presence of ZnO. *J. Environ. Chem. Eng.*, **3**, pp. 1107–1116.

50. Páez, M. R., Ochoa-Muñoz, Y., and Rodriguez-Páez, J. E. (2019). Efficient removal of a glyphosate-based herbicide from water using ZnO nanoparticles (ZnO-NPs). *Biocatal. Agric. Biotechnol.*, 22, pp. 101434.

51. Rumi Ishiki, R., Mitsugu Ishiki, H., and Takashima, K. (2005). Photocatalytic degradation of imazethapyr herbicide at TiO_2/H_2O interface. *Chemosphere*, **58**, pp. 1461–1469.

52. Ismail, A. A., Abdelfattah, I., Helal, A., Al-Sayari, S. A., Robben, L., and Bahnemann, D. W. (2016). Ease synthesis of mesoporous WO_3-TiO_2 nanocomposites with enhanced photocatalytic performance for photodegradation of herbicide imazapyr under visible light and UV illumination. *J. Hazard. Mater.* **307**, pp. 43–54.

53. Kumar, S., Nehra, M., Dilbaghi, N., Marrazza, G., Hassan, A. A., and Kim, K. H. (2019). Nano-based smart pesticide formulations: Emerging opportunities for agriculture. *J. Control. Release*, **294**, pp. 131–153.

54. Jampílek, J. and Kráľová, K. (2017). Nanopesticides: Preparation, targeting, and controlled release. In: *New Pesticides and Soil Sensors*, Academic Press, Cambridge, MA, USA, pp. 81–127.

55. Zhao, L., Huang, Y., Adeleye, A. S., and Keller, A. A. (2017). Metabolomics reveals $Cu(OH)_2$ nanopesticide-activated anti-oxidative pathways and decreased beneficial antioxidants in spinach leaves. *Environ. Sci. Technol.*, **51**, pp. 10184–10194.

56. Bourguet, D. and Guillemaud, T. (2016). *Sustainable Agriculture Reviews*, Lichtfouse, E., Ed., Springer International Publishing: Cham, Switzerland, p. 120.

57. Affam, A. C. and Chaudhuri, M. (2013). Degradation of pesticides chlorpyrifos, cypermethrin and chlorothalonil in aqueous solution by TiO_2 photocatalysis. *J. Environ. Manag*, **130**, pp. 160–165.

58. Hanh, N. T., Le Minh Tri, N., Van Thuan, D., Thanh Tung, M. H., Pham, T. D., Minh, T. D., Trang, H. T., Binh, M. T., and Nguyen, M. V. (2019). Monocrotophos pesticide effectively removed by novel visible light driven Cu doped ZnO photocatalyst. *J. Photochem. Photobiol. Chem.*, **382**, p. 111923.

59. Lamsal, K., Kim, S. W., Jung, J. H., Kim, Y. S., Kim, K. S., and Lee, Y. S. (2011). Application of silver nanoparticles for the control of *Colletotrichum* species in vitro and pepper anthracnose disease in field. *Mycobiology* **39**, pp. 194–199.

60. Jagana, D., Hegde, Y. R., and Lella, R. (2017). Green nanoparticles: A novel approach for the management of banana anthracnose caused by *Colletotrichum musae*. *Int. J. Curr. Microbiol. App. Sci.* **6**, pp. 1749–1756.

61. Giannousi, K., Avramidis, I., and Dendrinou-Samara, C. (2013). Synthesis, characterization and evaluation of copper based nanoparticles as agrochemicals against *Phytophthora infestans*. *RSC Adv.* **3**, pp. 21743–21752.

62. He, L., Liu, Y., Mustapha, A., and Lin, M. (2011). Antifungal activity of zinc oxide nanoparticles against *Botrytis cinerea* and *Penicillium expansum*. *Microbiol. Res.* **166**, pp. 207–215.

63. Wani, A. H. and Shah, M. A. (2012). A unique and profound effect of MgO and ZnO nanoparticles on some plant pathogenic fungi. *J. Appl. Pharmaceut. Sci.* **2**, pp. 40–44.

64. Sardella, D., Gatt, R., and Valdramidis, V. (2017). Physiological effects and mode of action of ZnO nanoparticles against postharvest fungal contaminants. In: Paper Presented at the 109th Annual Meeting of the American Phytopathological Society, San Antonio, TX, August.

65. Sharma, P., Sharma, A., Sharma, M., Bhalla, N., Estrela, P., Jain, A., Thakur, P., and Thakur, A. (2017). Nanomaterial Fungicides: In Vitro and In Vivo Antimycotic Activity of Cobalt and Nickel Nanoferrites on Phytopathogenic Fungi Global Challenges. 1. pp. 1700041.

66. Jo, Y. K., Kim, B. H., and Jung, G. (2009). Antifungal activity of silver ions and nanoparticles on phytopathogenic fungi. *Plant Dis.* **93**, pp. 1037–1043.

67. Elmer, W. and White, J. C. (2018). The future of nanotechnology in plant pathology. *Annu. Rev. Phytopathol.* **56**, pp. 111–133.

68. Elmer, W. H., Ma, C., and White, J. C. (2018). Nanoparticles for plant disease management. *Curr. Opin. Environ. Sci. Health* **6**, pp. 66–70.

69. Cai, L., Liu, C., Fan, G., Liu, C., and Sun, X. (2019). Preventing viral disease by ZnO NPs through directly deactivating TMV and activating plant immunity in Nicotiana benthamiana. *Environ. Sci. Nano* **6**, pp. 3653.

70. Cai, L., Cai, L., Jia, H., Liu, C., Wang, D., and Sun, X. (2020). Foliar exposure of Fe_3O_4 nanoparticles on *Nicotiana benthamiana*: Evidence for nanoparticles uptake, plant growth promoter and defense response elicitor against plant virus. *J. Hazard. Mat.* pp. 122415.

71. Abdelkhalek, A. and Al-Askar, A. A. (2020). Green synthesized ZnO nanoparticles mediated by Mentha Spicata extract induce plant systemic resistance against Tobacco mosaic virus. *Appl. Sci.* **10**, pp. 5054.

72. Ditta, A. and Arshad, M. (2016). Applications and perspectives of using nanomaterials for sustainable plant nutrition. *Nanotechnol. Rev.* **2**(5), pp. 209–229.

73. Meena, V. S., Maurya, B. R., Verma, J. P., and Meena, R. S. (2016). *Potassium Solubilizing Microorganisms for Sustainable Agriculture*, Springer International Publishing, Cham, Switzerland, pp. 331.

74. Bala, R., Kalia, A., and Dhaliwal, S. S. (2019). Evaluation of efficacy of ZnO nanoparticles as remedial zinc nanofertilizer for rice. *J. Soil Sci. Plant. Nutr.*, **19**, pp. 379–389.

75. Madusanka, N., Sandaruwan, C., Kottegoda, N., Sirisena, D., Munaweera, I., De Alwis, A., Karunaratne, V., and Amaratunga, G. A. J. (2017). Urea-hydroxyapatite-montmorillonite nanohybrid composites as slow release nitrogen compositions. *Appl. Clay Sci.*, **150**, pp. 303–308.

76. Rios, J. J., Garcia-Ibañez, P., and Carvajal, M. (2019). The use of biovesicles to improve the efficiency of Zn foliar fertilization. *Colloids Surf. B Biointerfaces*, **173**, pp. 899–905.

77. Miranda-Villagómez, E., Trejo-Tellez, L. I., Gómez-Merino, F. C., Sandoval-Villa, M., Sánchez-García, P., and Aguilar-Mendez, M. Á. (2019). Nanophosphorus fertilizer stimulates growth and photosynthetic activity and improves P status in rice. *Journal of Nanomaterials*.

78. Tarafdar, J. C., Raliya, R., and Rathore, I. (2012a). Microbial synthesis of phosphorus nanoparticles from tricalcium phosphate using *Aspergillus tubingensis* TFR-5. *J. Bionanosci.* **6**, pp. 84–89.

79. Tarafdar, J. C., Agrawal, A., Raliya, R., Kumar, P., Burman, U., and Kaul, R. K. (2012b). ZnO nanoparticles induced synthesis of polysaccharides and phosphatases by Aspergillus fungi. *Adv. Sci. Eng. Med.* **4**, pp. 1–5.

80. Kumaraswamy, R., Saharan, V., Kumari, S., Choudhary, R. C., Pal, A., Sharma, S. S., Rakshit, S., Raliya, R., and Biswas, P. (2021). Chitosan-silicon nanofertilizer to enhance plant growth and yield in maize (*Zea mays* L.). *Plant Physiol. Biochem.* **159**, pp. 53–66.

81. Salako, O., Kovo, A., Abdulkareem, A., Yusuf, S., Afolabi, E., and Auta, M. (2020). The effect of synthesized NPK loaded surfactant modified zeolite a based fertilizer in tomato (*Lycopersicum esclentum*) cultivation. *NSChE Journal*, **35**, pp. 50–50.

82. Manikandan, A. and Subramanian, K. (2016). Evaluation of zeolite based nitrogen nano-fertilizers on maize growth, yield and quality on inceptisols and alfisols. *Int J Plant Soil Sci*, **9**, pp. 1–9.

83. Askary, M., Amirjani, M. R., and Saberi, T. (2017). Comparison of the effects of nano-iron fertilizer with iron-chelate on growth parameters and some biochemical properties of *Catharanthus roseus*. *J. Plant Nutr.* **40**, pp. 974–982.

84. Rui, M., Ma, C., Hao, Y., Guo, J., Rui, Y., Tang, X., Zhao, Q., Fan, X., Zhang, Z., Hou, T., and Zhu, S. (2016). Iron oxide nanoparticles as a potential iron fertilizer for peanut (*Arachis hypogaea*). *Front. Plant Sci.* **7**, p. 815.

85. Das, C. K., Srivastava, G., Dubey, A., Roy, M., Jain, S., Sethy, N. K., Saxena, M., Harke, S., Sarkar, S., and Misra, K. (2016). Nano-iron pyrite seed dressing: A sustainable intervention to reduce fertilizer consumption in vegetable (beetroot, carrot), spice (fenugreek), fodder (alfalfa), and oilseed (mustard, sesamum) crops. *Nanotechnol. Environ. Eng.* **1**, p. 2.

86. Bastani, S., Hajiboland, R., Khatamian, M., and Saket-Oskoui, M. (2018). Nano iron (Fe) complex is an effective source of Fe for tobacco plants grown under low Fe supply. *J. Soil Sci. Plant Nutr.* **18**, pp. 524–541.

87. Mendel, R. R. and Hänsch, R. (2002). Molybdoenzymes and molybdenum cofactor in plants. *J. Exp. Bot.* **53**, pp. 1689–1698.

88. Elrys, A. S., Abdo, A. I. E., and Desoky, E. S. M. (2018). Potato tubers contamination with nitrate under the influence of nitrogen fertilizers and spray with molybdenum and salicylic acid. *Environ. Sci. Pollut. Res.* **25**, pp. 7076–7089.

89. Abbasifar, A., ValizadehKaji, B., and Iravani, M. A. (2020). Effect of green synthesized molybdenum nanoparticles on nitrate accumulation and nitrate reductase activity in spinach. *J. Plant Nutr.* **43**, pp. 13–27.

90. Osman, S. A., Salama, D. M., Abd El-Aziz, M. E., Shaaban, E. A., and AbdElwahede, M. S. (2020). The influence of MoO_3-NPs on agro-morphological criteria, genomic stability of DNA, biochemical assay, and production of common dry bean (*Phaseolus vulgaris* L.). *Plant Physiol. Biochem.* **151**, pp. 77–87.

91. Niamat, B., Naveed, M., Ahmad, Z., Yaseen, M., Ditta, A., Mustafa, A., Rafique, M., Bibi, R., and Minggang, X. (2019). Calcium-enriched animal manure alleviates the adverse effects of salt stress on growth, physiology and nutrients homeostasis of *Zea Mays* L. *MDPI-Plants* **8**(11), pp. 480, https://doi.org/10.3390/plants8110480.

92. Shahzad, H., Ullah, S., Iqbal, M., Bilal, H. M., Shah, G. M., Ahmad, S., Zakir, A., Ditta, A., Farooqi, M. A., and Ahmad, I. (2019). Effects of salinity sources and levels on growth, physiology and nutrient contents of maize crop. *Ital. J. Agron.* **14**, pp. 199–207, https://doi.org/10.4081/ija.2019.1326.

93. Hamid, S., Ahmad, I., Akhtar, M. J., Iqbal, M. N., Shakir, M., Tahir, M., Rasool, A., Sattar, A., Khalid, M., Ditta, A., and Zhu, B. (2021). *Bacillus subtilis* Y16 and biogas slurry enhanced potassium to sodium ratio and physiology of sunflower (*Helianthus annuus* L.) to mitigate salt stress. *Environ. Sci. Pollut. Res.* https://doi.org/10.1007/s11356-021-13419-2

94. Ullah, N., Ditta, A., Khalid, A., Mehmood, S., Rizwan, M. S., Mubeen, F., and Imtiaz, M. (2020). Integrated effect of algal biochar and plant growth

promoting rhizobacteria on physiology and growth of maize under deficit irrigations. *J. Soil Sci. Plant Nutr.* **20**, pp. 346–356. https://doi.org/10.1007/s42729-019-00112-0

95. Parveen, A., Ashraf, M. A., Hussain, I., Perveen, S., Rasheed, R., Mahmood, Q., Hussain, S., Ditta, A., Hashim, A., Al-Arjani, A.-B. F., Alqarawi, A. A., and Abd_Allah, E.-S. F. (2021). Promotion of growth and physiological characteristics in water stressed *Triticum aestivum* consequent to foliar-application of salicylic acid. *MDPI Water* **13**(9), p. 1316. https://doi.org/10.3390/w13091316.

96. Ahmad, H. T., Hussain, A., Aimen, A., Jamshaid, M. U., Ditta, A., Asghar, H. N., and Zahir, Z. A. (2021). Improving resilience against drought stress among crop plants through inoculation of plant growth-promoting rhizobacteria. In: Husen, A. and Jawaid, M. (Eds.). *Harsh Environment and Plant Resilience: Molecular and Functional Aspects*. Springer Cham. pp. 387–408. https://doi.org/10.1007/978-3-030-65912-7_16.

97. Ullah, N., Ditta, A., Imtiaz, M., Li, X., Jan, A. U., Mehmood, S., Rizwan, M. S., and Rizwan, M. (2021). Appraisal for organic amendments and plant growth-promoting rhizobacteria to enhance crop productivity under drought stress: A review. *J. Agron. Crop Sci.* **207**(5), pp. 783–802. https://doi.org/10.1111/jac.12502.

98. Ditta, A. (2018). Physiological and phenological responses of crop plants under heat stress. In: Ramakrishna, A. and Gill, S. S. (Eds.), *Metabolic Adaptations in Plants during Abiotic Stress*. Taylor and Francis, Boca Raton: CRC Press USA. pp. 55–63. https://www.taylorfrancis.com/books/e/9781351676830/chapters/10.1201%2Fb22206-5

99. Imtiaz, M., Ashraf, M., Rizwan, M. S., Nawaz, M. A., Rizwan, M., Mehmood, S., Yousaf, B., Yuan, Y., Mumtaz, M. A., Ditta, A., Ali, M., Mahmood, S., and Tu, S. (2018). Vanadium toxicity in chickpea (*Cicer arietinum* L.) grown in red soil: Effects on cell death, ROS and antioxidative systems. *Ecotoxicol. Environ. Safe.* **158**, pp. 139–144. https://doi.org/10.1016/j.ecoenv.2018.04.022.

100. Rizwan, M. S., Imtiaz, M., Zhu, J., Yousaf, B., Hussain, M., Ali, L., Ditta, A., Ihsan, M. Z., Huang, G., Ashraf, M., and Hu, H. (2021). Immobilization of Pb and Cu by organic and inorganic amendments in contaminated soil. *Geoderma*, **385**, p. 114803. https://doi.org/10.1016/j.geoderma.2020.114803.

101. Irshad, S., Xie, Z., Mehmood, S., Nawaz, A., Ditta, A., and Mahmood, Q. (2021). Insights into conventional and recent technologies for arsenic bioremediation: A systematic review. *Environ. Sci. Pollut. Res.* **28**(15), pp. 18870–18892. https://doi.org/10.1007/s11356-021-12487-8.

102. Jan, A. U., Hadi, F., Shah, A., Ditta, A., Nawaz, M. A., and Tariq, M. (2021). Plant growth regulators and EDTA improve phytoremediation potential and antioxidant response of *Dysphania ambrosioides* (L.) Mosyakin & Clemants in a Cd-spiked soil. *Environ. Sci. Pollut. Res.* https://doi.org/10.1007/s11356-021-13772-2.

103. Thind, S., Hussain, I., Ditta, A., Perveen, A., Rasheed, R., Ashraf, M. A., Hussain, S., and Mahmood, Q. (2021). Alleviation of Cd stress by silicon nanoparticles during different phenological stages of Ujala wheat variety. *Arab. J. Geosci.* **14**, 1028.

104. Ahmad, I., Tahir, M., Daraz, U., Ditta, A., Hussain, M. B., and Khan, Z. U. H. (2020). Responses and tolerance of cereal crops to metals and metalloids toxicity. In: Hassanuzzaman, M. (Ed.), *Agronomic Crops*. Springer, Singapore. pp. 235–264. https://doi.org/10.1007/978-981-15-0025-1_14.

105. Mehmood, S., Saeed, D. A., Rizwan, M., Khan, M. N., Aziz, O., Bashir, S., Ibrahim, M., Ditta, A., Akmal, M., Mumtaz, M. A., Ahmed, W., Irshad, S., Imtiaz, M., Tu, and Shaheen, A. (2018a). Impact of different amendments on biochemical responses of sesame (*Sesamum Indicum* L.) plants grown in lead-cadmium contaminated soil. *Plant Physiol. Biochem.* 132, pp. 345–355. https://doi.org/10.1016/j.plaphy.2018.09.019.

106. Mehmood, S., Rizwan, M., Bashir, S., Ditta, A., Aziz, O., Yong, L. Z., Dai, Z., Akmal, M., Ahmed, W., Adeel, M., Imtiaz, M., and Tu, S. (2018b). Comparative effects of biochar, slag and ferrous–Mn ore on lead and cadmium immobilization in soil. *Bull. Environ. Contamin. Toxicol.* **100**(2), pp. 286–292. https://doi.org/10.1007/s00128-017-2222-3.

107. Liu, Y. Z., Imtiaz, M., Ditta, A., Rizwan, M. S., Ashraf, M., Mehmood, S., Aziz, O., Mubeen, F., Ali, M., Elahi, N. N., Ijaz, R., Lelel, S., Shuang, C., and Tu, S. (2020). Response of growth, antioxidant enzymes and root exudates production towards As stress in *Pteris vittata* and *Astragalus sinicus* colonized by arbuscular mycorrhizal fungi. *Environ. Sci. Pollut. Res.* **27**, pp. 2340–2352. https://doi.org/10.1007/s11356-019-06785-5.

108. Sabir, A., Naveed, M., Bashir, M. A., Hussain, A., Mustafa, A., Zahir, Z. A., Kamran, M., Ditta, A., Núñez-Delgado, A., Saeed, Q., and Qadeer, A. (2020). Cadmium mediated phytotoxic impacts in *Brassica napus*: Managing growth, physiological and oxidative disturbances through combined use of biochar and *Enterobacter* sp. MN17. *J. Environ. Manag.* **265**, pp. 110522. https://doi.org/10.1016/j.jenvman.2020.110522.

109. Shri, P. U. and Pillay, V. (2017). Excess of soil zinc interferes with uptake of other micro and macronutrients in *Sorghum bicolor* (L.) plants. *Indian J. Plant. Physiol.*, **22**, pp. 304–308.

110. Mehmood, S., Imtiaz, M., Bashir, S., Rizwan, M., Irshad, S., Yuvaraja, G., Ikram, M., Aziz, O., Ditta, A., Rehman, S. U., Shakeel, Q., Mumtaz, M. A., Ahmed, W., Mahmood, S., Chen, D., and Tu, S. (2019). Leaching behaviour of Pb and Cd and transformation of their speciation in co-contaminated soil receiving different passivators. *Environ. Eng. Sci.* **36**(6), pp. 749–759. https://doi.org/10.1089/ees.2018.0503.

111. Ijaz, M., Rizwan, M. S., Sarfraz, M., Ul-Allah, S., Sher, A., Sattar, A., Ali, L., Ditta, A., and Yousaf, B. 2020. Biochar reduced cadmium uptake and enhanced wheat productivity in alkaline contaminated soil. *Int. J. Agri. Biol.* **24**, pp. 1633–1640. https://doi.org/10.17957/IJAB/15.1605.

112. Naveed, M., Bukhari, S. S., Mustafa, A., Ditta, A., Alamri, S., El-Esawi, M. A., Rafique, M., Ashraf, S., and Siddiqui, M. H. (2020). Mitigation of nickel toxicity and growth promotion in sesame through the application of a bacterial endophyte and zeolite in nickel contaminated soil. *Int. J. Environ. Res. Public Health.* **17**(23), pp. 8859. https://doi.org/10.3390/ijerph17238859.

113. Ullah, I., Ditta, A., Imtiaz, M., Mehmood, S., Rizwan, M., Rizwan, M. S., Jan, A. U., and Ahmad, I. (2020). Assessment of health and ecological risks of heavy metal contamination: A case study of agricultural soils in Thall, Dir-Kohistan. *Environ. Monit. Assess.* **192**, pp. 786. https://doi.org/10.1007/s10661-020-08722-3.

114. Ihsanullah, A., Al-Amer, A. M., Laoui, T., Al-Marri, M. J., Nasser, M. S., Khraisheh, M., and Atieh, M. A. (2016). Heavy metal removal from aqueous solution by advanced carbon nanotubes: Critical review of adsorption applications. *Sep. Purif. Technol.* **157**, pp. 141–161

115. Mustafa, G. and Komatsu, S. (2016). Toxicity of heavy metals and metal-containing nanoparticles on plants. *Biochimica et Biophysica Acta (BBA)-Proteins and Proteomics*, **1864**, pp. 932–944.

116. Lim, J. Y., Mubarak, N. M., Abdullah, E. C., Nizamuddin, S., Khalid, M., and Inamuddin (2018). Recent trends in the synthesis of graphene and graphene oxide based nanomaterials for removal of heavy metals—a review. *J. Ind. Eng. Chem.* **66**, pp. 29–44.

117. Huang, Y. and Keller, A. A. (2020). Remediation of heavy metal contamination of sediments and soils using ligand-coated dense nanoparticles. *PloS one*, **15**, p. e0239137.

118. Ahmad, R., Wolfbeis, O. S., Hahn, Y.-B., Alshareef, H. N., Torsi, L., and Salama, K. N. (2018). Deposition of nanomaterials: A crucial step in biosensor fabrication. *Materials Today Communications*, **17**, pp. 289–321.

119. Dar, F. A., Qazi, G., and Pirzadah, T. B. (2020). Nano-biosensors: NextGen diagnostic tools in agriculture. In: Hakeem, K. and Pirzadah, T. (Eds.), *Nanobiotechnology in Agriculture. Nanotechnology in the Life Sciences*. Springer.

120. Su, D., Li, H., Yan, X., Lin, Y., and Lu, G. 2020. Biosensors based on fluorescence carbon nanomaterials for detection of pesticides. *TrAC Trends in Analytical Chemistry*, 116126.

121. Walia, S. and Acharya, A. (2014). Fluorescent cadmium sulfide nanoparticles for selective and sensitive detection of toxic pesticides in aqueous medium. *J. Nanopart. Res.* **16**, p. 2778

122. Dar, A. I., Walia, S., and Acharya, A. (2016). Citric acid-coated gold nanoparticles for visual colorimetric recognition of pesticide dimethoate. *J. Nanopart. Res.* **18**, p. 233.

123. Muenchen, D. K., Martinazzo, J., De Cezaro, A. M., Rigo, A. A., Brezolin, A. N., Manzoli, A., De Lima Leite, F., Steffens, C., and Steffens, J. (2016). Pesticide detection in soil using biosensors and nanobiosensors. *Biointerface Research in Applied Chemistry*, **6**.

124. Srivastava, A. K., Dev, A., and Karmakar, S. (2017). Nanosensors for food and agriculture. *Nanoscience in Food and Agriculture* **5**. Springer.

125. Singh, G. and Kalia, A. (2019). Nano-enabled technological interventions for sustainable production, protection, and storage of fruit crops. *Nanoscience for Sustainable Agriculture*. Springer.

126. Lahiani, M. H., Chen, J., Irin, F., Puretzky, A. A., Green, M. J., and Khodakovskaya, M. V. (2015). Interaction of carbon nanohorns with plants: Uptake and biological effects. *Carbon*, **81**, pp. 607–619.

127. Mazumder, J. A., Khan, E., Perwez, M., Gupta, M., Kumar, S., Raza, K., and Sardar, M. (2020). Exposure of biosynthesized nanoscale ZnO to Brassica juncea crop plant: Morphological, biochemical and molecular aspects. *Scientific Reports*, **10**, pp. 1–13.

128. Kumar M., Shamsi T. N., Parveen R., and Fatima S. (2017). Application of nanotechnology in enhancement of crop productivity and integrated pest management. In: Prasad, R., Kumar, M., and Kumar, V. (Eds.), *Nanotechnology*. Springer, Singapore. https://doi.org/10.1007/978-981-10-4573-8_17.

129. Bumbudsanpharoke, N. and Ko, S. (2015). Nano-food packaging: An overview of market, migration research, and safety regulations. *J. Food Sci.* **80**, pp. R910–R923.

130. Prasad, R., Bhattacharyya, A., and Nguyen, Q. D. (2017). Nanotechnology in sustainable agriculture: Recent developments, challenges, and perspectives. *Front. Microbiol.* **8**, p. 1014.

131. Berekaa, M. M., 2015. Nanotechnology in food industry, advances in food processing, packaging and food safety. *Int. J. Curr. Microbiol. App. Sci.* **4**, pp. 345–357.

132. Devi, P. V., Duraimurugan, P., and Chandrika, K. (2019). *Bacillus thuringiensis*-based nanopesticides for crop protection. Nano-biopesticides today and future perspectives. Elsevier.

133. Sahu, P. K., Sao, R., Mondal, S., Vishwakarma, G., Gupta, S. K., Kumar, V., Singh, S., Sharma, D., and Das, B. K. (2020). Next generation sequencing based forward genetic approaches for identification and mapping of causal mutations in crop plants: A comprehensive review. *Plants,* **9**, pp. 1355.

134. Bramlett, M., Plaetinck, G., and Maienfisch, P. (2019). RNA-based biocontrols: A new paradigm in crop protection. *Engineering,* **6**, pp. 522–527.

135. Ghormade, V., Deshpande, M. V., and Paknikar, K. M. (2011). Perspectives for nano-biotechnology enabled protection and nutrition of plants. *Biotechnol. Adv.* **29**, pp. 792–803.

136. Bleotu, C., Matei, L., Dragu, L. D., Grigorescu, L., Diaconu, C. C., and Anton, G. (2018). Methods for plant genetic modification. *Genetically Engineered Foods.* Elsevier.

137. Harling, H., Martin-Ortigosa, S., Niessen, M., Streitner, C., Schumann, N., and Jongedijk, E. (2018). Methods and hybrids for targeted nucleic acid editing in plants. Google Patents.

138. Galbraith, D. W. (2007). Nanobiotechnology: Silica breaks through in plants. *Nat. Nanotechnol.* **2**, pp. 272–273.

139. Vijayakumar, P. S., Abhilash, O. U., Khan, B. M., and Prasad, B. L. V. (2010). Nanogold-loaded sharp-edged carbon bullets as plant-gene carriers. *Adv. Funct. Mater.* **20**, pp. 2416–2423.

140. Shang, Y., Hasan, M., Ahammed, G. J., Li, M., Yin, H., and Zhou, J. (2019). Applications of nanotechnology in plant growth and crop protection: A review. *Molecules,* **24**(14), pp. 2558.

141. Miller, J. B., Zhang, S., Kos, P., Xiong, H., Zhou, K., Perelman, S. S., Zhu, H., and Siegwart, D. J. (2017). Non-viral CRISPR/Cas gene editing in vitro and in vivo enabled by synthetic nanoparticle co-delivery of Cas9 mRNA and sgRNA. *Angew. Chem. Int. Ed.* **56**, pp. 1059–1063.

142. Farhang, B. (2009). Nanotechnology and applications in food safety. In: Barbosa-Canovas, G., Mortimer, A., Lineback, D., Spiess, W., Buckle, K., and Colonna, P. (Eds.), *Global Issues in Food Science and Technology.* Elsevier.

143. Scott, N. and Chen, H. (2003). Nanoscale science and engineering of agriculture and food systems. Report submitted to. Cooperative State research, Education and Extension Service (CSREES), U.S. Dept. of Agriculture. Available from http://www.nseafs.cornell.edu.

144. Mathew, A. P., Laborie, M. P., and Oksman, K. (2009). Cross-linked chitosan/chitin crystal nanocomposites with improved permeation selectivity and pH stability. *Biomacromolecules*, **10**, pp. 1627–1632.

145. Fraceto, L. F., Grillo, R., de Medeiros, G. A., Scognamiglio, V., Rea, G., and Bartolucci, C. (2016). Nanotechnology in agriculture: Which innovation potential does it have? *Front. Environ. Sci.* **4**, 20.

146. Servin, A. D. and White, J. C. (2016). Nanotechnology in agriculture: Next steps for understanding engineered nanoparticle exposure and risk. *NanoImpact*, **1**, pp. 9–12.

Chapter 7

Application of Nanocomposites and Nanomaterials for Treatment of Wastewater Containing Pollutants

Hossein Esmaeili
Department of Chemical Engineering, Bushehr Branch, Islamic Azad University, Bushehr, Iran
esmaeili.hossein@gmail.com, h.esmaeili@iaubushehr.ac.ir

Industrial and urban wastewater are important sources of environmental pollutants such as heavy metals, toxic dyes, nitrate, phosphate, and microorganisms. These pollutants are dangerous and cause several problems for human health. Therefore, these wastewaters should be refined. There are several procedures in the literature for the treatment of wastewater, including filtration, reverse osmosis, electrochemistry, advanced oxidation, membrane, and sorption method. Among them, the use of adsorbents has attracted many attentions because of the low costs, high efficiency, and simplicity. Activated carbon is among the most important sorbents, which has been used in previous studies. However, this adsorbent has several disadvantages. Recently, nanocomposites

Science and Applications of Nanoparticles
Edited by Waqar Ahmed and Ehsan Nourafkan

ISBN 978-981-4267-34-2 (Hardcover), 978-1-003-28029-3 (eBook)
www.jennystanford.com

and nanomaterials have been widely used due to the high specific area and high porosity. This chapter focuses on the application of nanoadsorbents and nanocomposites on the removal of contaminants such as heavy metals, dyes, hydrocarbons, and drugs from wastewater. Also, the reusability of the nanoadsorbents after the uptake process has been investigated, which indicates that nanoadsorbents have high potentials for the sorption of different contaminants after several reuse cycles. Moreover, the impact of effective factors such as pH, contact time, dosage of nanoadsorbent, concentration of pollutants, and temperature was thoroughly investigated on the removal efficiency.

7.1 Introduction

Water is one of the most significant natural sources in the universe, and it is necessary for creatures to survive [1, 2]. The global population is regularly growing and releasing contaminants, and industrial and urban wastewater is rising. In fact, agricultural wastes, industrial and human residues play a crucial role in sewage pollution. All these items produce different contaminants, change the water cycle, and raise global concerns, as their ultimate effect is on human health and wildlife. The presence of poisonous and heavy metals, hydrocarbons, bacteria, or other ions such as nitrates (NO^{3-}), sulfate (SO_4^{2-}), ammonia (NH^{4+}), and phosphate (PO_4^{3-}) in water or wastewater has a negative impact on the environment, which has been extensively investigated in previous studies [3].

Also, the pharmaceutical materials specified in the environment belong to several classes of human drugs such as antibiotics, anticonvulsants, anticancer agents, analgesics, beta-blockers, lipid regulators, contrast agents, and hormones, which can enter the sewage and cause several unfavorable impacts on human health [3]. Generally, there are several types of contaminants in the wastewater, including heavy metal ions, dyes, hydrocarbons (i.e., oil, gas condensate, etc.) and drug materials, and all of them have their own characteristics. Therefore, their removal from wastewater should be separately studied.

Removing these contaminants from wastewater is a crucial factor for human health. There are several ways for the treatment of urban and industrial wastewaters, including nanofiltration, microfiltration, reverse osmosis, ultrafiltration, electrochemical, advanced oxidation, ion exchange, chemical precipitation, chemical coagulation, and sorption method. Each of these methods have benefits and drawbacks [4].

Chemical precipitation is a method to eliminate contaminants from wastewater, and in this method, the solubility of contaminants is reduced at pH >8. In this method, chemicals react with contaminants to form an insoluble precipitate. These precipitations can be eliminated from the solution by the filtration process [5, 6].

Also, the ion-exchange approach applies natural minerals and adsorbents. These substances include active sites to separate pollutants from the solution into their surface by the ion-exchange mechanism [6]. This method can remove pollutants from the solution with a high removal efficiency [7]. However, the ion-exchange process has several disadvantages. For example, (1) the resins applied in this method are expensive; (2) this process needs high capital investment; and (3) sedimentation occurs in the membrane [8].

Another method to eliminate contaminants from the wastewater is the membrane technology, which contains filtration, microfiltration, ultrafiltration, nanofiltration, and reverse osmosis. This technology is almost applied to treat sea water. Also, this method is an expensive technology and cannot be used for any pollutants. Membranes, regardless of cost, cannot withstand some chemicals, pH values as well as microorganisms [9, 10].

Flocculation and coagulation processes are the aggregation of unstable particles for converting them to massive clots and then precipitate them. In this process, the addition of a flocculator may increase the formation of these clots. This technique is usually the last stage of solid–liquid separation [9, 11].

Moreover, the electrochemical method involves the transfer of electrons between contaminants and the electrodes used. This technique requires high investment and expensive devices. Other disadvantages include effluent neutralization and sludge disposal [9, 12].

Among these methods, the adsorption process is a low-cost, high-efficiency, simple system and design, abundance of adsorbents, and environmentally friendly. In this process, the adsorbate (pollutant) adsorbs on the surface of the adsorbent by chemical or physical mechanisms. Also, there are several low-cost and efficient adsorbents in the nature, whose surfaces can be modified by different materials to improve their active sites on the adsorbent surface as well as sorption capacity [13, 14].

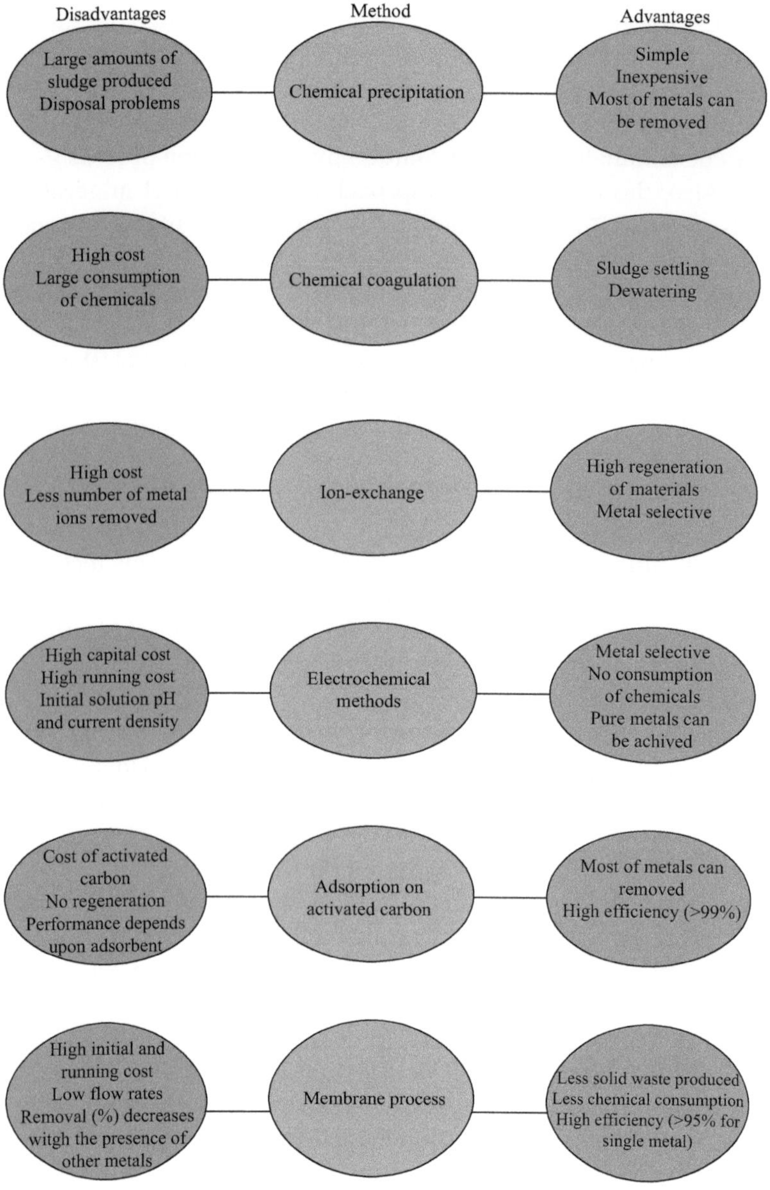

Figure 7.1 Advantages and disadvantages of different methods to remove pollutants from wastewater.

Recently, the use of nanoadsorbents and nanocatalysts has attracted many attentions. Nanomaterials have the high capability for removing contaminants due to the high activity, high specific area, and removal capacity [15, 16].

Figure 7.1 shows different methods for the removal of pollutants from wastewater as well as their advantages and disadvantages.

7.2 Nanomaterials Properties

Nanotechnology produces different types of materials at the nano-scale level. Nanoparticles (NPs) are an extensive class of materials, having at least one dimension less than 100 nm [17]. Scientists have found that the size of particles can affect their physiochemical properties. NPs are not simple molecules and consist of three layers, including surface layer, shell layer, and core. The surface layer can be formed by metal ions, surfactants, polymers, and other small molecules. Also, the shell layer can be a different material than the core, and the core is the central section of the NPs. NPs are classified into different categories, which depend on their size, morphology, and other properties. Based on physicochemical properties, NPs are classified as carbon- base, metal-base, semiconductor-base, ceramics-base, and polymeric-base [18].

Various techniques are utilized to measure the NPs size, including SEM or FESEM, TEM, DLS, XRD, and AFM. TEM, SEM, AFM, and XRD can determine the particles size better than other analyses; however, the DLS analysis can be applied to measure the NPs size at acutely low level. The high specific surface area (SSA) of nanomaterials proposes various applications of these materials and the BET analysis is the best method to measure the SSA of NPs. This analysis is based on the sorption/desorption procedure, and N_2 gas is applied for this analysis [18].

One of the main applications of NPs is their use in the treatment of effluents. To evaluate the impact of NPs in the environment, it is necessary to understand their reactivity, mobility, eco-toxicity, and stability. Nanoparticles have a critical role in the interface of solid/water because of high surface/mass ratio, and pollutants can be adsorbed on the surface of NPs [19]. The interaction of pollutants with nanoparticles depends on the NPs properties, including their size, surface area, porousness, components, morphology, and aggregate structure. The removal of toxic metal ions like Hg (II), Pb (II), Cd (II), Co (II), Ni (II), As (III), and As (V) from wastewater using NPs has attracted much attention because they have adverse impacts on human health. Fe_3O_4 NPs are one of these effective adsorbents for removing heavy metals [20]. Also, photodegradation by nanoparticles is an usual method, and many nanomaterials are used to do this. In a research, silica/NiO/ZnO NPs was used to photodegrade pollutants. The small size of NPs and high specific surface area simplified the photodegradation reaction [21].

7.3 Toxic Metals Removal from Wastewater

Toxic and heavy metal ions can mainly enter the water via mining, metallurgy, electroplating, smelting, chemical plants, paper industries, pesticides, agriculture, and household wastewater [22, 23]. Heavy metals like As(III), As(V), Cr(III), Cr(VI), Cu(II), Pb(II), Co, (II), Ni(II), Zn(II), Hg(II), etc. can accumulate in the food chains of humans and pose serious threats to human health [24–26]. These metal ions can damage the brain, kidneys, lungs, central nervous system, and other body parts [27]. Also, they can have negative impacts on the environment, when they are released into the environment and cannot be degraded by microorganism; vice versa, they accumulate via food chains. In addition, they are mostly toxic and carcinogenic [23, 28, 29]. These metal ions present in wastewater are categorized into three classes, which is presented in Fig. 7.2.

The maximum permissible limits for Ni, Cd, Pb, Zn, and Fe in drinking water are 0.2, 0.1, 0.05, 5, and 1 mg/L, respectively [30]. As per the WHO and FAO standards for drinking water recommended, a permissible limit for Cr (III) is less than 5 mg/L [31, 32]. Also, a permissible amount of 0.05 mg/L is considered for Cr (VI) in drinking water [33]. As per the WHO standards, the permissible limits for Mg, Na, and Cu in drinking water are 12 mg/L [34], 200 mg/L [35], and 3 mg/L [36], respectively. Also, the permissible limit for Co (II) in drinking water is not determined by the WHO [37]. Moreover, the allowable limit of Ag (II) in drinking water is 0.1 mg/L [38]. Based on the WHO, the maximum allowable values of Hg (II) and As (III) in drinking water are 0.001 and 0.01 mg/L, respectively [25].

Furthermore, based on the WHO, the allowable limit for U ion in drinking water is 0.03 mg/L [39].

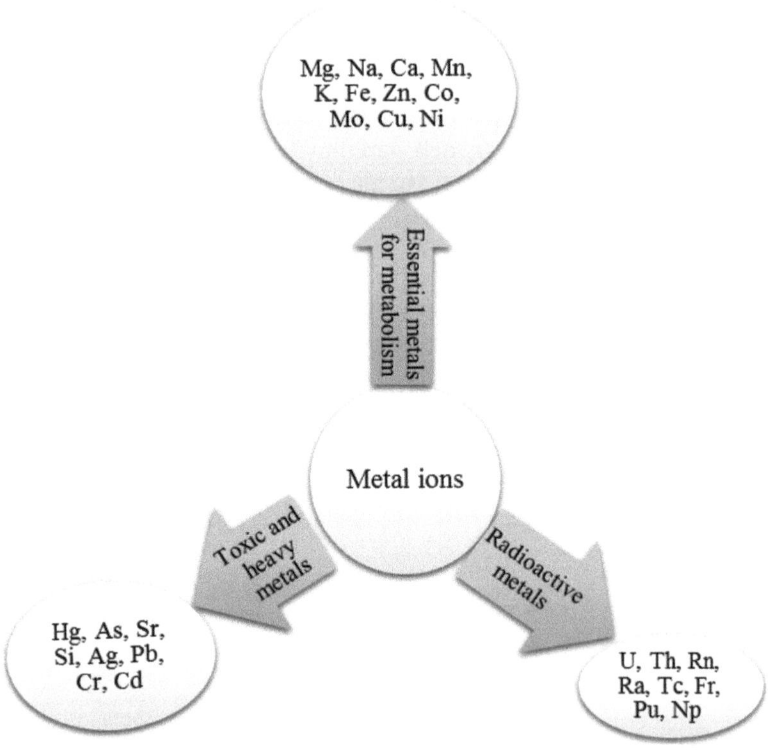

Figure 7.2 Different types of metal ions available in wastewater.

In the past years, porous metal-organic framework (MOF) nanomaterials have demonstrated their great supremacy in the elimination of hazardous materials from the wastewater because of their large porosity, high surface area, different pore structures, etc. [40]. Generally, nanomaterials have an external or internal dimension in the nanoscale range (1–100 nm) [41]. Due to the fine particles, NPs often show some special features such as high surface/volume ratio, high activity, and high SSA. These properties result in the extraordinary sorption capacity and high reactivity of the nanoadsorbents in removing contaminants from wastewater [23, 42]. There are several nanoadsorbents in the literature showing high removal efficiency of heavy metals, which include bentonite/

Fe_3O_4 nanocomposite [43], CaO/Fe_3O_4/SDS [31], graphene oxide/magnetic (GO/Fe_3O_4) [44], Fe_3O_4/talc nanocomposite [45], eggshell NPs [46], silver/quartz nanocomposite [47], cobalt-ferrite [48], MgO/Fe_3O_4 [49], polyaniline/ZnO [50], and magnetite NPs and magnetite-kaolinite nanocomposite [51]. The following table shows the removal efficiency of heavy metals using different nanomaterials from wastewater. As demonstrated, magnetic silica nanospheres, magnetite nanorods, eggshell nanoparticles, and ZnS nanocrystals showed the sorption capacity larger than 100 mg/g, and ZnS nanocrystals showed a remarkable sorption capacity (2000 mg/g).

Table 7.1 Removal efficiency and the maximum sorption capacity of several nanoadsorbents for the removal of heavy metals from aqueous media

Heavy Metals	Adsorbent	Conditions	Removal Efficiency (%)	q_{max} (mg/g)	Ref.
Cr (III)	CaO/Fe_3O_4/SDS nanocomposite			6.406	[31]
Copper	CMC-g-HAp/Fe_3O_4 nanocomposite	—	—	88.49	[22]
Nickel	CMC-g-HAp/Fe_3O_4 nanocomposite	—	—	86.95	[22]
Cu(II)	Magnetite nanorods	—	—	79.1	[52]
Fe (II)	Magnetite nanorods	—	—	127.01	[52]
Pb (II)	Magnetite nanorods	—	—	112.86	[52]
Zn (II)	Magnetite nanorods	—	—	107.27	[52]
Ni (II)	Magnetite nanorods	—	—	95.42	[52]
Cd (II)	Magnetite nanorods	—	—	88.39	[52]
Ni (II)	Nano-sized carbon in alginate beads	—	—	11.48	[53]
Ni (II)	Fe_3O_4/cyclodextrin polymer	—	—	13.2	[53]
Pb (II)	Montmorinollite@ Magnetite	—	—	38.15	[54]
As (V)	Montmorinollite@ Magnetite	—	—	19.10	[54]

Heavy Metals	Adsorbent	Conditions	Removal Efficiency (%)	q_{max} (mg/g)	Ref.
Ni (II)	3-Mercaptopropionic acid Fe_3O_4 NPs	Adsorbent dose = 0.1 g/L, pH = 6, temperature = 30°C, and ion concentration = 20 mg/L	96	42.01	[55]
Hg (II)	Magnetic silica nanospheres	Time = 20 min, pH = 4, dose = 5 g/L, temperature = 25°C, and ion concentration = 50 mg/L	98.8	303	[56]
Pb (II)	Magnetic silica nanospheres	Time = 20 min, pH = 4, adsorbent dose = 5 g/L, temperature = 25°C, and ion concentration = 50 mg/L	95.7	270.27	[56]
Hg (II)	Eggshell nanoparticle	pH = 9, temperature = 55°C, adsorbent dose = 1.5 g/L, ion concentration = 10 m/L, and time = 50 min	95	116.27	[29]
Cr (VI)	Magnetic multi-wall	Time = 600 min, pH = 3, and adsorbent dose = 0.1 g/L	100	16.23	[57]
Hg (II)	ZnS nanocrystals	Time = 5 min, and adsorbent dose = 10 g/L	99.99	2000	[58]

(*Continued*)

Table 7.1 (*Continued*)

Heavy Metals	Adsorbent	Conditions	Removal Efficiency (%)	q_{max} (mg/g)	Ref.
Ni (II)	Graphene nanosheets /δ-MnO_2	Time = 20 min, and adsorbent dose = 5 g/L	77.04	—	[59]
As (III)	Magnetic nanoparticles/zeolite	pH = 2.5, time = 15 min, and adsorbent dose = 0.5 g/L	95.6	44	[60]
Pb (II)	Zeolite	Time = 90 min and dose of adsorbent = 6 g/L	>80	88.31	[61]
Pb (II)	Acidified multi-walled carbon nanotubes (MWCNTs)	pH = 5, temperature = 25°C	—	49.71	[62]
Cu (II)	As-produced CNTs	pH = 6, temperature = 27°C	—	8.25	[63]
Cu (II)	NaOCl-modified CNTs	pH = 6, temperature = 27°C	—	47.39	[63]
Ni (II)	MWCNTs	pH = 5.4, temperature = 25°C	—	3.72	[64]
Ni (II)	NaClO-modified SWCNTs	Temperature = 25°C	—	47.76	[65]
Ni (II)	NaClO-modified MWCNTs	Temperature = 25°C	—	38.46	[65]
Pd (II)	Magnetic silica nanospheres	Time = 20 min, pH = 4, dose = 5 g/L, temperature = 25°C, ion concentration = 50 mg/L	96.4	256.41	[56]

7.4 Dye Removal from Wastewater

Dye is a different type of contaminant, which is extensively produced from industries such as dyeing, textile, printing, rubber, cosmetics, paper pulp, plastics, wool, coated paper, food, instant hair color, and paint industries [15, 66]. The presence of low concentration of dyes in the water is completely visible and unfavorable [67]. Synthetic dyes are the most common pollutants of water, which have an annual production of 10,000 tons, and 1–10% of which is discharged into the environment (water and soil). Therefore, the wastewaters of these industries can contaminate waters and soils irrigated. Dyes can cause several problems in the environment, because most of them are toxic, mutagenic, carcinogenic, and allergenic [68, 69]. Commercial dyes are classified in terms of their structure, applications, and particle charge when dissolved in water, including cationic dyes (basic dyes such as methylene blue), anionic dyes (such as methyl orange), and non-ionic dyes (dispersed dyes). Among them, cationic dyes are more toxic [15]. The intensity of colors in basic dyes is high, and they are even visible at very little concentrations [69]. Dyes can affect the photosynthetic activity because of reduced light penetration. Moreover, they are toxic to some aquatic life because they include metals, aromatics, etc. [70, 71]. Besides, dyes are mutagenic, carcinogenic, and teratogenic in different microbiological species. Furthermore, it may cause serious damage to humans, including kidney dysfunction, liver, brain, reproductive system, and central nervous system. Methylene blue (MB) dye is extensively used in the industry, which is not very toxic; however, it has harmful impacts [72]. This dye is not very toxic but can cause harmful effects and can lead to an enhanced heart rate, severe seizures, vomiting, jaundice, cyanosis, and necrosis in humans [68, 73]. Also, methyl violet (MV) dye is extensively applied in the textile to produce different pigments. Moreover, MV is carcinogenic, which can cause eye inflammation, corneal damage, and sensitivity to light. Inhalation of MV may irritate the respiratory tract and lead to vomiting, diarrhea, headache, and dizziness. Furthermore, prolonged exposure to MV can damage the mucous membrane and the gastrointestinal tract [15]. One of the most widely used dyes in the industry is methyl orange (MO), which is related to azo dyes and has a higher toxicity. MO is an anionic dye and is hazard for aquatic creatures, because of the presence of the azo group (-N=N-) in its structure [74]. In addition, brilliant blue (BB) dye is used in the leather and weaving industries. BB has been applied as a dietary supplement for dairy and confectionery products but is forbidden due to its toxic impacts on the humans and animals' health. BB is carcinogenic, and exposure to it results in disorders in the reproductive and nervous system and several diseases like severe allergies, itching, edema, and respiratory problems [75]. Effluent containing Congo Red (CR) is highly produced from textiles, dyeing, paper, printing, and plastic industries. CR is a carcinogen and a mutant and can be toxic for many organisms. Also, CR is an anionic dye [76]. Besides, the basic dye of Crystal Violet (CV) is applied for colored paper and ink for black ballpoint pens or inkjet printers. It is also applied to color a variety of products, including leather, detergents, antifreeze, and fertilizers [77]. Also, Safranin-T (ST) is a water-soluble dye, reddish brown and phenazine cationic dye. ST is used as a food dye in candy and cookies for diagnosing mucin and cartilage, which is considered to be carcinogenic. Because of its extensive use in the textile and food industries as colorant, the release of ST in wastewaters can contaminate these effluents in this way. The presence of ST dye in the wastewater can cause problems to respiratory systems, digestive, skin, and tract infections [78].

Several physical, chemical, and biological methods have been applied to eliminate dyes from wastewater. Among various methods, the adsorption process is an effective and low-cost method, which has been applied by several researchers [68]. Recently, the use of nanoadsorbents, nanocatalysts, and nanocomposites for the removal of dyes has attracted much attention. Table 7.2 provides the removal efficiency and the maximum adsorption capacity of various nanocomposites and nanoadsorbents for dye removal from wastewater. According to the table, some nanoadsorbents such as eggshell nanoparticle, calcium alginate/MWCNT, hydrogel-clay nanocomposite, poly vinyl alcohol/diethylenetriamine/ethylenediamine glutaraldehyde nanofiber, Fe_3O_4/AC nanocomposite, and cobalt-decorated SiO_2–Al_2O_3 indicated high sorption capacity (>100 mg/g) and among them cobalt nanoparticles-decorated SiO_2–Al_2O_3 had a sorption capacity of 2232 mg/g, which was an excellent sorption capacity. Also, the removal efficiency of these nanoadsorbents is more than 95%, which shows high removal efficiencies.

Table 7.2 The removal efficiency and the highest sorption capacity of different nanoadsorbents for the removal of dyes from synthetic wastewater

Nanoadsorbent	Dye Type	Conditions	Removal Efficiency (%)	q_{max} (mg/g)	Ref.
Eggshell nanoparticle	Methyl violet	pH = 9, temperature = 55°C, adsorbent dose = 1.5 g/L, ion concentration = 10 m/L, time = 50 min	96	123.45	[29]
Activated carbon/Zn/Alginate	Crystal violet	Temperature = 30°C, pH = 5, time = 50 min, adsorbent dose = 1 g, ion concentration = 60 mg/L	97	6.69	[79]
Calcium alginate/multi-walled carbon nanotubes	Methylene blue	—	—	606.1	[80]
Calcium alginate/multi-walled carbon nanotubes	Methyl orange	—	—	12.5	[80]
Cadmium sulfide/multi-walled carbon nanotubes/polymer nanocomposite	Methyl orange	Ion concentration = 15 mg/L, H_2O_2 concentration = 1.0 mM, time = 90 min, and photocatalyst dosage = 0.70 g/L	99	—	[81]
Chitosan/montmorillonite nanocomposites	Congo Red	—	—	54.52	[82]

(*Continued*)

Table 7.2 (*Continued*)

Nanoadsorbent	Dye Type	Conditions	Removal Efficiency (%)	q_{max} (mg/g)	Ref.
Hydrogel-clay nanocomposite	Safranine-T	—	—	484.2	[83]
Hydrogel-clay nanocomposite	Brilliant Cresyl Blue	—	—	494.2	[83]
Poly vinyl alcohol/ diethylenetriamine/ ethylenediamine glutaraldehyde nanofiber	Direct Red 23 (DR23)	—	—	370	[84]
Poly vinyl alcohol/ diethylenetriamine/ ethylenediamine glutaraldehyde nanofiber	Direct Blue (DB78)	—	—	400	[84]
Fe_3O_4/AC nanocomposite	Methylene blue	pH = 7, time = 30 min, dye concentration = 25 mg/L, temperature = 25°C	96.2	384.6	[85]
Cobalt-nanoparticles Decorated SiO_2–Al_2O_3	Methyl orange	pH = 7, temperature = 25°C, time = 60 min	99.96	2232	[86]

Nanoadsorbent	Dye Type	Conditions	Removal Efficiency (%)	q_{max} (mg/g)	Ref.
Si/Al	Methyl orange	—	—	23.09	[86]
Nickel (II) oxide NPs	Methyl orange	pH = 6, ion concentration = 20 mg/L, temperature = 30°C, time = 20 min	95.22	46.511	[87]
Functionalized TiO_2 nanocomposites	methyl orange	pH = 6.5, adsorbent dose = 0.01 g, ion concentration = 5 mg/L, time = 30 min	—	42.85	[88]
CO_3O_4 NPs	Methyl orange	pH = 6, adsorbent dose = 0.5 g, ion concentration = 50 mg/L, temperature = 30°C, time = 60 min	—	46.08	[89]
Ferric oxide biochar nanocomposites	Methyl orange	pH = 8, adsorbent dose = 5.0 g, ion concentration 50 mg/L, time = 30 min	-	20.53	[90]
MWCNTs	Methylene blue	—	—	42.3	[91]
MWCNTs	Neutral Red	—	—	77.5	[91]

7.5 Phosphate, Sulfate, and Nitrate Removal from Wastewater

Industrial and municipal wastewaters are one of the main contaminants of the environment. Among the inorganic pollutants, sulfate (SO_4^{2-}) is found in urban sewage as well as with various concentrations in industrial wastewaters. Also, its concentration is quickly enhancing in lakes and seas due to the drainage of urban and industrial wastewaters [92]. Excessive concentration of sulfate in the human stomach causes many problems for human health. Sulfate also causes pipe corrosion. In addition, it may cause unfavorable taste in water and diarrhea in human. The taste threshold changes between 250 mg per liter of sodium sulfate and 1000 mg per liter of calcium sulfate. The unfavorable taste of water is usually decreased at less than 250 mg/L. Also, the highest favorable concentration of sulfate ion in drinking water is about 250 mg/L and the highest allowable amount is 400 mg/L [93]. Damages caused by the release of sulfate are not direct because sulfate is not a toxic component. Nevertheless, the high concentration of sulfate unbalances the natural cycle of sulfur. Anion sulfate is found in water and industrial wastewater, including wastewater of chemical industry and drainage of acid mine. Consumption of drinking water containing a sulfate concentration of more than 600 mg/L usually leads to laxative impacts [94]. The sorption method can be selected as a suitable way for removing high concentration of sulfate ion from wastewater [95].

Also, phosphate ion (PO_4^{3-}) has an important duty in the growth of creatures. Phosphate may enter effluents via different ways such as agricultural industries, detergents, paints, and pharmaceuticals. Phosphorus is present in water as orthophosphate and polyphosphate [96]. The entrance of wastewater containing phosphate to the environment results in surface water eutrophication, toxicity, and enhanced oxygen consumption. To solve or decrease these problems, phosphate should be separated from wastewater. Phosphate ion concentration control at the output of municipal effluent treatment plants plays an important role in preventing the ablation of surface water [97]. USEPA has suggested an average phosphate concentration of 10 μg/L for lakes and rivers. This value is often considered the minimum concentration of phosphate to avoid the eutrophication process [98].

In addition, nitrate (NO_3^-) is a main nutrient needed for the plants' growth, and nitrogenous fertilizers are routinely used on agricultural lands around the world to enhance crop production [99]. High concentration of NO_3^- in drinking water more than the standard amount causes the blue syndrome in infants and cancer since nitrate is decreased to nitrite. Different sources of nitrate releasing into environment include mineral chemical fertilizers, animal manures, sanitary and industrial wastewater, plant wastes, domestic sewage, compost, nitrification of nitrogen in the atmosphere by nitrification bacteria, surface rainfall, and other sources of water contamination [100, 101]. The high amount of nitrate ion in drinking water affects human health and creates environmental hazards, which causes human cancer, eliminating biodiversity in the area, and eutrophication of surface waters [101]. According to the WHO and the European Union reports, the concentration of nitrate in drinking water should not exceed 50 mg NO_3^-/L, while the concentration of nitrate in drinking water according to the USEPA should not exceed 44 mg NO_3^-/L. Also, the allowable limit of nitrate in Australia is set at 50 mg NO_3^-/L for infants under 3 months of age and for children and adults over 3 months. Moreover, South Africa has set an allowable value of 20 mg NO_3^-/L [99, 102].

One of the main methods to attenuate these types of pollutants from wastewater is the use of nanomaterials, which have been recently applied. Table 7.3 gives some important nanomaterials such as nanoadsorbents for removing these contaminants from wastewater. As shown in this table, magnetic multi-walled carbon nanotubes, magnetic multi-walled carbon nanotubes, and MgO/$Fe_2(MoO_4)_3$ have demonstrated the highest adsorption capacity of nitrate, sulfate, and phosphate ions, respectively.

Table 7.3 The removal efficiency and the highest sorption capacity of phosphate, sulfate, and nitrate from wastewater using different nanoadsorbents

Sorbent	Pollutant	Conditions	Removal Efficiency (%)	q_{max} (mg/g)	Ref.
MgO/Fe_3O_4	Phosphate	Temperature = 25°C, pH = 9, time = 20 min, adsorbent dose = 5 g/L, and ion concentration = 1 mg/L	98.9	2.95	[97]
$MgO/Fe_2(MoO_4)_3$	Phosphate	Temperature = 45°C, pH = 6, time = 80 min, adsorbent dose = 2 g/L, ion concentration = 10 mg/L	98.38	30.21	[16]
HFO@ triethylamine-functionalized polystyrene resin	Phosphate	—	—	18	[23]
Magnetic multi-walled carbon nanotubes	Sulfate	pH = 5.96, adsorbent dose = 0.1 g/L, temperature = 25°C, time = 90 min	93.28	56.94	[94]
Ba–SnO	Sulfate	—	83±2	—	[103]
$Ba–SnCl_2$	Sulfate	—	76±2	—	[103]
$Ba–SnCl_4$	Sulfate	—	79±2	—	[103]

Sorbent	Pollutant	Conditions	Removal Efficiency (%)	q_{max} (mg/g)	Ref.
HFO@ triethylamine-functionalized polystyrene resin	Nitrate	—	—	45	[23]
Magnetic multi-walled carbon nanotubes	Nitrate	pH = 4, adsorbent dose = 0.1 g/L, temperature = 25°C, time = 90 min	97.15	183.11	[100]
Chitosan-polystyrene-Zn nanocomposite	Nitrate	Initial concentration = 10 mg/L, pH = 3, time = 30 min, adsorbent dose = 20 g/L	90	0.541	[104]
ZnO NPs	Nitrate	Ion concentration = 10 mg/L, pH = 3, time = 30 min, adsorbent dose = 20 g/L	100	4.184	[104]
TiO_2/SiO_2 nanocomposites	Nitrate	pH = 6.81	92.84	86.59	[105]
TiO_2/nZVI nanocomposite	Nitrate	Adsorbent dosage = 0.982 g/L, pH = 4.185, time = 150.091 min	98.226	—	[106]
Nano-alumina	Nitrate	Temperature = 25°C, pH = 4.4	—	4	[107]
Fe/Pt NPs	Nitrate	—	97	103	[108]

7.6 Drugs Removal from Wastewater

Drugs and pharmaceutical residues are environmental contaminants, and their presence in water is an environmental problem, and extensive studies on them have been done over the past decades. There are various types of drugs in wastewater, and among them, non-steroidal anti-inflammatory drugs (NSAIDs) are a group of drugs with several impacts such as antipyretic, analgesic, and anti-inflammatory [109, 110].

Naproxen (Nap), Diclofenac (Dic), and Ibuprofen (Ibu) are part of the NSAIDs family, which are abundantly found in hospital wastewater. The presence of these contaminants in the wastewater is dangerous for human health. Naproxen belongs to the group arylacetic acid, and its presence in water can have a significant toxic impact on the ecosystem. Also, the products of its photodegradation are more toxic than Naproxen. Moreover, Dic is identified in wastewater, because of its high consumption and resistance to biological decomposition. Due to their stability and hydrophilicity, they can be easily seen in water. Furthermore, Ibuprofen is found in the wastewater at a dose of 24.6 μg/L. The consumption of Ibu is high and has low solubility and high mobility in aqueous solution. Therefore, separation of NSAIDs from the hospital wastewater is essential [109].

The residuals of these compounds are extremely perilous to human health and aquatic environment because of their toxicity, mutagenicity, and genotoxicity. The main sources of cytostatic components discharged to wastewater treatment plants (WWTPs) are urban and hospital wastewaters. Conventional methods are not suitable for efficient removal of many organic pollutants such as pharmaceuticals from wastewater. These methods must be adapted to the local contamination sources, and in some cases, the tertiary wastewater treatment should be selected. There are several methods in the literature, which are used for the efficient removal of anticancer drugs from wastewater such as chemical, biological, and physicochemical processes [111].

Table 7.4 The removal efficiency and the highest sorption capacity of drugs from wastewater using various nanoadsorbents

Nanoadsorbent	Drug	Conditions	Removal Efficiency (%)	q_{max} (mg/g)	Ref.
Copper NPs	Ibu	Temperature = 25°C, pH = 4.5, 10.0 mg Cu NPs, time = 60 min	74.4	33.9	[109]
Copper NPs	Nap	Temperature = 25°C, pH = 4.5, 10.0 mg Cu NPs, time = 60 min	86.9	33.9	[109]
Copper NPs	Dic	Temperature = 25°C, pH = 4.5, 10.0 mg Cu NPs, time = 60 min	91.4	36.0	[109]
Multi-walled carbon nanotubes	Dic	—	—	19.9	[115]
Multi-walled carbon nanotubes	Triclosan	—	—	19.7	[115]
Magnetic multi-walled carbon nanotubes	Nap	Temperature = 25°C and pH = 3	67.2	20.75	[116]
Nanographene	Ibu	Temperature = 50°C, time = 60 min, adsorbent dose = 1 g/L, and drug concentration = 20.0 mg/L	82.7	11.9	[117]
Nanographene	Nap	Temperature = 50°C, time = 60 min, adsorbent dose = 1 g/L, and drug concentration = 20.0 mg/L	97.3	17.8	[117]

(Continued)

Table 7.4 (*Continued*)

Nanoadsorbent	Drug	Conditions	Removal Efficiency (%)	q_{max} (mg/g)	Ref.
Nanographene	Dic	Temperature = 50°C, time = 60 min, adsorbent dose = 1 g/L, and drug concentration = 20.0 mg/L	99.3	19.3	[117]
Nanographene	Ketoprofen (KETO)	Temperature = 50°C, time = 60 min, adsorbent dose = 1 g/L, and drug concentration = 20.0 mg/L	94.9	16.6	[117]
Nanoporous carbon	Ciprofloxacin	Time = 25 min, pH = 6	—	416.7	[118]
Carbon nanofibers	Ciprofloxacin	Time = 120 min, and pH = 6	—	10.36	[119]
Carbon nanotubes	Sulfamethoxazole	Time = 60 min and pH = 7	—	33.58	[120]
Carbon nanotubes	Sulfamethoxazole	Time = 120 min and pH = 4	—	12	[121]
Green synthesized magnetite NPs	Piperacillin, tazobactam, sulfamethoxazole, tetracycline, trimethoprim, ampicillin and erythromycin	Time = 30 min and pH = 5.5	>90%	—	[122]

Nanoadsorbent	Drug	Conditions	Removal Efficiency (%)	q_{max} (mg/g)	Ref.
γ-Fe_2O_3/NiO nanocomposite	Ibuprofen	—	—	323	[123]
Magnetic NPs coated zeolite	Ibu, Nap, and gemfibrozil	Time = 10 min and pH = 4	95	—	[124]
γ-Fe_2O_3/NiO nanocomposites	17β-Estradiol	Time = 12 min and pH = 6	—	35.232	[123]
Core–shell CNTs@MIPs nanocomposites	Estrone, Estriol, and diethylstilbestrol	—	—	113.5	[125]
Graphite oxide/poly(acrylic acid) grafted chitosan nanocomposite	Dorzolamide	pH = 3	—	334	[126]
MWCNTs	Olaquindox	pH = 3 and time = 2 min	—	99.7	[127]
XAD-7 resin	Erythromycin	—	—	0.82 m mol/g	[128]
XAD-16 resin	Erythromycin	—	—	0.34 m mol/g	[128]
IRA-410 resin	Erythromycin	—	—	0.91 m mol/g	[128]
Zinc oxide @carbon nanofibers	Amoxicillin	—	—	156	[129]

Hospital wastewater contains a wide variety of micro-pollutants due to research works, laboratory, diagnosis, and patient disposal. Chemicals that appear in the hospital wastewater usually include all types of drugs and their metabolites, X-ray agents, disinfectants, halogenated solvents, heavy and toxic metals, etc. These wastewaters are commonly discharged into general sewage networks without pretreatment with high drug contamination and join the municipal sewage with an extensive variety of chemicals that must be treated together in WWTPs [112].

WWTPs pollutants are usually human and veterinary drugs, surfactants, personal care drugs, pesticides, and different industrial additives [113]. Due to the probable dangers to human health, the presence of drugs at low concentration has caused a great deal of concern [114]. The removal efficiency and the maximum sorption capacity of several nanomaterials for the removal of drugs from wastewater are provided in Table 7.4. As demonstrated in Table 7.4, some nanomaterials such as nanoporous carbon, γ-Fe_2O_3/NiO nanocomposite, core–shell CNTs@MIPs nanocomposites, graphite oxide/poly(acrylic acid) grafted chitosan nanocomposite, and zinc oxide @carbon nanofibers showed very high sorption capacities (greater than 100 mg/g).

7.7 Hydrocarbon Removal from Wastewater

Oil spill pollution is one of the most important challenges in the world and has a great negative effect on the environment, because it is considered one of the most hazardous contaminants and the life of marine creatures is at high risk. Oil can be present in the wastewater as tiny droplets. Industrial progression and human activities have increased oily wastewater. There are many sources of wastewater containing oil, such as petrochemical plants, crude oil spill, power plants, and household residuals (e.g., waste cooking oil) [130]. Also, it is a blend of various organic and inorganic components [131]. Organic toxic compounds (e.g., oil and grease) cause environmental harms to aquatic organisms, plants, animals, and they also have mutagenic and carcinogenic impacts on human. Moreover, oil and grease decrease the biological activity in wastewater, which results in a decrease in the concentration of dissolved oxygen in the wastewater and eventually, it causes environmental damage [132]. Therefore, the discharge of oil-contaminated wastewater will be one of the most important environmental problems in the oil industry [131]. Hydrocarbons mostly occur in wastewater when the fine droplets of oil or gas condensate are dispersed in water and form an emulsion. This type of emulsion (oil in water or O/W) is commonly detected in refinery wastewater containing a small portion of oil droplets dispersed in the water, which must be separated [133, 134].

The presence of oil in the water blocks the pipes in treatment units, which must be cleaned and sometimes replaced. Therefore, it increases the cost of maintenance and inspection [132]. Also, the presence of oil in the water has other adverse impacts such as corrosion in pipes, leakage during processing, and sabotage of oil infrastructure. Annually, about 20 million tons of oil are dumped into the ocean [130].

It is very important to eliminate oil content from effluents before discharging them into the environment [130]. There are several procedures to eliminate oil from emulsion, which include thermal, chemical, mechanical, and physical methods. The use of chemical methods is a common way to destabilize or separate oil from emulsion [68, 135]. In the chemical method, an adsorbent or demulsifier can be applied for separating the oil from the emulsion. Recently, the use of adsorbents has attracted much attention. Many studies are being conducted using advanced and cost-effective adsorbents for effectively removing oil from oily emulsions [130, 136]. Various adsorbents like polyethylene, chalk powder, sawdust, polyurethane, nylon fibers, and activated carbon have been investigated for removing oil. However, removing oil contaminants from wastewater using these adsorbents is expensive [137]. Also, the biological method is extensively applied; however, this method is usually slow and limited because of non-biodegradable and toxic contaminants, which lead to the toxicity of microorganisms. The physical procedures like filtration or microfiltration can be applied to separate the pollutants; however, these methods produce high amounts of sludge, which are toxic and difficult to remove. As mentioned above, a fundamental and powerful technology is required for the treatment of industrial wastewaters. Nanotechnology can be applied as an effective and potential technique to remediate industrial wastewater and other environmental issues. Nanotechnology can be categorized into three main classes: nanoadsorbents, nanocatalysts, and nanomembranes. A great number of works have been recently done on the removal of contaminants from industrial wastewater using nanoadsorbents. Nanoadsorbents have high active sites and high specific surface area, which lead to high sorption capacity. Materials that can be applied to produce nanoadsorbents include activated carbon, clay, silica, metal oxides, and other materials in the form of composite. Also, other nanomaterials like metal oxides and semiconductors have been extensively used by scientists in the treatment of industrial wastewater. There are various types of nanocatalysts to remove the contaminants from wastewater, including electrocatalysts, catalysts obtained by the Fenton method, and catalysts with antimicrobial properties. Moreover, the treatment of wastewater under pressure has been used in the nanomembrane technology as an ideal method to enhance the quality of water [138].

NPs have been recently considered as powerful adsorbents. The smaller size of NPs enhances their SSA, which increases the sorption capacity of NPs for the adsorbing oil [138]. There are a few works in the literature on removing and separating oil from wastewater, which are provided in Table 7.5. According to the results presented in Table 7.5, GO and MWCNTs-silica aerogel showed excellent sorption capacities for the removal of oil from wastewater.

7.8 Reusability of Nano-Sorbents

The reusability of adsorbents plays a critical role for practical applications, economic and cost effectiveness of the process in wastewater treatment [144]. For the recovery and reuse of nanoadsorbents, different regenerating agents like acids, alkalis, and EDTA have been used in previous studies although the reusability of the used adsorbents and the recovery of heavy metal ions is an important task [145].

There are a few works on the reusability of nanoadsorbents in the treatment of wastewater, and most of them have been done on the reusability of conventional adsorbents. Comparing different studies, several adsorbents such as manganese dioxide@ sand, Fe_3O_4@ sand, nano-zerovalent iron, and Fe(III) oxide/hydroxide NPs were recovered and reused using alkaline chemicals such as NaOH after the sorption of heavy metals [145].

Table 7.5 The removal efficiency and the maximum sorption capacity of nanomaterials for the removal of oils from wastewater

Adsorbent	Type of Emulsion	Conditions	Removal Efficiency (%)	q_{max} (mg/g)	Ref.
Hydrophobic nanosilica	Gasoline in water	Low pH	99	—	[136]
Graphene oxide (GO)	—	4 min	>90	1335	[139]
Silica	—	—	100	169.4	[131]
wood sawdust coated Fe_3O_4 functionalized stearic acid	Motor oil-water	5 min and 0.1 g nanoadsorbent	—	28.32	[140]
Wood sawdust coated Fe_3O_4 functionalized stearic acid	Crude oil	10 min and 0.1 g nanoadsorbent	—	41.22	[140]
Styrene/divinyl benzene/ potassium persulfate, sorbitan monooleate	Oily wastewater	75 min, 300 rpm, 0.79 g/L of sorbent dose	99.75	—	[141]
Pristine CNTs	1,2-dichlorobenzene in water	—	—	30.8	[142]
Graphitized CNTs	1,2-dichlorobenzene in water	—	—	28.7	[142]
Hierarchical porous MWCNTs-silica aerogel	Diesel in water	—	—	24420	[143]

Table 7.6 The results of desorption and regeneration of nanoadsorbents in the removal of pollutants from wastewater

Nanoadsorbent	Pollutants	Removal efficiency (%)	Desorption agent	Desorption/regenerating efficiency (%)	Ref.
Nano-zero-valent iron	As (V) and As (III)	100% As (III) and 100% As (V)	NaOH	95.5% (n = 8) for As (III) and 72.7% (n = 8) for As (V)	[146]
Nano-TiO_2	Se (IV)	>96 %	0.1 M NaOH	>95% (n = 1)	[147]
Titanium dioxide NPs	Pb (II), Cu (II), and Zn (II)	—	0.01 M $NaNO_3$ Metal-free solution	98% of Pb (II), 99% of Cu (II) and 100% of Zn (II) were desorbed at pH = 2 and after 24 h	[148]
TiO_2 NPs	Pb (II), Cu (II) and Zn (II)	94% $n = 4^*$	EDTA	92% (n = 4)	[149]
Fe(III) oxide/hydroxide nanoparticle	Cr(VI)	—	NaOH and $BaCl_2$	95–97% of the adsorbed Cr (VI) was desorbed	[150]
Orange peel powder/ Fe_3O_4 NPs	Cd (II)	—	0.1 M HNO_3	98% of the adsorbed Cd^{2+} was desorbed	[151]
Maghemite NPs	Cr (VI)	97.3%	NaOH	87.7% (n = 6)	[152]
Maghemite NPs	Cr (VI)	97.3%	NaAC	19.7% (n = 6)	[152]
Maghemite NPs	Cr (VI)	97.3%	$NaHCO_3$	73.5% (n = 6)	[152]
Maghemite NPs	Cr(VI)	97.3%	Na_2CO_3	69.6% (n = 6)	[152]
Maghemite NPs	Cr (VI)	97.3%	Na_3PO_4	82.9% (n = 6)	[152]

*n = number of cycles

7.9 Effective Factors on Wastewater

There are several factors affecting the removal of contaminants from wastewater, including temperature, contact time, pH, dose of nanoadsorbent, and concentration of pollutants in the wastewater [42]. It will be economical if the sorption process performs at low temperatures, for example, 25°C [15]. If the sorption efficiency increases with increasing temperature, the sorption process will be endothermic, otherwise it is exothermic [15, 153]. Also, the solution pH is a critical factor on the sorption of pollutants from wastewater. At low pH values, the aqueous solution has positive charges (H^+) and the aqueous media will have negative charges (OH^-) at high pH values [68]. Therefore, the sorption process depends on the charge on the pollutant. Some pollutants have negative charge, and other pollutants have positive charge. In this case, a competition will create between the charge in the solution and the charge on the pollutant. Similar charges repel each other, while the opposing charges attract each other [68].

Moreover, the contact time is another important parameter on the sorption of pollutants from effluents, because it affects the economic process. One of the features of using nanosorbents is the sorption process in a short time [15]. Furthermore, adsorbent dosage is one of the most critical factors in the removal of pollutants from wastewater, because it determines the maximum sorption capacity of an adsorbent for a specified amount of the adsorbate [154]. The high SSA of nanoadsorbents causes a small dose of them to be used in the sorption process [155].

Besides, the initial ion concentration is one of the key factors on the sorption capacity and the best sorption efficiency occurs at low ion concentrations, because there are many active sites on the sorbent surface to adsorb ions and almost all ions can be adsorbed on the sorbent surface. But by increasing the ion concentration, the active sites on the surface of the adsorbent will be filled and all ions cannot be adsorbed on the sorbent surface [46, 49].

7.10 Conclusion

In this chapter, the effects of various types of nanomaterials on the removal of pollutants from wastewater have been studied. The studied pollutants include heavy metals, dyes, phosphorous, nitrate, sulfate, types of drugs, and oil. Each of these contaminants have different behaviors, and removing them are very different. There are various types of nanomaterials in the literature such as metal oxide NPs, magnetic NPs, metal oxide/magnetic nanocomposite, etc., which can be effective on the removal of contaminants with removal efficiencies higher than 95%. For example, magnetic silica nanospheres, magnetite nanorods, eggshell nanoparticles, and ZnS nanocrystals showed the highest removal efficiency of heavy metal ions. Also, calcium alginate/MWCNT, hydrogel-clay nanocomposite, Fe_3O_4/AC nanocomposite, and cobalt-decorated SiO_2–Al_2O_3 indicated high efficiencies for dyes removal. Moreover, MWCNT/Fe_3O_4 and MgO/$Fe_2(MoO4)_3$ have demonstrated the highest sorption capacity of nitrate, sulfate, and phosphate ions and nanoporous carbon, γ-Fe_2O_3/NiO nanocomposite, CNTs@MIPs nanocomposites, graphite oxide/poly (acrylic acid) grafted chitosan nanocomposite, and ZnO@carbon nanofibers showed very high sorption capacities for drugs removal from wastewater. Besides, GO and MWCNTs-silica aerogel showed the maximum sorption capacity of oil from wastewater emulsions.

References

1. Abshirini, Y., Foroutan, R., and Esmaeili, H. (2019). Cr (VI) removal from aqueous solution using activated carbon prepared from Ziziphus spina–christi leaf, *Mater. Res. Express*, **6**, p. 045607.
2. Yang, W., Shi, X., Wang, J., Chen, W., Zhang, L., Zhang, W., Zhang, X., and Lu, J. (2019). Fabrication of a novel bifunctional nanocomposite with improved selectivity for simultaneous nitrate and phosphate removal from water, *ACS Appl. Mater. Interfaces*, **11**, pp. 35277–35285.
3. Deblonde, T., Cossu-Leguille, C., and Hartemann, P. (2011). Emerging pollutants in wastewater: A review of the literature, *Int. J. Hyg. Envir. Heal.*, **214**, pp. 442–448.
4. Foroutan, R., Mohammadi, R., Farjadfard, S., Esmaeili, H., Saberi, M., Sahebi, S., Dobaradaran, S., and Ramavandi, B. (2019). Characteristics and performance of Cd, Ni, and Pb bio-adsorption using Callinectes sapidus biomass: Real wastewater treatment, *Environ. Sci. Pollut. Res.*, **26**, pp. 6336–6347.

5. Brbooti, M., Abid, B., and Najah, M. (2011). Removal of heavy metals using chemicals precipitation, *J. Eng. Technol.*, **29**, pp. 595–612.
6. Hashim, M. A., Mukhopadhyay, S., Sahu, J. N., and Sengupta, B. (2011). Remediation technologies for heavy metal contaminated groundwater, *J. Environ. Manage.*, **92**, pp. 2355–2388.
7. Hubicki, Z. and Koodynsk, D. (2016) Selective removal of heavy metal ions from waters and waste waters using ion exchange methods. In: *Ion Exchange Technologies*, pp. 193–240.
8. Kononova, O. N., Bryuzgina, G. L., Apchitaeva, O. V., and Kononov, Y. S. (2015). Ion exchange recovery of chromium(VI) and manganese(II) from aqueous solutions, *Arab. J. Chem.*, **12**, pp. 2713–2720.
9. Judd, S. and Jefferson, B. (2003). *Membranes for Industrial Wastewater Recovery and Re-use*, Elsevier.
10. Mutamim, N. S. A., Noor, Z. Z., Hassan, M. A. A., and Olsson, G. (2012). Application of membrane bioreactor technology in treating high strength industrial wastewater: A performance review, *Desalination*, **305**, pp. 1–11.
11. El Samrani, A. G., Lartiges, B. S., and Villiéras, F. (2008). Chemical coagulation of combined sewer overflow: Heavy metal removal and treatment optimization, *Water Res.*, **42**, pp. 951–960.
12. Fu, F. and Wang, Q. (2011). Removal of heavy metal ions from wastewaters: A review, *J. Environ. Manage.* **92**, pp. 407–418.
13. Renu, B., Agarwal, M., and Singh, K. (2017). Methodologies for removal of heavy metal ions from wastewater: An overview, *Interdiscip. Environ. Rev.*, **18**, p. 124.
14. Gulnaz, O., Sahmurova, A., and Kama, S. (2011). Removal of Reactive Red 198 from aqueous solution by *Potamogeton crispus*, *Chem. Eng. J.*, **174**, pp. 579–585.
15. Boushehrian, M. M., Esmaeili, H., and Foroutan, R. (2020). Ultrasonic assisted synthesis of Kaolin/$CuFe_2O_4$ nanocomposite for removing cationic dyes from aqueous media, *J. Environ. Chem. Eng.*, p. 103869.
16. Esmaeili, H., Foroutan, R., Jafari, D., and Aghil Rezaei, M. (2020). Effect of interfering ions on phosphate removal from aqueous media using magnesium oxide@ ferric molybdate nanocomposite, *Korean J. Chem. Eng.*, **37**, pp. 804–814.
17. Laurent, S., Forge, D., Port, M., Roch, A., Robic, C., Vander Elst, L., and Muller, R. N. (2008). Magnetic iron oxide nanoparticles: Synthesis, stabilization, vectorization, physicochemical characterizations, and biological applications, *Chem. Rev.*, **108**, pp. 2064–2110.

18. Khan, I., Saeed, K., and Khan, I. (2019). Nanoparticles: Properties, applications and toxicities. *Arab. J. Chem.*, **12**, pp. 908–931.

19. Takmil, F., Esmaeili, H., Mousavi, S. M., and Hashemi, S. A. (2020). Nano-magnetically modified activated carbon prepared by oak shell for treatment of wastewater containing fluoride ion, *Adv. Powder Technol.*, **31**, pp. 3236–3245.

20. Mueller, N. C. and Nowack, B. (2008). Exposure modeling of engineered nanoparticles in the environment, *Environ. Sci. Technol.*, **42**, pp. 4447–4453.

21. Rogozea, E. A., Petcu, A. R., Olteanu, N. L., Lazar, C. A., Cadar, D., and Mihaly, M. (2017). Tandem adsorption-photodegradation activity induced by light on NiO-ZnO p–n couple modified silica nanomaterials, *Mater. Sci. Semicond. Process.*, **57**, pp. 1–11.

22. Foroutan, R., Ahmadlouydarab, M., Ramavandi, B., and Mohammadi, R. (2018). Studying the physicochemical characteristics and metals adsorptive behavior of CMC-g-HAp/Fe_3O_4 nanobiocomposite, *J. Environ. Chem. Eng.* **6**, pp. 6049–6058.

23. Yang, J., Hou, B., Wang, J., Tian, B., Bi, J., Wang, N., Li, X., and Huang, X. (2019). Nanomaterials for the removal of heavy metals from wastewater, *Nanomaterials*, **9**, p. 424.

24. Esmaeili, H. and Tamjidi, S. (2020). Ultrasonic-assisted synthesis of natural clay/Fe_3O_4/graphene oxide for enhance removal of Cr (VI) from aqueous media, *Environ. Sci. Pollut. Res.*, **27**, pp. 31652–31664.

25. Teimouri, A., Esmaeili, H., Foroutan, R., and Ramavandi, B. (2018). Adsorptive performance of calcined Cardita bicolor for attenuating Hg (II) and As (III) from synthetic and real wastewaters, *Korean J. Chem. Eng.*, **35**, pp. 479–488.

26. Foroutan, R., Esmaeili, H., Rishehri, S. D., Sadeghzadeh, F., Mirahmadi, S., Kosarifard, M., and Ramavandi, B. (2017). Zinc, nickel, and cobalt ions removal from aqueous solution and plating plant wastewater by modified Aspergillus flavus biomass: A dataset, *Data Brief*, **12**, pp. 485–492.

27. Reglero M. M., Taggart M. A., Lidia M. G., and Rafael M. (2009). Heavy metal exposure in large game from a lead mining area: Effects on oxidative stress and fatty acid composition in liver, *Environ. Pollut.* **157**, pp. 1388–1395.

28. Esmaeili, H., Mousavi, S. M., Hashemi, S. A., Chiang, W. H., and Abnavi, S. A. (2020). Activated carbon@ MgO@ Fe_3O_4 as an efficient adsorbent for As (III) removal, *Carbon Lett.* **31**, pp. 851–862.

29. Foroutan, R., Mohammadi, R., Farjadfard, S., Esmaeili, H., Ramavandi, B., and Sorial, G. A. (2019). Eggshell nano-particle potential for methyl violet and mercury ion removal: Surface study and field application, *Adv. Powder Technol.,* **30**, pp. 2188–2199.

30. Nazir, R., Khan, M., Masab, M., Rehman, H. U., Rauf, N. U., Shahab, S., Ameer, N., Sajed, M., Ullah, M., Rafeeq, M., and Shaheen, Z. (2015). Accumulation of heavy metals (Ni, Cu, Cd, Cr, Pb, Zn, Fe) in the soil, water and plants and analysis of physico-chemical parameters of soil and water collected from Tanda Dam Kohat, *Int. J. Pharm. Sci. Res.,* **7**, p. 89.

31. Tamjidi, S. and Esmaeili, H. (2019). Chemically modified CaO/Fe_3O_4 nanocomposite by sodium dodecyl sulfate for Cr (III) removal from water, *Chem. Eng. Technol.*, **42**, pp. 607–616.

32. Ahmadi, A., Foroutan, R., Esmaeili, H., and Tamjidi, S. (2020). The role of bentonite clay and bentonite clay@ $MnFe_2O_4$ composite and their physico-chemical properties on the removal of Cr (III) and Cr (VI) from aqueous media, *Environ. Sci. Pollut. Res.*, **27**, pp. 14044–14057.

33. Rani, N., Singh, B., and Shimrah, T. (2017). Chromium (VI) removal from aqueous solutions using Eichhornia as an adsorbent, *J. Water Reuse Desal.*, **7**, pp. 461–467.

34. Kumar, P. S., Korving, L., van Loosdrecht, M. C., and Witkamp, G. J. (2019). Adsorption as a technology to achieve ultra-low concentrations of phosphate: Research gaps and economic analysis, *Water Res. X*, p. 100029.

35. Manjula, P. and Warrier, C. U. (2019). Evaluation of water quality of Thuthapuzha Sub-basin of Bharathapuzha, Kerala, India, *Appl. Water Sci.*, **9**, p. 70.

36. Parveen, R., Abbasi, A. M., Shaheen, N., and Shah, M. H. (2020). Accumulation of selected metals in the fruits of medicinal plants grown in urban environment of Islamabad, Pakistan, *Arab. J. Chem.*, **13**, pp. 308–317.

37. Chennaiah, J. B., Rasheed, M. A., and Patil, D. J. (2014). Concentration of heavy metals in drinking water with emphasis on human health, *Int. J. Plant Anim. Environ. Sci.*, **4**, pp. 205–214.

38. Varjani, S. J., Agarwal, A. K., Gnansounou, E., and Gurunathan, B. (2018) *Bioremediation: Applications for Environmental Protection and Management* (No. BOOK). New York, NY: Springer.

39. Bajwa, B. S., Kumar, S., Singh, S., Sahoo, S. K., and Tripathi, R. M. (2017). Uranium and other heavy toxic elements distribution in the drinking

water samples of SW-Punjab, India, *J. Radiat. Res. Appl. Sc.*, **10**, pp. 13–19.

40. Khan, N. A., Hasan, Z., and Jhung, S. H. (2013). Adsorptive removal of hazardous materials using metal-organic frameworks (MOFs): A review, *J. Hazard. Mater.*, **244**, pp. 444–456.

41. Buzea, C. and Pacheco, I. (2017) *Nanomaterial and Nanoparticle: Origin and Activity*. Springer International Publishing; Cham, Switzerland.

42. Tamjidi, S., Esmaeili, H., and Moghadas, B. K. (2019). Application of magnetic adsorbents for removal of heavy metals from wastewater: A review study, *Mater. Res. Express*, **6**, p. 102004.

43. Ahmadi, F. and Esmaeili, H. (2018). Chemically modified bentonite/ Fe_3O_4 nanocomposite for Pb (II), Cd (II), and Ni (II) removal from synthetic wastewater, *Desalin. Water Treat.*, **110**, pp. 154–167.

44. Mousavi, M., Hashemi, A., Arjmand, O., Amani, A. M., Babapoor, A., Fateh, M. A., Fateh, H., Mojoudi, F., Esmaeili, H., and Jahandideh, S. (2018). Erythrosine adsorption from aqueous solution via decorated graphene oxide with magnetic iron oxide nano particles: Kinetic and equilibrium studies. *Acta Chim. Slov.*, **65**, pp. 882–894.

45. Kalantari, K., Ahmad, M. B., Masoumi, H. R. F., Shameli, K., Basri, M., and Khandanlou, R. (2014). Rapid adsorption of heavy metals by Fe_3O_4/ talc nanocomposite and optimization study using response surface methodology, *Int. J. Mol. Sci.*, **15**, pp. 12913–12927.

46. Foroutan, R., Oujifard, A., Papari, F., and Esmaeili, H. (2019). Calcined *Umbonium vestiarium* snail shell as an efficient adsorbent for treatment of wastewater containing Co (II), *3 Biotech*, **9**, p. 78.

47. El-Tawil, R. S., El-Wakeel, S. T., Abdel-Ghany, A. E., Abuzeid, H. A., Selim, K. A., and Hashem, A. M. (2019). Silver/quartz nanocomposite as an adsorbent for removal of mercury (II) ions from aqueous solutions, *Heliyon*, **5**, p. 02415.

48. Khoshkerdar, I. and Esmaeili, H. (2019). Adsorption of Cr (III) and Cd (II) ions using mesoporous cobalt-ferrite nanocomposite from synthetic wastewater, *Acta Chim. Slov.*, **66**, pp. 208–216.

49. Abshirini, Y., Esmaeili, H., and Foroutan, R. (2019). Enhancement removal of Cr (VI) ion using magnetically modified MgO nanoparticles, *Mater. Res. Express*, **6**, p. 125513.

50. Ahmad, R. (2019). Polyaniline/ZnO nanocomposite: A novel adsorbent for the removal of Cr (VI) from aqueous solution. In: *Advances in Composite Materials Development*, IntechOpen.

51. Lasheen, M. R., El-Sherif, I. Y., Sabry, D. Y., El-Wakeel, S. T., and El-Shahat, M. F. (2016). Adsorption of heavy metals from aqueous solution by magnetite nanoparticles and magnetite-kaolinite nanocomposite: Equilibrium, isotherm and kinetic study, *Desalin. Water Treat.,* **57**, pp. 17421–17429.

52. Karami, H. (2013). Heavy metal removal from water by magnetite nanorods, *Chem. Eng. J.,* **219**, pp. 209–216.

53. Jung, W., Jeon, B. H., Cho, D. W., Roh, H. S., Cho, Y., Kim, S. J., and Lee, D. S. (2015). Sorptive removal of heavy metals with nano-sized carbon immobilized alginate beads, *J. Ind. Eng. Chem.*, **26**, pp. 364–369.

54. Irawan, C., Nata, I. F., and Lee, C. K. (2019). Removal of Pb (II) and As (V) using magnetic nanoparticles coated montmorillonite via one-pot solvothermal reaction as adsorbent, *J. Environ. Chem. Eng.*, **7**, p. 103000.

55. Venkateswarlu, S., Kumar, S. H., and Jyothi, N. (2015). Rapid removal of Ni (II) from aqueous solution using 3-Mercaptopropionic acid functionalized bio magnetite nanoparticles, *Water Resour. Ind.* **12**, pp. 1–7.

56. Vojoudi, H., Badiei, A., Bahar, S., Ziarani, G. M., Faridbod, F., and Ganjali, M. R. (2017). A new nano-sorbent for fast and efficient removal of heavy metals from aqueous solutions based on modification of magnetic mesoporous silica nanospheres, *J. Magn. Magn. Mater.*, **441**, pp. 193–203.

57. Huang, Z. N., Wang, X. L., and Yang, D. S. (2015). Adsorption of Cr (VI) in wastewater using magnetic multi-wall carbon nanotubes, *Water Sci. Eng.*, **8**, pp. 226–232.

58. Qu, Z., Yan, L., Li, L., Xu, J., Liu, M., Li, Z., and Yan, N. (2014). Ultraeffective ZnS nanocrystals sorbent for mercury (II) removal based on size-dependent cation exchange, *ACS App. Mater. Int.*, **6**, pp. 18026–18032.

59. Varma, S., Sarode, D., Wakale, S., Bhanvase, B. A., and Deosarkar, M. P. (2013). Removal of nickel from wastewater using graphene nanocomposite, *Int. J. Chem. Phys. Sci.*, **2**, pp. 132–139.

60. Attia, S. T. M., Hu, X. L., and Yin, D. Q. (2014). Synthesised magnetic nanoparticles coated zeolite (MNCZ) for the removal of arsenic (As) from aqueous solution, *J. Experiment. Nanosci.*, **9**, pp. 551–560.

61. Visa, M. (2016). Synthesis and characterization of new zeolite materials obtained from fly ash for heavy metals removal in advanced wastewater treatment, *Powder Technol.*, **294**, pp. 338–347.

62. Wang, H. J., Zhou, A. L., Peng, F., Yu, H., and Chen, L. F. (2007). Adsorption characteristic of acidified carbon nanotubes for heavy metal Pb(II) in aqueous solution, *Mater. Sci. Eng. A*, **466**, pp. 201–206.

63. Wu, C. H. (2007). Studies of the equilibrium and thermodynamics of the adsorption of Cu^{2+} onto as-produced and modified carbon nanotubes, *J. Colloid Interface Sci.*, **311**, pp. 338–346.

64. Yang, S. T., Li, J. X., Shao, D. D., Hu, J., and Wang, X. K. (2009). Adsorption of Ni(II) on oxidized multi-walled carbon nanotubes: Effect of contact time, pH, foreign ions and PAA, *J. Hazard. Mater.*, **166**, pp. 109–116.

65. Lu, C. Y., Liu, C. T., and Su, F. S. (2009). Sorption kinetics, thermodynamics and competition of Ni^{2+} from aqueous solutions onto surface oxidized carbon nanotubes, *Desalination*, **249**, pp. 18–23.

66. Anastopoulos, I., Hosseini-Bandegharaei, A., Fu, J., Mitropoulos, A. C., and Kyzas, G. Z. (2018). Use of nanoparticles for dye adsorption, *J. Dispers. Sci. Technol.*, **39**, pp. 836–847.

67. Acar, I., Bal, A., and Güçlü, G. (2012). Adsorption of basic dyes from aqueous solutions by depolymerization products of post-consumer PET bottles, *CLEAN–Soil, Air, Water*, **40**, pp. 325–333.

68. Esmaeili, H. and Foroutan, R. (2019). Adsorptive behavior of methylene blue onto sawdust of sour lemon, date palm, and eucalyptus as agricultural wastes, *J. Dispers. Sci. Technol.*, **40**, pp. 990–999.

69. Yagub, M. T., Sen, T. K., Afroze, S., and Ang, H. M. (2014). Dye and its removal from aqueous solution by adsorption: A review, *Adv. Colloid Interface Sci.*, **209**, pp. 172–184.

70. Lazar, T. (2005). Color chemistry: Synthesis, properties, and applications of organic dyes and pigments. *Color Res. Appl.*, **30**, pp. 313–314.

71. Sharafzad, A., Tamjidi, S., and Esmaeili, H. (2020). Calcined lotus leaf as a low-cost and highly efficient biosorbent for removal of methyl violet dye from aqueous media, *Int. J. Environ. Anal. Chem.* **101** (15), pp. 2761–2784.

72. Wu, Y., Zhang, L., Gao, C., Ma, J., Ma, X., and Han, R. (2009). Adsorption of copper ions and methylene blue in a single and binary system on wheat straw, *J. Chem. Eng. Data.*, **54**, pp. 3229–3234.

73. Hameed, B. H. and Ahmad, A. A. (2009). Batch adsorption of methylene blue from aqueous solution by garlic peel, an agricultural waste biomass, *J. Hazard. Mater.* **164**, pp. 870–875.

74. Karri, R. R., Tanzifi, M., Yaraki, M. T., and Sahu, J. N. (2018). Optimization and modeling of methyl orange adsorption onto polyaniline nano-

adsorbent through response surface methodology and differential evolution embedded neural network, *J. Environ. Manage.*, **223**, pp. 517–529.

75. Mittal, A. (2006). Use of hen feathers as potential adsorbent for the removal of a hazardous dye, Brilliant Blue FCF, from wastewater, *J. Hazard. Mater.*, **128**, pp. 233–239.

76. Wanyonyi, W. C., Onyari, J. M., and Shiundu, P. M. (2014). Adsorption of Congo red dye from aqueous solutions using roots of *Eichhornia crassipes*: Kinetic and equilibrium studies, *Energy Procedia*, **50**, pp. 862–869.

77. Laskar, N. and Kumar, U. (2018). Adsorption of crystal violet from wastewater by modified bambusa tulda, *KSCE J. Civ. Eng.*, **22**, pp. 2755–2763.

78. Gupta, V. K., Jain, R., Saleh, T. A., Nayak, A., Malathi, S., and Agarwal, S. (2011). Equilibrium and thermodynamic studies on the removal and recovery of safranine-T dye from industrial effluents, *Sep. Sci. Technol.*, **46**, pp. 839–846.

79. Duraipandian, J., Rengasamy, T., and Vadivelu, S. (2017). Experimental and modeling studies for the removal of crystal violet dye from aqueous solutions using eco-friendly *Gracilaria corticata* seaweed activated carbon/Zn/alginate polymeric composite beads, *J. Polym. Environ.*, **25**, pp. 1062–1071.

80. Sui, K., Li, Y., Liu, R., Zhang, Y., Zhao, X., Liang, H., and Xia, Y. (2012). Biocomposite fiber of calcium alginate/multi-walled carbon nanotubes with enhanced adsorption properties for ionic dyes, *Carbohydr. Polym.* **90**, pp. 399–406.

81. Zhu, H. Y., Yao, J., Jiang, R., Fu, Y. Q., Wu, Y. H., and Zeng, G. M. (2014). Enhanced decolorization of azo dye solution by cadmium sulfide/multi-walled carbon nanotubes/polymer composite in combination with hydrogen peroxide under simulated solar light irradiation, *Ceram. Int.* **40**, pp. 3769–3777.

82. Wang, L. and Wang, A. (2007). Adsorption characteristics of Congo Red onto the chitosan/montmorillonite nanocomposite, *J. Hazard. Mater.*, **147**, pp. 979–985.

83. Kaşgöz, H. and Durmus, A. (2008). Dye removal by a novel hydrogel-clay nanocomposite with enhanced swelling properties, *Polym. Adv. Technol.*, **19**, pp. 838–845.

84. Mahmoodi, N. M., Mokhtari-Shourijeh, Z., and Ghane-Karade, A. (2017). Synthesis of the modified nanofiber as a nanoadsorbent and its dye

removal ability from water: Isotherm, kinetic and thermodynamic, *Water Sci. Technol.*, **75**, pp. 2475–2487.

85. Ghasemi, M., Mashhadi, S., and Azimi-Amin, J. (2018). Fe_3O_4/AC nanocomposite as a novel nano adsorbent for effective removal of cationic dye: Process optimization based on Taguchi design method, kinetics, equilibrium and thermodynamics, *J. Water Environ. Nanotechnol.*, **3**, pp. 321–336.
86. Arshadi, M., Faraji, A. R., and Mehravar, M. (2015). Dye removal from aqueous solution by cobalt-nano particles decorated aluminum silicate: kinetic, thermodynamic and mechanism studies, *J. Colloid Interface Sci.*, **440**, pp. 91–101.
87. Baig, U., Uddin, M. K., and Gondal, M. A. (2020). Removal of hazardous azo dye from water using synthetic nano adsorbent: Facile synthesis, characterization, adsorption, regeneration and design of experiments, *Colloids Surf. A Physicochem. Eng.*, **584**, p. 124031.
88. Ahmad, A., Razali, M. H., Mamat, M., Mehamod, F. S. B., and Anuar Mat Amin, K. (2017). Adsorption of methyl orange by synthesized and functionalized-CNTs with 3-aminopropyltriethoxysilane loaded TiO_2 nanocomposites, *Chemosphere*, **168**, p. 474.
89. Uddin, M. K. and Baig, U. (2019). Synthesis of Co_3O_4 nanoparticles and their performance towards methyl orange dye removal: Characterisation, adsorption and response surface methodology, *J. Clean. Prod.*, **211**, pp. 1141–1153.
90. Chaukura, N., Murimba, E. C., and Gwenzi, W. (2016). Synthesis, characterisation and methyl orange adsorption capacity of ferric oxide–biochar nano-composites derived from pulp and paper sludge, *Appl. Water Sci.*, **7**, pp. 2175–2186.
91. Qu, S., Huang, F., Yu, S., Chen, G., and Kong, J. (2008). Magnetic removal of dyes from aqueous solution using multi-walled carbon nanotubes filled with Fe_2O_3 particles. *J. Hazard. Mater.* **160**, pp. 643–647.
92. Rahmati, M., Yeganeh, G., and Esmaeili, H. (2019). Sulfate ion removal from water using activated carbon powder prepared by *Ziziphus spina-christi* lotus leaf, *Acta Chim. Slov.*, **66**, pp. 888–898.
93. Pol, L. W. H., Lens, P. N., Stams, A. J., and Lettinga, G. (1998). Anaerobic treatment of sulphate-rich wastewaters, *Biodegradation*, **9**, pp. 213–224.
94. USEPA (2003). Drinking water advisory: consumer acceptability advice and health effects analysis on sulfate, USEPA.
95. Cao, W., Dang, Z., Zhou, X. -Q., Yi, X.-Y., Wu, P. -X., Zhu, N.-W., and Lu, G. -N. (2011) Removal of sulphate from aqueous solution using modified

rice straw: Preparation, characterization and adsorption performance, *Carbohydr. Polym.*, **85**, pp. 571–577.

96. Sarvestani, F. S., Esmaeili, H., and Ramavandi, B. (2016). Modification of *Sargassum angustifolium* by molybdate during a facile cultivation for high-rate phosphate removal from wastewater: Structural characterization and adsorptive behavior, *3 Biotech*, **6**, p. 251.

97. Habiby, S. R., Esmaeili, H., and Foroutan, R. (2019). Magnetically modified MgO nanoparticles as an efficient adsorbent for phosphate ions removal from wastewater, *Sep. Sci. Technol.*, **55**, pp. 1910–1921.

98. Kumar, P. S. (2019). Assessment of corrosion and scaling potential of the groundwater in the Thanjavur district using hydrogeochemical analysis and spatial modeling techniques, *SN Applied Sciences*, **1**, p. 395.

99. Loganathan, P., Vigneswaran, S., and Kandasamy, J. (2013). Enhanced removal of nitrate from water using surface modification of adsorbents: A review, *J. Environ. Manage.*, **131**, pp. 363–374.

100. Alimohammadi, V., Sedighi, M., and Jabbari, E. (2016). Response surface modeling and optimization of nitrate removal from aqueous solutions using magnetic multi-walled carbon nanotubes, *J. environ. Chem. Eng.*, **4**, pp. 4525–4535.

101. Yang, L., Xu, P., Yang, M., and Bai, H. (2017). The characteristics of steel slag and the effect of its application as a soil additive on the removal of nitrate from aqueous solution, *Environ. Sci. Pollut. Res.*, **24**, pp. 4882–4893.

102. Battas, A., Gaidoumi, A.E., Ksakas, A., and Kherbeche, A. (2019). Adsorption study for the removal of nitrate from water using local clay, *Sci. World J.*, **2019**.

103. Johnson, I. E., Chatterjee, S., Hall, G. B., Burton, S. D., Campbell, E. L., Conroy, M. A., Du, Y., Fujimoto, M. S., Varga, T., Kruger, A. A., and Levitskaia, T. G. (2018). Inorganic Ba–Sn nanocomposite materials for sulfate sequestration from complex aqueous solutions, *Environ. Sci. Nano*, **5**, pp. 890–903.

104. Keshvardoostchokami, M., Babaei, S., Piri, F., and Zamani, A. (2017). Nitrate removal from aqueous solutions by ZnO nanoparticles and chitosan-polystyrene–Zn nanocomposite: Kinetic, isotherm, batch and fixed-bed studies, *Int. J. Biol. macromol.*, **101**, pp. 922–930.

105. Mohammadi, M., Sedighi, M., and Alimohammadi, V. (2019). Modeling and optimization of Nitrate and total Iron removal from wastewater by TiO_2/SiO_2 nanocomposites, *Int. J. Nano Dimens.*, **10**, pp. 195–208.

106. Hejri, Z., Hejri, M., Omidvar, M., and Morshedi, S. (2019). Synthesis of TiO_2/nZVI nanocomposite for nitrate removal from aqueous solution, *Int. J. Ind. Chem.*, **10**, pp. 225–236.

107. Bhatnagar, A., Kumar, E., and Sillanpää, M. (2010). Nitrate removal from water by nano-alumina: Characterization and sorption studies, *Chem. Eng. J.*, **163**, pp. 317–323

108. Daud, M., Khan, Z., Ashgar, A., Danish, M. I., and Qazi, I. A. (2015). Comparing and optimizing nitrate adsorption from aqueous solution using Fe/Pt bimetallic nanoparticles and anion exchange resins, *Journal of Nanotechnology*, **2015**.

109. Husein, D. Z., Hassanien, R., and Al-Hakkani, M. F. (2019). Green-synthesized copper nano-adsorbent for the removal of pharmaceutical pollutants from real wastewater samples, *Heliyon*, **5**, p. 02339.

110. Imbimbo, B. P., Solfrizzi, V., and Panza, F. (2010). Are NSAIDs useful to treat Alzheimer's disease ormild cognitive impairment? *Front. Aging Neurosci.* **2**, p. 19.

111. Zhang, J., Chang, V. W. C., Giannis, A., and Wang, J. Y. (2013). Removal of cytostatic drugs from aquatic environment: A review, *Sci. Total Environ.*, **445–446**, pp. 281–298.

112. Macías-García, A., García-Sanz-Calcedo, J., Carrasco-Amador, J. P., and Segura-Cruz, R. (2019). Adsorption of paracetamol in hospital wastewater through activated carbon filters, *Sustainability*, **11**, p. 2672.

113. Petrie, B., Barden, R., and Kasprzyk-Hordern, B. (2015). A review on emerging contaminats in wastewaters and the environment: Current knowledge, understudied areas and recommendations for future monitoring, *Water Res.*, **72**, pp. 3–27.

114. De García, S. O., Pinto, G. P., Encina, P. G., and Mata, R. I. (2013). Consumption and occurrence of pharmaceutical and personal care products in the aquatic environment in Spain, *Sci. Total Environ.*, **444**, pp. 451–465.

115. Hu, X., Cheng, Z., Sun, Z., and Zhu, H. (2017). Adsorption of Diclofenac and Triclosan in aqueous solution by purified multi-walled carbon nanotubes, *Pol. J. Environ. Stud.*, **26**.

116. İlbay, Z., Şahin, S., Kerkez, Ö., and Bayazit, Ş. (2015). Isolation of naproxen from wastewater using carbon-based magnetic adsorbents, *Int. J. Environ. Sci. Technol.*, **12**, pp. 3541–3550.

117. Al-Khateeb, L. A., Hakami, W., and Salam, M. A. (2017). Removal of non-steroidal anti-inflammatory drugs from water using high surface area

nanographene: Kinetic and thermodynamic studies, *J. Mol. Liq.*, **241**, pp. 733–741.

118. Li, S., Zhang, X., and Huang, Y. (2017). Zeolitic imidazolate framework-8 derived nanoporous carbon as an effective and recyclable adsorbent for removal of ciprofloxacin antibiotics from water, *J. Hazard. Mater.*, **321**, pp. 711–719.

119. Li, X., Wang, W., Dou, J., Gao, J., Chen, S., Quan, X., and Zhao, H. (2016). Dynamic adsorption of cip-rofloxacin on carbon nanofibers: Quantitative measurement by in situ fluorescence, *J. Wat. Proc. Eng.*, **9**, pp. 14–20.

120. Caldeira, M., Leandro, T., Alves, V., Sérgio, G., and deAquino, F. (2015). Activated carbons from agricultural byproducts (pine tree and coconut shell), coal, and carbon nanotubes as adsorbents for removal of sulfamethoxazole from spiked aqueous solutions: Kinetic and thermodynamic studies, *Ind. Crop. Prod.*, **74**, pp. 111–121.

121. Chen, B., Sun, W., Wang, C., and Guo, X. (2017). Size-dependent impact of inorganic nanoparticles on sulfamethoxazole adsorption by carbon nanotubes, *Chem. Eng. J.*, **316**, pp. 160–170.

122. Stan, M., Lung, I., Soran, M., Leostean, C., Popa, A., Stefan, M., Lazar, M. D., Opris, O., Silipas, T., and Porav, A. S. (2017). Removal of antibiotics from aqueous solutions by green synthesized magnetite nanoparticles with selected agro-waste extracts, *Proc. Saf. Environ. Prot.*, **107**, pp. 357–372.

123. Fakhri, A. and Behrouz, S. (2015). Improved uptake of steroid hormone from aqueous solution using γ-Fe_2O_3/NiO nanocomposites, *J. Ind. Eng. Chem.*, **26**, pp. 61–66.

124. Attia, T. M. S., Hu, X. L., and Qiang, Y. D. (2013). Synthesized magnetic nanoparticles coated zeolite for the adsorption of pharmaceutical compounds from aqueous solution using batch and column studies, *Chemosphere*, **93**, pp. 2076–2085.

125. Pan, B., Lin, D., Mashayekhi, H., and Xing, B. (2008). Adsorption and hysteresis of bisphenol A and 17α-ethinyl estradiol on carbon nanomaterials, *Environ. Sci. Technol.*, **42**, pp. 5480–5485.

126. Kyzas, G. Z., Bikiaris, D. N., Seredych, M., Bandosz, T. J., and Deliyanni, E. A. (2014). Removal of dorzolamide from biomedical wastewaters with adsorption onto graphite oxide/poly(acrylic acid) grafted chitosan nanocomposite, *Bioresour. Technol.*, **152**, pp. 399–406.

127. Zhang, L., Xu, T., Liu, X., Zhang, Y., and Jin, H. (2011). Adsorption behavior of multi-walled carbon nanotubes for the removal of olaquindox from aqueous solutions, *J. Hazard. Mater.*, **197**, pp. 389–396.

128. Ribeiro, M. H. and Ribeiro, I. A. (2003). Modelling the adsorption kinetics of erythromycin onto neutral and anionic resins, *Bioproc. Biosyst. Eng.*, **26**, pp. 49–55.

129. Chaba, J. M. and Nomngongo, P. N. (2019). Effective adsorptive removal of amoxicillin from aqueous solutions and wastewater samples using zinc oxide coated carbon nanofiber composite, *Emerg. Contam.*, **5**, pp. 143–149.

130. Sulyman, M., Sienkiewicz, M., Haponiuk, J., and Zalewski, S. (2018). New approach for adsorptive removal of oil in wastewater using textile fibers as alternative adsorbent, *Acta Sci. Agric.*, **2**, pp. 1–6.

131. Franco, C. A., Martínez, M., Benjumea, P., Patiño, E., and Cortés, F. B. (2014). Water remediation based on oil adsorption using nanosilicates functionalized with a petroleum vacuum residue, *Adsorp. Sci. Technol.*, **32**, pp. 197–207.

132. Abd El-Gawad, H. S. (2014). Oil and grease removal from industrial wastewater using new utility approach, *Advances in Environmental Chemistry*, **2014**.

133. Esmaeili, H., Esmaeilzadeh, F., and Mowla, D. (2014). Effect of surfactant on stability and size distribution of gas condensate droplets in water, *J. Chem. Eng. Data*, **59**, pp. 1461–1467.

134. Esmaeili, H., Esmaeilzadeh, F., and Mowla, D. (2018). Destabilization and separation of gas condensate from wastewater using different surfactant demulsifiers, *Tenside Surfact. Det.*, **55**, pp. 153–161.

135. Esmaeili, H., Esmaeilzadeh, F., and Mowla, D. (2019). Mathematical modeling of destabilizing gas condensate droplets in water emulsions using the population balance method, *Tenside Surfact. Det.*, **56**, pp. 119–125.

136. Syed, S., Alhazzaa, M. I., and Asif, M. (2011). Treatment of oily water using hydrophobic nano-silica, *Chem. Eng. J.*, **167**, pp. 99–103.

137. Banerjee, S. S., Joshi, M. V., and Jayaram, R. V. (2006). Treatment of oil spills using organo-fly ash, *Desalination*, **195**, pp. 32–39.

138. Anjum, M., Miandad, R., Waqas, M., Gehany, F., and Barakat, M. A. (2019). Remediation of wastewater using various nano-materials, *Arab. J. Chem.*, **12**, pp. 4897–4919.

139. Diraki, A., Mackey, H. R., McKay, G., and Abdala, A. (2019). Removal of emulsified and dissolved diesel oil from high salinity wastewater by adsorption onto graphene oxide, *J. Environ. Chem. Eng.*, **7**, pp. 103106.

140. Soliman, E. M., Ahmed, S. A., and Fadl, A. A. (2020). Adsorptive removal of oil spill from sea water surface using magnetic wood sawdust as a

novel nano-composite synthesized via microwave approach, *J. Environ. Health Sci. Eng.*, **18**, pp. 79–90.

141. El Naggar, A. M., Noor El-Din, M. R., Mishrif, M. R., and Nassar, I. M. (2015). Highly efficient nano-structured polymer-based membrane/sorbent for oil adsorption from O/W emulsion conducted of petroleum wastewater, *J. Dispers. Sci. Technol.*, **36**, pp. 118–128.

142. Peng, X., Li, Y., Luan, Z.; Di, Z., Wang, H., Tian, B., and Jia, Z. (2003). Adsorption of 1,2-dichlorobenzene from water to carbon nanotubes, *Chem. Phys. Lett.*, **376**, pp. 154–158.

143. Huang, J., Liu, H., Chen, S., and Ding, C. (2016). Hierarchical porous MWCNTS-silica aerogel synthesis for high-efficiency oily water treatment, *J. Environ. Chem. Eng.*, **4**, pp. 3274–3282.

144. Naushad, M., Alqadami, A. A., AlOthman, Z. A., Alsohaimi, I. H., Algamdi, M. S., and Aldawsari, A. M. (2019). Adsorption kinetics, isotherm and reusability studies for the removal of cationic dye from aqueous medium using arginine modified activated carbon, *J. Mol. Liq.*, **293**, p. 111442.

145. Lata, S., Singh, P. K., and Samadder, S. R. (2015). Regeneration of adsorbents and recovery of heavy metals: A review, *Int. J. Environ. Sci. Technol.*, **12**, pp. 1461–1478.

146. Zhu, H., Jia, Y., Wu, X., and Wang, H. (2009). Removal of arsenic from water by supported nano zero-valent iron on activated carbon, *J. Hazard. Mater.*, **172**, pp. 1591–1596.

147. Zhang, L., Liu, N., Yang, L., and Lin, Q. (2009). Sorption behavior of nano-TiO_2 for the removal of selenium ions from aqueous solution, *J. Hazard. Mater.* **170**, pp. 1197–1203.

148. Hu, J. and Shipley, H. J. (2012). Evaluation of desorption of Pb(II), Cu(II) and Zn(II) from titanium dioxide nanoparticles, *Sci. Total Environ.*, **431**, pp. 209–220.

149. Hu, J. and Shipley H. J. (2013). Regeneration of spent TiO_2 nanoparticles for Pb(II), Cu(II), and Zn(II) removal, *Environ. Sci. Pollut. Res.*, **20**, pp. 5125–5137.

150. Zelmanov, G. and Semiat, R. (2011). Iron Fe(III) oxide/hydroxide nanoparticles-based agglomerates suspension as adsorbent for chromium Cr(IV) removal from water and recovery, *J. Sep. Purif. Technol.*, **80**, pp. 330–337.

151. Gupta, V. K. and Nayak, A. (2012). Cadmium removal and recovery from aqueous solutions by novel adsorbents prepared from orange peel and Fe_2O_3 nanoparticles, *Chem. Eng. J.*, **180**, pp. 81–90.

152. Hu, J., Chen, G., and Lo, I. M. C. (2005). Removal and recovery of Cr(VI) from wastewater by maghemite nanoparticles, *Water Res.*, **39**, pp. 4528–4536.

153. Esmaeili, H., Esmaeilzadeh, F., and Mowla, D. (2019). Effect of salinity, pH, and temperature on stability of gas condensate in water emulsions using different surfactants, *Iran. J. Chem. Chem. Eng.*, **38**, pp. 151–166.

154. Baik, S., Zhang, H., Kim, Y. K., Harbottle, D., and Lee, J. W. (2017). Enhanced adsorption capacity and selectivity towards strontium ions in aqueous systems by sulfonation of CO_2 derived porous carbon, *RSC adv.*, **7**, pp. 54546–54553.

155. Abbasi, S., Foroutan, R., Esmaeili, H., and Esmaeilzadeh, F. (2019). Preparation of activated carbon from worn tires for removal of Cu (II), Ni (II) and Co (II) ions from synthetic wastewater, *Desalin. Water Treat.*, **141**, pp. 269.

Chapter 8

Application of Nanotechnology in Cosmetics

Adriana de Melo,[a] Marcelo Lancellotti,[b] and Daisy Machado[c]

[a]*Centro Regional Universitário de Espirito Santo do Pinhal, UniPinhal, Av. Hélio Vergueiro Leite s/n – Jardim Universitário I Espírito Santo do Pinhal, São Paulo 139990-000, Brasil*
Unimogi, Av. Padre Jaime 2600 – Jardim Serra Dourada Mogi Guaçu, São Paulo 13844-070, Brasil
Etec Dr Caolimo da Motta e Silva, Rodovia ESP 346, Km 204 s/n – Morro Azul Espírito Santo do Pinhal, São Paulo 139900-000, Brasil
[b]*Faculdade de Ciências Farmacêuticas, Universidade Estadual de Campinas, UNICAMP, Rua Cândido Portinari 200 – Cidade Universitária Zeferino Vaz Campinas, São Paulo 13083-871, Brasil*
[c]*Universidade São Francisco, USF, Avenida São Francisco de Assis 218 - Cidade Universitária, Bragança Paulista, São Paulo 12916-900, Brasil*
koymelo@yahoo.com.br

Nanotechnology has led to a significant growth of the cosmetic industry, which is one of the most revolutionary technologies of the 21st century. Its application and innovation have grown exponentially in many areas. However, some concerns still exist about its use and

---

Science and Applications of Nanoparticles
Edited by Waqar Ahmed and Ehsan Nourafkan

ISBN 978-981-4267-34-2 (Hardcover), 978-1-003-28029-3 (eBook)
www.jennystanford.com

potential risks to public health, which led the competent authorities to put certain conditions to harmonize and control the introduction of nanocosmetics in the market. Some concerns hold to the type of nanomaterials used, their stability, cutaneous absorption, route of exposure, and how they are formulated in cosmetics products. Taking into account these aspects, it is fundamental that the production of nanocosmetics contains several analyses that ensure their quality, safety, and efficacy, similar to drugs.

8.1 Introduction

According to the European Commission [2020], a cosmetic is "any substance or preparation that will be put in contact with various external parts of the human body (epidermis, hair, nails, lips, and external genital organs) or with the teeth and mucosa of the oral cavity for the sole purpose, or mainly to clean, perfume, change its appearance and correct body odor and protect or maintain in good condition."

The cosmetic field is based on the assumption that smaller particles are more easily absorbed into the skin and repair the damage more effectively [2]. Nanotechnology includes the knowledge and use of materials that react or work at scale, contributing to the development of structures, devices, or systems with innovative properties and functions [3].

Nanotechnology has seen tremendous development during recent years. According to the PubMed search, there are currently more than 110,000 published journal articles using the word "nanotechnology;" in 2000, there were 53 published articles, and in 2020, there were 11,548 published articles. This increase, by nearly 22,000%, shows the importance that this area has acquired in the last 20 years [4]. If this research is more specific, using "cosmetic nanotechnology," the result present in the same research platform is 2757 published journal articles, with 65 published articles found in 2005 and 187 published articles in 2020, being that in 2011 and 2012 had a peak with more than 250 published articles per year [5].

The cosmetic industry was one of the first to implement the use of nanotechnology in product development. However, nanotechnology has also dominated the fields of medicine, due to the ability to cross barriers, localized accumulation in tumors, and increased solubility of drugs [6, 7]. One reason for the high proportion could be that this industry focuses on the final products. Most of these applications happen by topical administration, making the skin a target of great study for cosmetics and medicine.

The skin is an organ with complex physiological processes, and this is where the first visible signs of aging arise. This process is greatly influenced by factors such as sun exposure, genetics, stress, and the environment. Consequently, skincare has gained importance in terms of physical science, since knowledge about the anatomic physiology of the skin is increasingly needed, for the development of methods that maintain or improve the health and appearance of the skin. The new cosmetic products include mechanisms of action against the aging processes, through active substances incorporated in carrier nanoparticles [8].

This chapter will address the nanoparticles with the most significant impact on the cosmetic area, their properties, characteristics, and the regulatory perspective for cosmetics with nanoparticles (nanocosmetics), in order to evaluate their safety and efficacy.

8.2 Definition of Nanoparticles

A nanoparticle can be identified as an individual particle whose size is within a range between 1 and 100 nm (Fig. 8.1). However, this interval is not required and can be extended. Also, the European Union considers that a nanoparticle should also be "an insoluble or persistent bio particle and manufactured voluntarily, with one or more external structures, or with an internal dimension," and therefore can be a system for transporting active substances, to improve, for example, penetration into the skin of them [9]. The nanoparticle has several forms, from spherical to dendritic, and this form will depend on the application to which it is intended [10].

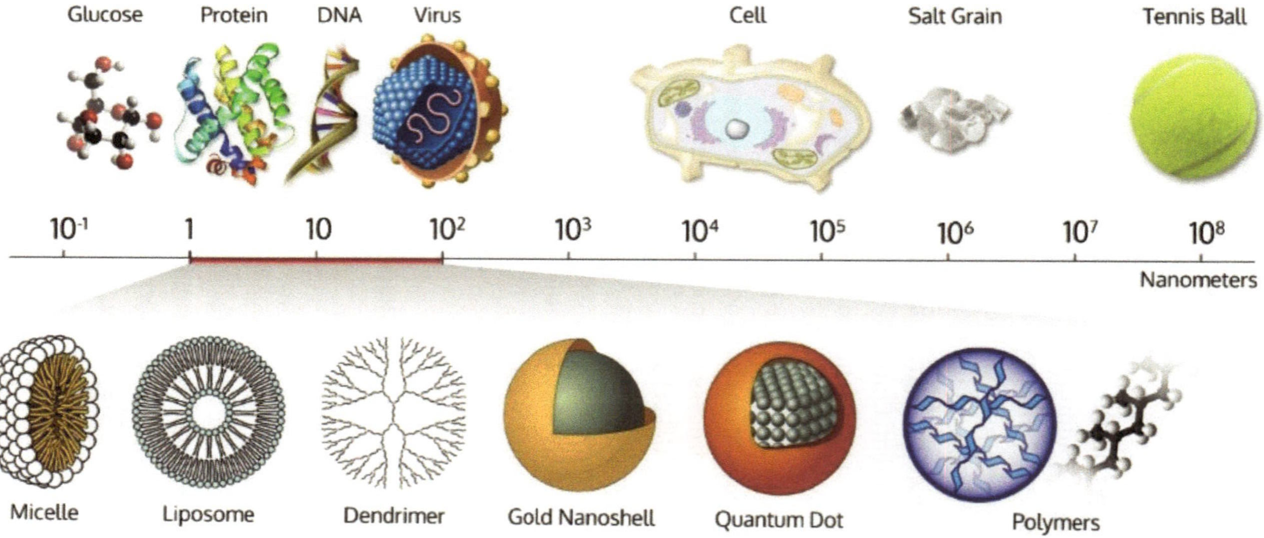

Figure 8.1 Example of the size of the nanoparticles (Adapted from Ref. [11]).

8.3 General Characteristics of Nanoparticles

Nanoparticles can be produced with different materials, including polymers, lipids, metals, among others. Besides, they can also be manufactured in a wide variety of ways, in order to create particles that function as principles at the same time, or that enclose them inside [12]. Therefore, nanoparticles are characterized by a sizeable volume–surface ratio, a large number of particles per unit of weight, and because they have confinement or quantum effects, that is, the nanoparticles have properties dependent on their size [13]. These characteristics of nanoparticles showed different properties in the initial state. Some characteristics that change when a particle enters the nanoscale range are melting temperature, solubility, and optical and magnetic properties [14].

A titanium oxide nanoparticle can illustrate what happens on the nanoscale. To the naked eye, titanium oxide is white, but its nanoparticles are transparent. This is because the movement of the electrons of titanium atoms is limited; therefore, titanium nanoparticles react to light differently to normal-scale titanium particles. This property is quite advantageous in the new generation of sunscreens that contain titanium oxide nanoparticles as a UV filter since they do not leave the skin whitish [12, 13]. However, nanoparticles can acquire many properties and, thus, can have numerous applications in cosmetics.

8.4 Nanoparticles for Cosmetic Formulation

To proceed with the use of nanomaterials in the formulation of cosmetics, they must bring benefits concerning the material in its conventional form, whether in the stability, application, or performance of the product. Some of the features that they may improve are the penetration and stability of some active substances, such as unsaturated fatty acids, vitamins or antioxidants, incorporated into nanoparticles, increase the effectiveness of some UV filters on the skin, sustained release of active substances for more prolonged effect [15], and make the product more aesthetically appealing to consumers [8]. Table 8.1 shows different nanomaterials and their action.

Table 8.1 Examples of nanomaterials and their actions [14]

Class	Material	Action
Active metal and metal oxide	Ag	Antibacterial
	$Ag/Ti/Ti_xO_y$	Haircare
	Au	Conjugated silk microfiber
Carbon	Fullereno	Free radical inhibition
Silic	SiO_2	Tightens skin, delivery of active ingredients
	$SiO2/Me_xO_y$ $Mica/ZnO/ Ti_xO_y$	Sunscreen
Vesicular lipid nanocarriers	Lipossome	Delivery of active ingredients
Synthetic polymers	Nanocapsule	Delivery of active ingredients

In order to choose the best nanoparticle for their transport, it is also essential to know the anatomic physiology of the skin; factors such as skin hydration, transepidermal water loss (TEWL), and wrinkles and other imperfections can influence the efficacy and permeability of nanoparticles [14]. For example, solid lipid nanoparticles (SLNs), depending on their size, can form a film on the surface of the skin, which has an occlusive effect (a protective barrier that decreases TEWL). This effect improves skin hydration and thus influences the percutaneous absorption of active substances [16].

8.5 Types of Nanoparticles Used in Cosmetics

Many nanoparticles have been developed in order to fill the gaps that conventional cosmetic formulations have, either in the modulation of the skin barrier or in the production of new systems for the transport of active substances of interest [15–17]. Each of them has properties that differentiate them depending on the functions they will perform or optimize.

8.5.1 Non-metallic Nanoparticles

8.5.1.1 Liposome

Liposomes were the first nanosystems to be introduced in cosmetics and arose in the 1960s when Alex Bangham, through an experienced laboratory, found that the hydration of a lipid film formed by phospholipids led to the formation of closed vesicles similar to those found in membrane cell phones. Bangham et al. called these lipid structures liposomes [16, 18]. Liposomes are spherical vesicles of nanometric sizes, made up of single or multiple concentric lipid bilayers of phospholipids and cholesterol, which are separated by internal aqueous compartments. They may differ in size, composition, surface electrical charge, and structure. In the presence of an excess of water, molecules can organize themselves and form one (unilamellar vesicles) or more (multilamellar vesicles) concentric bilayers [19–22]. The unilamellar vesicles can also be classified according to their size as small unilamellar vesicles, large unilamellar vesicles, and giant unilamellar vesicles [23, 24]. Liposomes have a charge on the surface, which can be positive, negative, or neutral depending on their chemical composition (a type of phospholipids, active ingredients, and other lipids used). These systems can encapsulate hydrophilic molecules in their internal aqueous compartment and lipophilic, amphiphilic, or hydrophilic molecules charged in the bilayers of their membranes, through electrostatic or hydrophobic interactions. Such versatile characteristics allow the use of liposomes as vehicles for transporting and releasing ingredients active in cosmetics [19–23].

In the recent years, the use of liposomes in cosmetics has been widely applied, taking into account the advantages they present [19–21]:

- Obtaining a controlled release of active ingredients, with a subsequent increase in their effectiveness and decrease in the number of product applications.
- Liposomes are biodegradable, biocompatible, and non-immunogenic systems, which, by decreasing their toxicity, prevent local and/or systemic adverse effects from occurring.
- Reducing instability and decomposition of sensitive active ingredients.

- Composition similar to cell membranes, allowing them to act as skin moisturizers. This hydration is also achieved due to the occlusive effect, which translates into the formation of a film on the surface of the skin, which allows the maintenance of its integrity and prevents the loss of transepidermal water.
- Able to carry hydrophilic molecules in the aqueous core and lipophilic molecules trapped inside the bilayer.
- Possibility of vectorization to specific target sites, preventing active ingredients from reaching the deepest layers of the skin, i.e., avoiding their systemic absorption.

These aspects, along with their nontoxic nature and long circulation time, make them a really efficient nanocarrier for cosmeceuticals delivery [5, 20]. In this way, liposomes are used to convey active ingredients in cosmetics because they can act as transporters, releasing the encapsulated molecules directly into the skin (Fig. 8.2); reservoir for controlled release; and barrier to prevent systemic absorption. The incorporation and distribution of active ingredients in liposomes depend on the physicochemical characteristics of the former and the lipid and structural composition of the latter [21, 24].

Nanoliposome-based delivery of antioxidants for skin wrinkles in antiaging treatment is a possibility.

Despite all the advantages they present, the application of liposomes in cosmetics may present drawbacks, such as [21–23]:

- High production costs associated with the use of phospholipids.
- Low physicochemical stability over time.
- Difficulties in effectively protecting chemically visible assets.
- Risk of immediate release of encapsulated active ingredients.
- The possibility of encapsulating tiny quantities of active ingredients.
- Difficulty in transposing production methods to an industrial scale.

The cosmetic industry has mainly used liposomes, intending to increase the trans-size of cosmetic ingredients in the skin [26]. They are small vesicles, spherical, with a diameter between 50 and 150 nm, which contain one or more membranes similar to phospholipid bilayers, and have an aqueous nucleus inside. Generally, the main components of liposome membranes are phosphatidylcholine [18]. To stabilize the system, cholesterol can also be incorporated, making it more rigid (suitable only for local cutaneous transport), or a surfactant can be integrated into the membrane, such as sodium collagen, providing flexibility and ultra-deformability to liposomes (called in this case transforms, having the ability to penetrate deeper into the skin) [27].

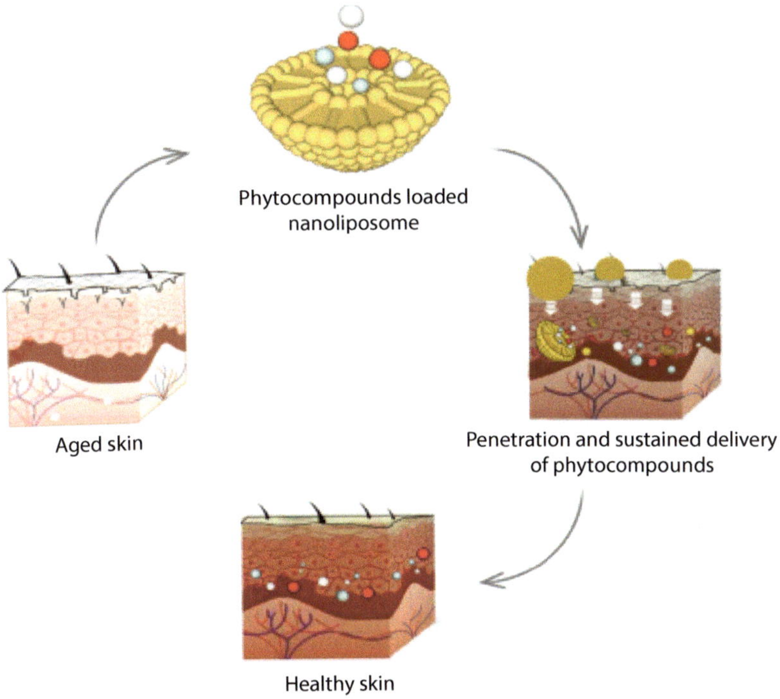

Figure 8.2 Liposome delivery of antioxidants for aged skin to antiaging treatment [25].

8.5.1.2 Applications of liposomes in cosmetics

Liposomes are found in many cosmetic formulations in order to transport active substances to the epidermis. Also, they can be incorporated into cosmetics to increase the number of lipids in the stratum corneum and to moisturize the skin [18].

Liposomes provide protection from the external environment and improve the physicochemical properties of the encapsulated molecules, such as fat-soluble vitamins and antioxidants, because liposomes are used in skincare products as moisturizers and antiaging agents [28].

Liposomes are used to carry UV filters, transporting them more effectively on the skin surface. UVB filters such as octyl methoxycinnamate (OMC) are incorporated into liposomes to increase their safety and efficacy in the formulation and increase retention in the stratum corneum (Fig. 8.3).

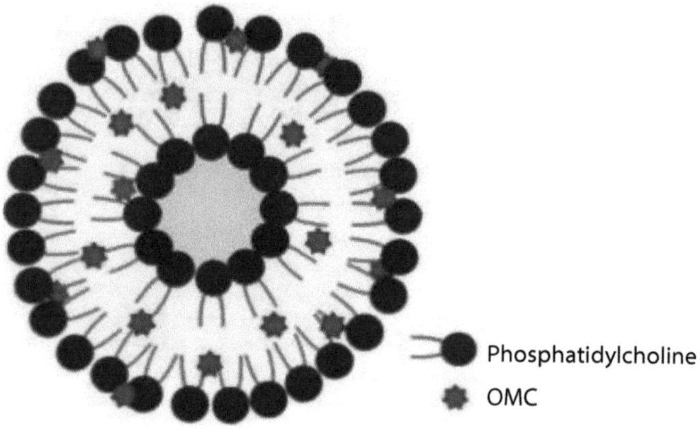

Figure 8.3 Schematic structure of the liposome containing OMC (Adapted from Ref. [29]).

Formulation containing liposomes/OMC showed more excellent safety in histocompatibility tests, lower toxicity, and more significant OMC deposition in the stratum corneum.

Photolytic enzymes, extracted from *Anacystis nidulans* marine plant, are encapsulated in liposomes, associated with sun protection products, and released during the irradiation of light, so these liposomes were able to prevent UV-induced DNA damage [5, 30].

Antioxidants such as carotenoids, CoQ10, and lycopene and active components like vitamins A, E, and K have been incorporated into liposomes, which are used to amplify their physical and chemical stability when dispersed in water [28].

Vitamin C, known as ascorbic acid, is an antioxidant and used by cosmeceuticals as an antiaging product due to stimulated collagen production. However, factors such as temperature, light, and storage can degrade vitamin C in non-active products; therefore, its restriction in handling and use. Thus, nanotechnology helps in the distribution and adequate vitamin C supply [5, 30]. One study showed that vitamin-C-loaded liposomes coated with pectin reported better skin storage and permeation stability when compared to free vitamin C [31]. In another study, an improvement in antioxidant activity was observed by the high trapping efficiency of vitamin C in liposomes, which left it with greater stability [32].

Diseases such as eczema and burns are treated with compounds that have a healing property on the skin, such as *Aloe vera* pulp, which is composed of polysaccharides that demonstrate such properties [21, 31]. The ability to stimulate the synthesis of collagen and elastin fibers determines the healing, antiaging, and moisturizing properties of *Aloe vera* [32, 33]. In a study, liposomes loaded with *Aloe vera* leaf gel extract showed increased skincare bioavailability [34]. In addition, coencapsulation of *Aloe vera* with curcumin enhances phytocompound delivery to the skin and protects antioxidant activity [25].

Some examples of commercially available liposomal cosmetic products are included in Table 8.2 with product name, functional component, manufacturer of the product, and uses.

Table 8.2 Marketed liposomal cosmetic formulation [25, 34]

Product Name	Functional Component	Marketed by	Uses
Revitalift Double Lifting	Pro retinol A	Lancome (L'Oreal)	Antiwrinkle cream for intense lifting action
Efect du Soleil	Tanin liposomal	Lancome (L'Oreal)	Antioxidant
Capture totale	Longoza flower extract, collagen	Dior	Antiaging
C-Vit Liposomal Serum	Vitamin C liposomal	Sesderma	Highly antioxidant, antiwrinkles and brightening serum, which restores vitality of skin

8.5.1.3 Nanoemulsions

Nanoemulsions are stable nanodispersions with droplets in diameter between 10 and 500 nm. These systems consist of two immiscible phases, in which a liquid (the dispersed phase) is dispersed into another phase (the continuous phase), which can be oily, aqueous, and one or more surfactants. Depending on the type of surfactant used and the proportion of the two phases, it can be prepared as oil-in-water (O/W) or water-in-oil (W/O) emulsion [18]. These have advantages over conventional emulsions, such as smaller droplets (between 5 and 200 nm), better distribution on the skin, excellent contact surface, better transport and release of active substances, better occlusive effect, excellent stability, and aesthetically more appealing on the skin [35].

8.5.1.4 Applications of nanoemulsions in cosmetics

The use of nanoemulsions in the formulation of cosmetics has an added value considering the reduced size of droplets since this can enhance the bioavailability of active substances that disperse in both hydrophilic and lipophilic phases [29].

Shampoos are not formulated only with the function of cleaning the hair; it is also essential that they hydrate and soften the hair. First, it was difficult to establish a formula that could clean and soften the hair at the same time since the silicone oil, the main component of the softeners, is dragged with the eliminated particles in the cleaning.

One study proceeded to the preparation of nanoemulsions, where it would be tested whether surfactants facilitate the deposition of silicone oil on the surface of hair. The nanoemulsion was obtained by mixing Span 80, Tween 80, and n-butanol, the format of an emulsifier phase. Then, dimethyl silicone oil was added as a lipophilic phase, and in the end, water was added under continuous agitation. After testing the nanoemulsion's stability, the shampoo was applied, with the nanoemulsion incorporate, in the hair (the process is presented in Fig. 8.4). As the nanoemulsion has two surfactants (Span 80 and Tween 80), with the ability to create a hydrophilic–lipophilic balance similar to that found in the emulsive hydrolipidic layer of the hair, the absorption of silicone oil is enhanced [36].

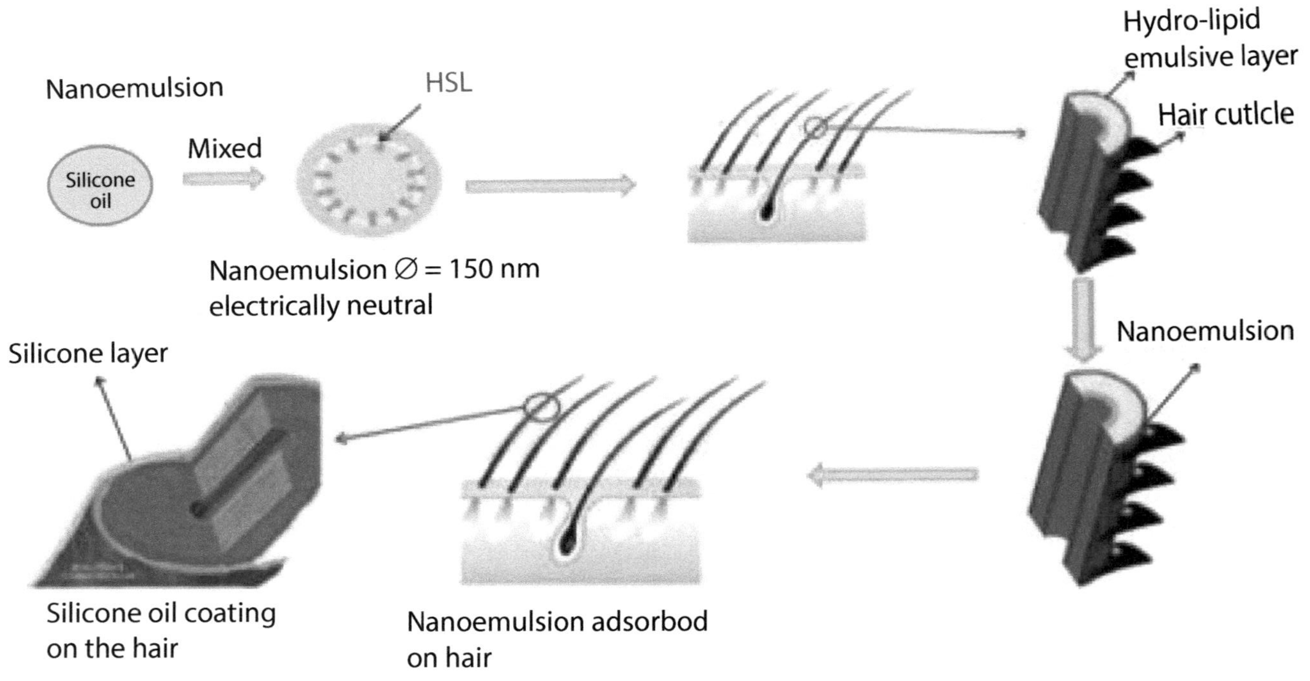

Figure 8.4 Illustrative scheme of the nanoemulsion absorption process in the emulsive hydrolipidic layer of the hair (Adapted from Ref. [36]).

Several plant-derived bioactive compounds are used in nanoemulsions, such as flavonoids and polyphenols. A quercetin-rich nanoemulsion study, prepared from the ethanolic extract of *Achyrocline satureioides*, showed an increase in the retention of bioactive compounds in the skin with antioxidant activities [18, 37]. In another study, genistein, a soy bioactive compound, associated with tocomin showed enhanced skin protection from UV damage [38]. Pomegranate peel contains antioxidant polyphenols, which have photoprotection properties, so when these polyphenols were associated in nanoemulsions, they increase their concentration and skin penetration [39].

Some examples of commercially available nanoemulsion cosmetic products are included in Table 8.3 with product name, functional component, manufacturer of the product, and uses.

Table 8.3 Marketed nanoemulsion cosmetic formulation [40]

Product Name	Functional Component	Marketed by	Uses
Coco Mademoiselle Fresh Moisture Mist	Tocopherol, ascorbic acid	Chanel	Prolongs fragrance effect
Nanovital Vitanics Crystal Moisture Cream	Arbutin, ginseng	Vitacos Cosmetics	Skin moisturizing, elastic, and lightening effects
Skinlastin	Acetyl Hexapeptide-8 (Argireline), Lanolin	Proctor & Gamble	Antiaging, antiwrinkles

8.5.1.5 Nanocapsules

Nanocapsules are polymeric, colloidal nanoparticles with a liquid nucleus of lipid nature, surrounded by an external polymeric layer, and a surfactant layer in the interphase. They can have a size between 10 and 1000 nm [41]. Several polymers have been used in the production of nanocapsules, such as poly-lactic acid (PLA), so these have been of great interest to researchers as they are biodegradable and biocompatible [18].

8.5.1.6 Complications of nanocapsules in cosmetics

Nanocapsules have been incorporated into cosmetics to prevent unpleasant odors, protect sensitive active substances, and avoid possible incompatibilities between the ingredients of the formulation [41].

Tests have been conducted on the stability and efficacy of nanocapsules in deodorant formulations, focusing on the transport and sustained release of perfumes. These fragrances have little solubility, and also the aromatics are unstable molecules, making their use limited. Thus, the encapsulation of these can make the formulation more effective [41].

Two formulations of nanocapsules were produced in the study. Some contained chlorobenzene, a highly hydrophobic aromatic compound (structurally similar to other compounds used in perfumes) used in the industry in bathroom deodorants and other cleaning products; and the others contained fluorescein, an easily detectable dye (Fig. 8.5). Several of the trials ensured the biological preservation of the skin and analyzed the release profile of the substantiation contained in the nanocapsules. The results showed that those containing fluorescein had a rapid release in 6 h and a slower phase up to 24 h. In the case of those with chlorobenzene (more hydrophobic aromatic compound), a more constant and sustained profile was obtained, prolonging the release of the fragrance until 48 h. However, this means that there is a potential application of nanocapsules in the perfume industry, where compounds structurally similar to chlorobenzene are used [41].

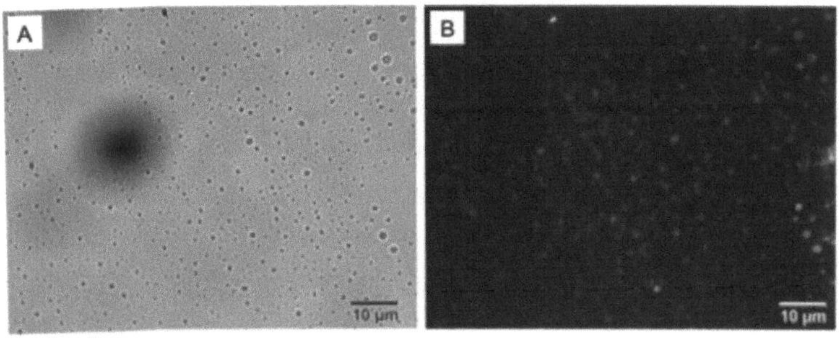

Figure 8.5 PLA nanocapsules in an optical microscope (A); and fluorescence microscope containing fluorescein (Adapted from Ref. [41]).

8.5.1.7 Solid lipid nanoparticles (SLNs)

SLNs emerged as an improved version of the nanosystems mentioned above since they have a uniform, stable, hydrophilic shell and lipophilic nucleus (Fig. 8.6), which reduces the ability to move molecules of the incorporated active substances and thus allows to control or limit the flow of these during storage [18]. The diameter of these nanoparticles is usually between 40 and 1000 nm. The lipids most used for core construction are acylglycerols, fatty acids, steroids, and waxes. To improve dispersion stability, lipid nuclei are submerged in solutions with surfactants that organize a layer around them. Hydrophobic molecules may be incorporated in the solid core of SLNs leading to an enhanced permeation, stability, target specificity, low toxicity, low degradation, as well as a controlled release [42]. The occlusive behavior is the responsible reason for the application of nanocarriers, such as SLNs, in cosmeceutical formulations, since this behavior enables the formation of a skin superficial film, reducing the transepidermal water loss and improving skin hydration [43].

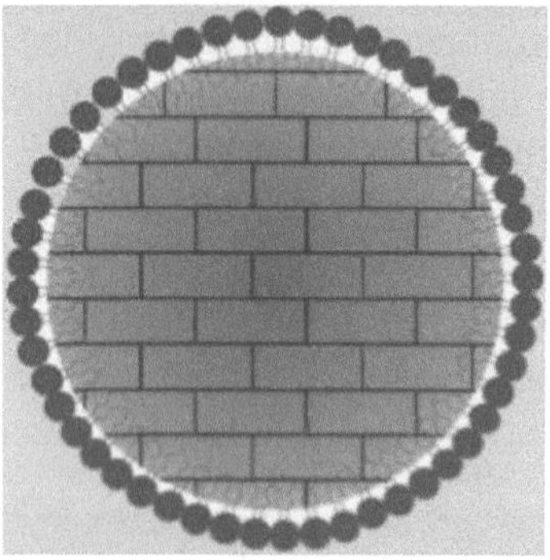

Figure 8.6 A schematic structure of SLN (Adapted from Ref. [42]).

Despite the good luck of commitment of SLNs, there are some disadvantages in their use, such as limitations in the release of active substances and insufficient load capacity of active substances, due to their limited solubility in the lipid nucleus. To solve these problems, the second generation of SLNs, called nanostructure lipid carriers (NLCs), emerged, which differ from their precursors in the addition of liquid lipids to form a liquid/solid lipid nucleus. This alteration allowed the NLC matrix to present structural defects, not allowing the formation of perfect lipid crystals and facilitating the release of active substances. The rearrangement leaves spaces that increase the load capacity of the nanoparticle [16].

8.5.1.8 Applications of SLNs in cosmetics

SLNs are widely used in numerous applications due to the excellent performance in maintaining active substances on the skin surface, or in deeper layers. As they easily organize a regular and cohesive film on the skin, the sun provokes an occlusive effect that leads to proper hydration of the skin [44].

One study formulated SLNs containing retinol (vitamin A alcohol) and SLNs containing retinol palmitate (vitamin A ester), being incorporated into a hydrogel and O/W cream, to test the influence on the penetration of active substances, compared with conventional formulations. The results obtained were positive for the formulas with SLNs, where their use altered the release profile of retinol and retinol palmitate and improved the properties of the stratum corneum [3, 15, 28].

Auraptene is an antioxidant coumarin that exhibits anti-inflammatory effects and has poor water solubility and low skin permeation. To improve physicochemical limitations, SLNs containing auraptene were prepared, characterized, and used [45]. SLNs have UV-resistant properties and act as physical sunscreens on their own, thus improving photoprotection when they are combined with molecular sunscreen [46]. SLNs as carrier for 3,4,5-trimethoxybenzoylchitin, and vitamin E sunscreens have been developed to enhance UV protection [47].

Some examples of commercially available SLN cosmetic products are Allure Body Cream, Allure Parfum Bottle, and Allure Eau the Parfum Spray all marketed by Chanel [40]. The first products containing NLC were introduced in the cosmetic market, namely, NanoRepair Q10 cream and NanoRepair Q10 serum, Dr. Rimpler GmbH, Germany, offering increased skin penetration. Table 8.4 shows examples of commercially available NLC cosmetic products.

Table 8.4 Marketed NLC cosmetic formulation [40, 48]

Product Name	Functional Component	Marketed by	Uses
Iope Supervital Extra Moist Softener	Flavonoids	Amore Pacific	Antiaging
Intensive Serum Nanorepair Q10	Coenzyme Q10-loaded nanopearls	Dr. Rimpler	Antiwrinkle serum, fights sign of aging
Swiss Cellular White Illuminating Eye Essence	Ascorbyl glucoside, tocotrienols	La prairie	Removes under eye darkness and discoloration

8.5.1.9 Dendrimers

Dendrimers are well-defined polymeric structures, similar to a tree (Fig. 8.7), and of dimensions between 2 and 20 nm [49]. They are symmetrical around the nucleus, often with three-four generations of high density in the terminal groups of the periphery, and both in the nucleus as in the branches are deposited particles. These nanocarriers are produced from a starting nitrogen element, to which a number of other elements, like carbons, are added in a step-by-step polymerization growth process, which leads to successive branches [5]. Also, they can contain both hydrophilic and lipophilic substances.

8.5.1.10 Applications of dendrimers in cosmetics

As they have a compact structure, reasonable structural control, and high solubility both in aqueous and apolar medium, dendrimers are excellent transporting nanosystems used in products such as hair gels, shampoos, sunscreens, and anti-acne cosmetics [50]. A carbosiloxane dendrimer was created that can provide exceptional resistance to water and skin sebum, adhesive properties, and a shiny appearance [3, 15, 28].

Dendrimers are being used as nanotechnology-based cosmeceuticals for various applications in haircare and skincare. Companies like L'Oreal, Te Dow Company, Wella, and Unilever have several patents for application of dendrimers in cosmeceuticals [40].

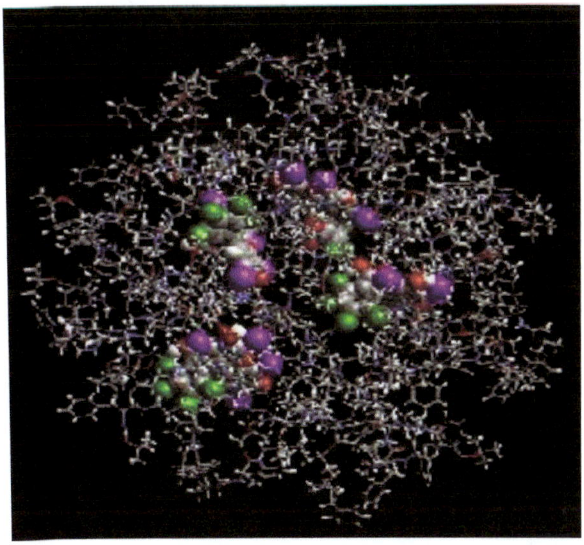

Figure 8.7 Illustrative scheme of a dendrimer (Adapted from [49]).

Peptide dendrimers were selected due of their ability to reduce toxicity and biodegradability. Polyamidoamine dendrimers have the ability to increase asset delivery and the penetration action of ingredients [5, 40]. *In vitro* skin penetration of indomethacin, from the polyamidoamine dendrimer formulations, was enhanced compared to the pure drug suspension. *In vivo* absorption from a transdermal patch was found to increase 2.7% with the 0.2% dendrimer formulation [51]. A carbosiloxane-based dendrimer in cosmetic formulation claimed that it can provide better resistance to sebum and water [52].

8.5.1.11 Niosomes

Niosomes are liposomes formed by non-ionic surfactants (e.g., tweens, span, brijs, alkyl amides, sorbitan ester, crown ester, polyoxyethylene alkyl ether) of the aqueous nucleus, biodegradable, and having more stability than liposomes. Niosomes can be multilamellar or unilamellar vesicles, and non-ionic surfactants are structurally similar to the phospholipids used in liposomes, since these surfactants are also composed of a lipophilic tail and a hydrophilic head. They have dimensions between 100 nm and 2000 nm [18, 28].

Surfactants consist of two parts: one or more lipophilic components, and a hydrophilic head (Fig. 8.8), with an ester or amide bond between the two regions. Like liposomes, niosomes can form multilamellar structures and are suitable for the delivery of both lipophilic and/or hydrophilic molecules; but they are more stable than liposomes, their production cost is lower than liposomes, and they have higher skin penetration, entrapment efficiency, and yield [53]. On account of the non-ionic composition of niosomes, they are considered biodegradable, biocompatible, non-irritant, and non-immunogenic [54].

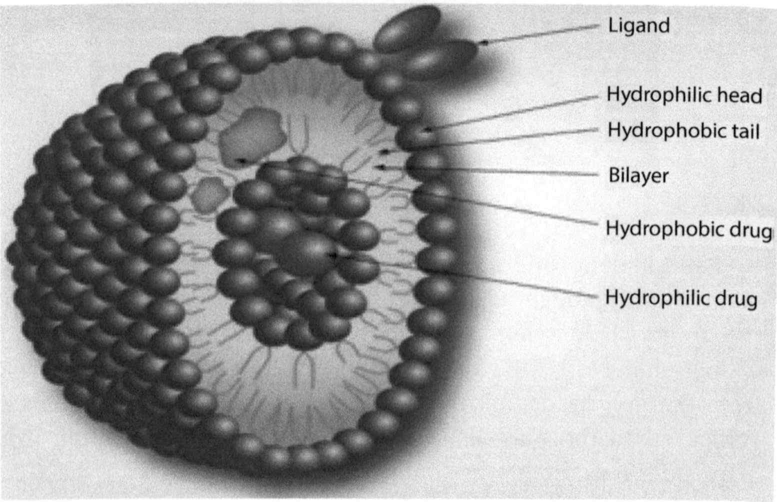

Figure 8.8 Niosome structure (Adapted from Ref. [53]).

8.5.1.12 Applications of niosomes in cosmetics

In cosmetics, niosomes have been used in the same applications as liposomes, since they reduce the possibilities of skin irritation, improve skin absorption and penetration, and increase the deposition of the substances they incorporate in the region of application [56]. Niosomes were first developed by L'Oreal in 1970 by the research and development of synthetic liposomes.

In investigations, an elastic niosome has been produced after the use of ethanol and non-ionic surfactants such as Tween 61 and

Span 60. Ethanol is an efficient permeation enhancer and is placed in niosomes to interact with the terminal portion of the lipids of the stratum corneum, reducing the melting point of these and, however, increasing the fluidity of the skin lipids and the permeability of cell membranes. Thus, it allowed a faster and more effective transport of the substances through the skin pores, because although they are of smaller diameter, the nanoparticle can deform and cross them [55].

Junyaprasert et al. formulated niosomes using Span 60 and Tween 60 (2:1), loaded with ellagic acid, an antioxidant, and a potential skin-whitening agent. The observation of *in vitro* skin permeation test revealed that the ellagic acid releasing from niosomes is a function of vesicle size, the amount of loaded ellagic acid, and the type of solubilizers (which could act as a permeation enhancer) [57]. These niosomes showed greater entrapment efficiency and higher amount of ellagic acid delivery to the deeper skin layers when compared to an ellagic acid solution.

Another study was developed to evaluate the activity of elastic niosomes compared to regular ones in topical antiaging formulations. These have been loaded with gallic acid, which is a potent antioxidant in emulsions and/or lipid systems. The results pointed to slightly higher storage stability for non-elastic niosomes, while elastic niosones obtained better results in terms of transdermal absorption [58].

Curcumin is the active compound in turmeric or *Curcuma longa*, which is used in skincare as a crude extract or tuber powder. It shows a wide range of beneficial properties, including antiaging, moisturizing, antioxidant, antineoplastic, antibacterial, anti-inflammatory, and wound-healing properties [57, 59]. Niosomes were also reported to improve curcumin bioavailability and have shown to be a more efficient nanocarrier in protecting curcumin from deterioration when compared to other nanocarriers employing surfactants [60]. In addition to the individual compounds, coadministration of curcumin with resveratrol in the nanosized niosomal formulations enhanced skin beauty with increased antioxidant activity [61].

Some examples of commercially available niosomal cosmetic products are included in Table 8.5 with product name, functional component, manufacturer of the product, and uses.

Table 8.5 Marketed niosomal cosmetic formulation [40, 48]

Product Name	Functional Component	Marketed by	Uses
Niosome	Pro retinol A glyceropolyether with moisturizers functional component	Lancome (L'Oreal)	Foundation cream, clear and white skin tone
Nactosome	Vitamins	Lancome (L'Oreal)	Antioxidant, antiwrinkles
Eusu Niosome Makam Pom Whitening Facial Cream	Makam Pom extract enriched vitamin C	Eusu	Skin whitening
Anti-Age Response Cream	Ginseng extract and mineral salts	Simply Man Match	Treatment of wrinkles

8.5.2 Metal Nanoparticles

8.5.2.1 Silver nanoparticles

These metallic nanoparticles have unique properties, which have provided them with immense applications in the industry. Silver nanoparticles have been known for hundreds of years. Silver in its metallic state is not reactive, but in contact with the moisture of the skin, or the fluids of a wound, it becomes ionized. Ionized silver is quite reactive and can bind to bacteria and cause changes in the bacterial cell wall, leading to the death of the bacteria (Fig. 8.9). Inside the bacterial cell, the silver ion increases the production of reactive oxygen species, due to their interaction with the respiratory chain, which inhibits bacterial cell growth. Besides, silver nanoparticles can also bind to the phosphorous and sulfur groups of DNA or bacterial RNA, thus preventing the replication and translation of the bacterium [62].

8.5.2.2 Applications of silver nanoparticles in cosmetics

Silver nanoparticles represent an added value for cosmetics due to their antibacterial and antifungal properties. Products such as deodorants, mascaras, antiaging creams, toothpastes, soaps, and lotions have used these nanoparticles in their formulations [63].

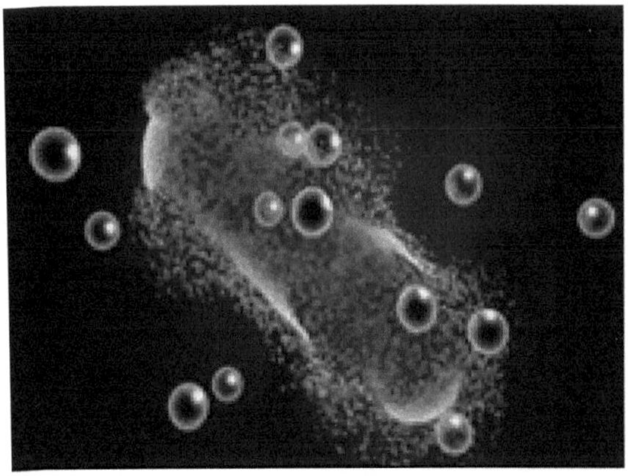

Figure 8.9 Illustration of a bacterium to be destroyed by silver nanoparticles (Adapted from Ref. [62]).

A fresh, nano-antibacterial deodorant was developed using silver nanoparticles, which end up combining relatively low production costs with strong antibacterial properties, no toxicity, and no adverse effects [64]. Silver nanoparticles showed acceptable preservation efficacy against bacteria and fungi and did not penetrate human skin [65].

8.5.3 Carbon Nanoparticles

Carbon nanomaterials such as graphene, carbon nanotubes, and fullerenes are some of the most thoroughly studied nanomaterials. This makes carbon the second most common nanomaterial after silver in commercial products within all areas of use [14].

8.5.3.1 Fullerenes (Buckyballs)

Fullerenes are made up of a series of rings of carbon atoms, which give them a three-dimensional spherical shape (Fig. 8.10), similar to a soccer ball (therefore called *Buckyballs*) [15]. They are incredibly hydrophobic nanoparticles, which limits their use in aqueous solutions. However, with the help of surfactants, or changes to the surface, their solubility in the aqueous medium increases, enabling their application in cosmetic formulations [66].

Figure 8.10 Structure of a fuller free C_{60} (Adapted from Ref. [66]).

8.5.3.2 Applications of fullerenes in cosmetics

The penetration degree of fullerenes seems to be dependent on the vehicle type. *In vivo* experiments showed that fullerenes penetrated deeper into the stratum corneum layers when applied with toluene, cyclohexane, or chloroform instead of mineral oil [67]. With tremendous antioxidant power and properties against oxygen-free radicals, fullerenes have been the subject of research for applications in antiaging creams or sunscreens, being incorporated into hair and skincare [66, 68]. Their antioxidant effect may be reportedly used in the promotion of hair growth, if oxidative stress is the cause of hair loss [69]. Another investigation reported a possible inhibitory activity of melanogenesis on the part of fullerenes, making them of great interest as active substances, for the formulation of skin bleaching products. Fullerene can act like a skin protector since it inhibits keratinocyte differentiation [70, 71]. Fullerenes have greater penetration into the epidermis without causing skin irritation, which makes them a good option for the development of topical formulations to treat acne vulgaris [72]. Fullerene-based photodynamic therapy was used for skin-based disease treatment with the bioactive compounds, such as curcumin [73].

8.6 Advantages and Disadvantages of Nanoparticles in Cosmetics

The main uses of nanotechnology in the formulation of cosmetics are as UV filters and as vehicles in the transport of active substances, due to the reduced dimensions and the greater area of contact with the application zone [3].

Additionally, there are other advantages:

- They enable a controlled release of active substances, prolonging the effects, such as perfumes.
- Being multifunctional structures, they simplify the manufacture of cosmetic products, avoiding incompatibilities between ingredients, such as antiaging creams, since they usually have more complex formulas.
- Nanoparticles preserve sensitive active substances such as vitamins and antioxidants.

However, the use of nanoparticles can bring risks and disadvantages:

- The reduced size of nanoparticles increases their chemical reactivity, increasing the number of ROS that can induce cytotoxicity in the skin.
- Also because the particles are small, they can be inhaled and deposited in the lungs, which can later evolve into health problems such as pulmonary fibrosis or even carcinogenesis [74].

8.7 Final Considerations

The European Union has been working to improve cosmetic regulations. There are many products with active substances and nanoparticles that should be kept under further evaluation. However, there is still a lack of knowledge of biological systems and the properties of nanoparticles, which allow a better prediction of their effects on health and the environment. However, in other countries, such as China, priority is being given to economic growth and progress in innovation to the detriment of consumer safety, and therefore the number of patents related to nanotechnological applications has increased in the Chinese market. In the future, the right balance must be struck between regulations ensuring the quality and safety of nanocosmetics, and technological innovation [3].

Nowadays, nanotechnology improves the safety, efficacy, stability, and aesthetic appearance of cosmetic products, which leads to a positive response on the part of those consumed.

A step forward is being taken in the case of nanoemulsions, as companies are eliminating polymer formulations such as PEG, and the number of consumers who prefer to use cosmetics with natural ingredients has increased. Also, reports from the manufacturers themselves indicate that the sensitivity of the skin to touch has improved and thus the performance of the product has increased [3].

In addition to the regulatory aspects, a public opinion is important for the development of this vast area of science and industry. The emergence of reports confirming the quality, efficacy, and safety of cosmetics from the consumers' perspective will encourage the scientific community to research new nanoparticles and also to improve the existing ones. This scientific community integrates multidisciplinary professionals, such as pharmacists, who have made a significant contribution to cosmetic innovation, increasingly directed to products that, in addition to providing a better look to consumers, can also act on more complex aspects, such as the treatment of dermatological diseases.

References

1. European Commission – DocsRoom. (2020). Ongoing MDR and IVDR guidance plan. http://ec.europa.eu/DocsRoom/documents/13021/attachments/1/translations/
2. Jeevanandam, J., Barhoum, A., Chan, Y. S., Dufresne, A., and Daquah, M. K. (2018). Review on nanoparticles and nanostructured materials:

History, sources, toxicity and regulations, *Beilstein J Nanotechnol*, **9**, pp. 1050–1074.

3. Chaudhri, N., Soni, G. C., and Prajapati, S. K. (2015). Nanotechnology: An advance tool for nano-cosmetics preparation, *Int J Pharma Res Rev*, **4**, pp. 28–40.
4. Farokhzad, O. C. and Langer, R. (2009). Impact of nanotechnology on drug delivery, *ACS Nano*, **3**, pp. 16–20.
5. Santos, A. C., Morais, F., Simões, A., Pereira, I., Sequeira, J. A. D., Pereira-Silva, M., Veiga, F., and Ribeiro, A. (2019). Nanotechnology for the development of new cosmetic formulations, *Expert Opin Drug Deliv*, **16**, pp. 313–330.
6. Johnson, S. (2012). *Encyclopedia of Applied Ethics* **3**, ed. Chadwich, R., "Nanotechnology," (Academic Press, San Diego) pp. 182–185.
7. Subramani, K., Elhissi, A., Subbiah, U., and Ahmed, W. (2019). *Nanobiomaterials in Clinical Dentistry*, eds. Subramani, K and Ahmed, W., Chapter 1 "Introduction to nanotechnology," (Elsevier, Amsterdam) pp. 3–18.
8. Mu, L. and Sprando, R. L. (2010). Application of nanotechnology in cosmetics, *Pharm Res*, **27**, pp. 1746–1749.
9. DeLouise, L. A. (2012). Applications of nanotechnology in dermatology, *J Invest Dermatol*, **132**, pp. 964–975.
10. Golubovic-Liakopoulos, N., Simon, S. R., and Shah, B. (2011). Nanotechnology use with cosmeceuticals, *Semin Cutan Med Surg*, **30**, pp. 176–180.
11. WichLab. Projects overview. https://www.wichlab.com/research.
12. Zamboni, W. C., Torchilin, V., Patri, A. K., Hrkach, J., Stern, S., Lee, R., Nel, A., Panaro, N. J., and Grodzinski, P. (2012). Best practices in cancer nanotechnology: Perspective from NCI nanotechnology alliance, *Clin Cancer Res*, **18**, pp. 3229–3241.
13. Papakostas, D., Rancan, F., Sterry, W., Blume-Peytavi, U., and Vogt, A. (2011). Nanoparticles in dermatology, *Arch Dermatol Res*, **303**, pp. 533–550.
14. Mihranyan, A., Ferraz, N., and Stromme, M. (2012). Current status and future prospects of nanotechnology in cosmetics, *Prog Mater Sci*, **57**, pp. 875–910.
15. Duarah, S., Pujari, K., Durai, R. D., and Narayanan, V. H. (2016). Nanotechnology based cosmeceutical: A review, *Int J Appl Pharm*, **8**, pp. 8–12.

16. Puglia, C. and Bonina, F. (2012). Lipid nanoparticles as novel delivery systems for cosmetics and dermal pharmaceuticals, *Expert Opin Drug Deliv*, **9**, pp. 429–441.

17. Padamwar, M. N. and Pokharkar, V. B. (2006). Development of vitamin loaded topical liposomal formulation using factorial design approach: Drug deposition and stability, *Int J Pharm*, **320**, pp. 37–44.

18. Wu, X. and Guy, R. H. (2009). Applications of nanoparticles in topical drug delivery and in cosmetics, *J Drug Deliv Sci Technol*, **19**, pp. 371–384.

19. Kalra, J. and Bally, M. B. (2013). *Fundamentals of Pharmaceutical Nanoscience*, eds. Uchegbu, I., Schätzlein, A., Cheng, W., and Lalatsa, A. Chapter 3 "Liposomes," (Springes, New York), pp. 27–63.

20. Akbarzadeh, A., Rezaei-Sadanady, R., Davaran, S., Joo, S. W., Zarghami, N., Hanifehpour, Y., Samiei, M., Kouhi, M., and Nejati-Koshki, K. (2013). Liposome: Classification, preparation, and applications, *Nanoscale Res Lett*, **22**, p. 102.

21. Torchilin, V. P. (2005). Recent advances with liposomes as pharmaceutical carrier, *Nat Rev Drug Discov*, **4**, pp. 145–160.

22. Allen, T. M. and Cullis, P. R. (2013). Liposomal drug delivery systems: From concept to clinical applications, *Adv Drug Deliv Rev*, **65**, pp. 36–48.

23. Eckert, G. P., Chang, S., Eckmann, J., Copanaki, E., Hagl, S., Hener, U., Müller, W. E., and Kögel, D. (2011). Liposome-incorporated DHA increases neuronal survival by enhancing non-amyloidogenic APP processing, *Biochim Biophys Acta*, **1808**, pp. 236–243.

24. Sharma, B. and Sharma, A. (2012). Future prospect of nanotechnology in development of anto-ageing formulations, *Int J Pharm Sci*, **4**, pp. 57–66.

25. Ganesan, P. and Choi, D. K. (2016). Current application of phytocompound-based nanocosmeceuticals for beauty and skin therapy, *Int J Nanomed*, **11**, pp. 1987–2007.

26. Katz, L. M., Dewan, K., and Bronaugh, R. L. (2015). Nanotechnology in cosmetics, *Food Chem Toxicol*, **85**, pp. 127–137.

27. Ascenso, A., Raposo, S., Batista, C., Cardoso, P., Mendes, T., Praça, F. G., Bentley, M. V. L. B., and Simões, S. (2015). Development, characterizations, and skin delivery studies of related ultradeformable vesicles: Transfersomes, ethosomes, and transethosomes, *Int J Nanomedicine*, **10**, pp. 5937–5851.

28. Arif, T., Nisa, N., Amin, S. S., Shoib, S., Mushtaq, R., and Shawl, M. R. (2015) Therapeutic and diagnostic applications of nanotechnology in dermatology and cosmetics, *J Nanomed Biotherapeutic Discov*, **5**, pp. 1–10.

29. Mota, A. C. V., Freitas, Z. M. F., Ricci Júnior, E., Dellamora-Ortiz, G. M., Santos-Oliveira, R., Ozzetti, R. A., Vergnanini, A. L., Ribeiro, V. L., Silva, R. S., and Santos, E. P. (2013). In vivo and in vitro evaluation of octyl methoxycinnamate liposomes, *Int J Nanomed*, **8**, pp. 4689–4701.

30. Ahmadi Ashtiani, H. R., Bishe, P., Lashgari, N., Nilforoushzadeh, M. A., and Zare S. (2016). Liposomes in cosmetics, *J Skin Stem Cell*, **3**, p. e65815.

31. Zhou, W., Liu, W., Zou, L., Liu, W., Liu, C., Liang, R., and Chen, J. (2014). Storage stability and skin permeation of vitamin C liposomes improved by pectin coating, *Colloids Surf B Biointerfaces*, **117**, pp. 330–337.

32. Yang, S., Liu, C., Liu, W., Yu, H., Zheng, H., Zhou, W., and Hu, Y. (2013). Preparation and characterization of nanoliposomes entrapping medium-chain fatty acids and vitamin C by lyophilization, *Int J Mol Sci*, **14**, pp. 19763–19773.

33. Radha, M. H. and Laxmipriya, N. P. (2014). Evaluation of biological properties and clinical effectiveness of Aloe vera: A systematic review, *J Tradit Complement Med*, **5**, pp. 21–26.

34. Takahashi, M., Kitamoto, D., Asikin, Y., Takara, K., and Wada, K. (2009). Liposomes encapsulating Aloe vera leaf gel extract significantly enhance proliferation and collagen synthesis in human skin cell lines, *J Oleo Sci*, **58**, pp. 643–650.

35. Yukuyama, M. N., Ghisleni, D. D. M., Pinto, T. J. A., and Bou-Chacra, N. A. (2016). Nanoemulsion: Process selection and application in cosmetics — a review, *Int J Cosmet Sci*, **38**, pp. 13–24.

36. Hu, Z., Liao, M., Chen, Y., Cai, Y., Meng, L., Liu, Y., Lv, N., Liu, Z., and Yuan, W. (2012). A novel preparation method for silicone oil nanoemulsions and its application for coating hair with silicone, *Int J Nanomed*, **7**, pp. 5719–5724.

37. Zorzi, G. K., Caregnato, F., Moreira, J. C. F., Teixeira, H. F., and Carvalho, E. L. S. (2016). Antioxidant effect of nanoemulsions containing extract of *Achyrocline satureioides* (Lam) D.C. – Asteraceae, *AAPS PharmSciTech*, **17**, pp. 844–850.

38. Brownlow, B., Nagarai, V. J., Nayel, A., Joshi, M., and Elbayoumi, T. (2015). Development and in vitro evaluation of vitamin E enriched

nanoemulsion vehicles loaded with genistein for chemoprevention against UVB-induced skin damage, *J Pharm Sci*, **104**, pp. 3510–3523.

39. Baccarin, T. and Lemos-Senna, E. (2017). Potential application of nanoemulsion for skin delivery of pomegranate peel polyphenols, *AAPS PharmSciTech*, **18**, pp. 3307–3314.

40. Kaul, S., Gulati, N., Verma, D., Mukherjee, S., and Nagaich, U. (2018). Role of nanotechnology in cosmeceuticals: A review of recent advances, *J Pharm (Cairo)*, **2018**, p. 3420204.

41. Hosseinkhani, B., Callewaert, C., Vanbeveren, N., and Boon, N. (2015). Novel biocompatible nanocapsules for slow release of fragrances on the human skin, *N Biotechnol*, **32**, pp. 40–46.

42. Morales, J. O., Valdes, K., Morales, J., and Oyarzun-Ampuero, F. (2015). Lipid nanoparticles for the topical delivery of retinoids and derivatives, *Nanomedicine (Lond)*, **10**, pp. 253–269.

43. Garcês, A., Amaral, M. H., Sousa Lobo, J. M., and Silva, A. C. (2018). Formulations based on solid lipid nanoparticles (SLN) and nanostructures lipid carriers (NLC) for cutaneous use: a review, *Eur J Pharm Sci*, **112**, pp. 159–167.

44. Jenning, V., Schäfer-Korting, M., and Gohla, S. (2000). Vitamin A – loaded solid lipid nanoparticles for topical use: Drug release properties, *J Control Release*, **66**, pp. 115–126.

45. Daneshmand, S., Jaafari, M. R., Movaffagh, J., Malaekeh-Nikouei, B., Iranshahi, M., Moghaddam, A. S., Najaran, Z. T., and Golmohammadzadeh, S. (2018). Preparation, characterization, and optimization of auraptene-loaded solid lipid nanoparticles as a natural anti-inflammatory agent: In vivo and in vitro evaluations, *Colloids Surf B Biointerfaces*, **164**, pp. 332–339.

46. Arora, N., Agarwa, S., and Murthy, R. S. (2012). Latest technology advances in cosmaceuticals, *Int J Pharm Sci Drug Res*, **4**, pp. 168–182.

47. Patidar, A., Singh Thakur, D., Kumar, P., and Verma, J. (2010). A review on novel lipid based nanocarriers, *Int J Pharm Sci*, **2**, pp. 30–35.

48. Ahmad, U., Ahmad, Z., Khan, A. A., Akhtar, J., Singh, S. P., and Ahmad, F. J. (2018). Strategies in development and delivery of nanotechnology based cosmetic products, *Drug Res (Stuttg)*, **68**, pp. 545–552.

49. Kaur, M. (2014). Dendrimers: Synthesis, its dosage forms and advantage over linear polymers, *Mintage J Pharm Med Sci*, **3**, pp. 15–19.

50. Iimura, T. and Furukawa, H. (2010). US10047199B2. Copolymer having carbosiloxane dendrimer structure, and composition and

cosmetic containing the same, Google Patents. https://patents.google.com/patent/US10047199B2/en.

51. Chauhan, A. S., Sridevo, S., Chalasani, K. B., Jain, A. K., Jain, S. K., Jain, N. K., and Diwan, P. V. (2003). Dendrimer-mediated transdermal delivery: Enhanced bioavailability of indomethacin, *J Control Release*, **90**, pp. 335–343.

52. Iimura, T. and Hori, S. (2014). WO2014030771. Copolymer having carbosiloxane dendrimer structure and composition and cosmetic containing the same, Patentscope. https://patentscope.wipo.int/search/en/detail.jsf?docId=WO2014030771.

53. Chandu, V. P., Arunachalam, A., Jeganath, S., Yamini, K., Tharangini, K., and Chaitanya, G. (2012). Niosomes: A novel drug delivery system, *Int J Novel Trends Pharm Sci*, **2**, pp. 25–31.

54. Uchechi, O., Ogbonna, J. D. N., and Attama, A. A. (2014) *Application of Nanotechnology in Drug Delivery*, ed. Sezer, A. D., Chapter 6 "Nanoparticles for Dermal and Transdermal Drug Delivery," (IntechOpen, London) pp. 193–235.

55. Manosroi, A., Jantrawut, P., Akazawa, H., Akihisa, T., Manosroi, W., and Manosroi, J. (2011). Transdermal absorption enhancement of gel containing elastic niosomes loaded with gallic acid from *Termilalia chebula* galls, *Pharm Biol*, **49**, pp. 553–562.

56. Manosroi, A., Jantrawut, P., and Monosroi, J. (2008). Anti-inflammatory activity of gel containingnovel elastic niosomes entrapped with diclofenac diethlylammonium, *Int J Pharm*, **360**, pp. 156–163.

57. Junyaprasert, V. B., Singhsa, P., Suksiriworapong, J., and Chantasart, D. (2012). Physicochemical properties and skin permeation of Span 60/Tween 60 niosomes of ellagic acid, *Int J Pharm*, **423**, pp. 303–311.

58. Rai, M., Yadav, A., and Gade, A. (2009). Silver nanoparticles as a new generation of antimicrobials, *Biotechnol Adv*, **27**, pp. 76–83.

59. Mahmood, K., Zia, K. M., Zuber, M., Salman, M., and Anjum, M. N. (2015). Recent developments in curcumin and curnumin based polymeric materials for biomedical applications: a review, *Int J Biol Macromol*, **81**, pp. 877–890.

60. Mandal, S., Banerjee, C., Ghosh, S., Kuchlyan, J., and Sarkar, N. (2013). Modulation of the photophysical properties of curcumin in nonionic surfactant (Tween-20) forming micelles and niosomes: A comparative study of different microenvironments, *J Phys Chem B*, **117**, pp. 6957–6968.

61. Tavano, L., Muzzalupo, R., Picci, N., and Cindio, B. (2014). Co-encapsulation of lipophilic antioxidants into niosomal carriers: Percutaneous permeation studies for cosmeceutical applications, *Colloids Surf B Biointerfaces*, **114**, pp. 144–149.

62. Kon, K. Destruction of bacterium by silver nanoparticles. Illustration 343969862. Shutterstock. https://www.shutterstock.com/pt/image-illustration/destruction-bacterium-by-silver-nanoparticles-illustration-343969862.

63. Fujian Province Nan'An City Bangdeng Shoes Co Ltd. (2013). CN104224593A. Cool-type nanometer antibacterial deodorant and preparation method thereof, Google Patents. https://patents.google.com/patent/CN104224593A/en

64. Lohani, A., Verma, A., Joshi, H., Yadav, N., and Karki, N. (2014). Nanotechnology-based cosmeceuticals, *ISRN Dermatol*, **2014**, pp. 843687.

65. Kokura, S., Handa, O., Takagi, T., Ishikawa, T., Naito, Y., and yoshikawa, T. (2010). Silver nanoparticles as a safe preservative for use in cosmetics, *Nanomedicine*, **6**, pp. 570–574.

66. Lens, M. (2011). Recent progress in application of fullerenes in cosmetics, *Recent Pat Biotechnol*, **5**, pp. 67–73.

67. Xia, X. R., Monteiro-Riviere, N. A., and Riviere, J. E. (2010). Skin penetration and kinetics of pristine fullerenes (C60) topically exposed in industrial organic solvents, *Toxicol Appl Pharmacol*, **242**, pp. 29–37.

68. Niwa, N., Takada, H., and Matsubayashi, K. (2005). WO2006011633A1. Skin melanin-controlling agent, Google Patents. https://patents.google.com/patent/WO2006011633A1/en

69. Mousavi, S. Z., Nafisi, S., and Maibach, H. I. (2017). Fullerene nanoparticle in dermatological and cosmetic applications, *Nanomedicine*, **13**, pp. 1071–1087.

70. Dolez, P. I. (2015) *Nanoengineering*, ed Dolez, P. I., Chapter 1.1 "Nanomaterials definitions, classifications, and applications," (Elsevier, Amsterdam) pp. 3–40.

71. Melo, A., Amadeu, M. S., Lancellotti, M., de Holanda, L. M., and Machado, D. (2015). The role of nanomaterials in cosmetics: National and international legislative aspects, *Química Nova*, **38**, pp. 599–603.

72. Inui, S., Aoshima, H., Nishiyama, A., and Itami, S. (2011). Improvement of acne vulgaris by topical fullerene application: Unique impact on skin care, *Nanomedicine*, **7**, pp. 238–241.

73. Zhang, Z., Chen, X., Rao, W., Long, F., Yan, L., and Yin, Y. (2015). Preparation of novel curcumin-imprinted polymers based on magnetic multi-walled carbon nanotubes for the rapid extraction of curcumin from ginger powder and kiwi fruit root, *J Sep Sci*, **38**, pp. 108–114.

74. Wacker, M. G., Proykova, A., and Santos, G. M. L. (2016). Dealing with nanosafety around the globe – regulation vs. innovation, *Int J Pharm*, **509**, pp. 95–106.

Chapter 9

Oral Lipid-Based Carriers: Overcoming the Challenges Associated with Conventional Treatments of Non-Small Cell Lung Cancer

James Glover,[a] Sakib Yousaf,[b] and Iftikhar Khan[a]
[a]*School of Pharmacy and Biomolecular Sciences, Liverpool John Moores University, Liverpool L3 3AF, United Kingdom*
[b]*School of Pharmacy and Biomedical Sciences, University of Central Lancashire, Preston PR1 2HE, United Kingdom*
i.khan@ljmu.ac.uk, iftikharkhans@yahoo.com

Non-small cell lung cancer is the second most diagnosed cancer attributing to over 1 million annual deaths globally. The majority of patients diagnosed with this disease will undergo a chemotherapeutic regimen during the course of their illness and consequently severe adverse events can arise. The aim of this review is to overcome the common challenges associated with oral dosage forms of anticancer treatments that possess poor drug bioavailability. The literature

Science and Applications of Nanoparticles
Edited by Waqar Ahmed and Ehsan Nourafkan

ISBN 978-981-4267-34-2 (Hardcover), 978-1-003-28029-3 (eBook)
www.jennystanford.com

revealed that various lipid-based drug-delivery systems are viable vesicles to entrap these chemotherapeutic agents for uptake into the intestinal lymphatics, which can bypass first-pass metabolism, thus greatly improving oral bioavailability. However, the balance between the parameters of oral bioavailability, drug concentration, cytotoxicity, and surface functionalization requires further research before oral dosage forms of chemotherapies are approved.

9.1 Introduction

Cancer, a disease characterized by non-regulated cell proliferation, is one of the leading causes of death today. The second most commonly diagnosed cancer is lung cancer, which is behind that of non-melanocytic skin cancer. However, lung cancer exhibits the highest death rate of any malignancy attributing 1.59 million deaths globally per year, over 1 million of which are smokers or former smokers according to the World Health Organization (WHO) [1, 2]. Lung cancer can be broadly classified into two categories: non-small cell lung cancer (NSCLC) making up 84% of all identified lung cancers and small cell lung cancer (SCLC) accounting for approximately 15% of lung cancer cases (Fig. 9.1). However, the frequency of SCLC diagnosis among men and women has continued to decline over the past 20 years [2].

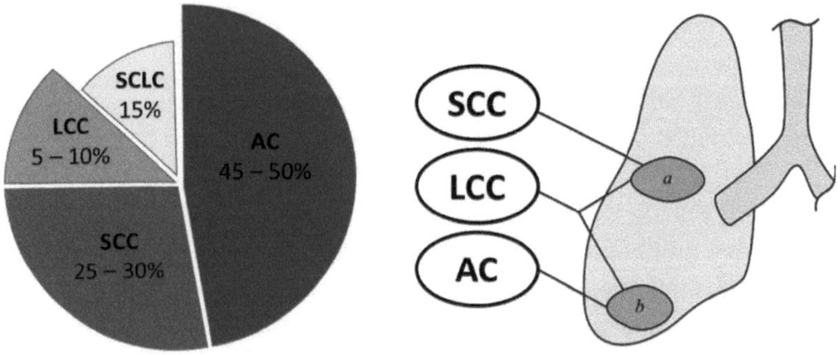

Figure 9.1 The frequency of different lung cancer diagnoses and location in the lung where each histological type of NSCLC metastasizes in the lungs. *A* = Central lung region, *b* = Peripheral region, Adenocarcinoma (AC), Large Cell Carcinoma (LCC), Squamous Cell Carcinoma (SCC).

Currently, the Food and Drug Administration (FDA) has approved many anticancer drugs such as doxorubicin, paclitaxel, and raloxifene, which are heavily impacted by first-pass metabolism via Cytochrome P450 enzymes. These therapies often exhibit low oral bioavailability due to poor aqueous solubility and high P-glycoprotein effluxing, resulting in variable pharmacokinetics following administration of an oral dosage form [3–5]. Therefore, intravenous (I/V) delivery is often required for a patient to witness any beneficial therapeutic effect. However, IV delivery can provoke hypertensive reactions from the formulation excipients and potentially cause an increase in dose-dependent toxicity from the active pharmaceutical ingredient.

According to the Biopharmaceutical Classification System (BCS), class II and class IV drugs are defined as poor water-soluble drugs. More specifically, class II drugs have low solubility and high permeability, and class IV drugs have both low solubility and low permeability. These drug molecules are discovered through novel techniques such as high-throughput screening and computational modeling. However, currently more than 40% of new pharmacologically active chemical entities are lipophilic and exhibit poor water solubility [6]. Thus, many drugs fail to make it to the advanced stages of clinical development due to limitations in solubility and bioavailability. This poses the question to pharmaceutical companies as to what measures can be explored to allow these drugs to have sufficient bioavailability in the body following administration of an oral dosage form.

Upon analysis, it was discovered that approximately 6% of all approved new molecular entities from 1975 to 2013 utilized solubilization technologies with lipid-based drug-delivery systems being the most common technique used [7]. At present, there are over 25 different drug molecules used in over 35 lipid-based delivery systems for oral dosage forms, which are approved by the FDA, most of which are delivered in a soft capsule [8, 9].

It is apparent that the intestinal lymphatic system is an auspicious route of administration for anticancer therapeutics, and that phospholipid-derived lipid-based formulations such as liposomes, nanostructured lipid carriers (NLCs), and solid lipid nanoparticles (SLNs) are viable vesicles for access to the intestinal lymphatics. These vesicles offer a wide variety of uses due to differences in vehicle composition as well as physiochemical properties that can be administered and have the potential to target specific disease states in the lymphatics such as human immunodeficiency virus (HIV) [10]. Oral lipid-based formulations provide a new potential outlook on anticancer therapy due to their ability to entrap lipophilic drug compounds and consequently improve oral bioavailability. The possible benefits of using these drug-delivery systems in the future include higher patient compliance, and ease of drug administration, resulting in reduced patient hospitalizations required for treatment.

This review will highlight how oral lipid-based therapies bypass hepatic first-pass metabolism via uptake into the intestinal lymphatic system and explore different phospholipid-derived vesicles that can be used with FDA-approved lipophilic drug compounds to allow greater affinity to the intestinal lymphatic system, thus improving oral bioavailability. Furthermore, current FDA-approved treatments for NSCLC are associated with drawbacks and lipid-based oral dosage forms are new potential vesicular delivery systems in order to help and overcome these challenges.

9.2 Non-Small Cell Lung Cancer (NSCLC) Pathophysiology

Adenocarcinoma (AC) developed in the bronchial mucosal glands, squamous cell carcinoma (SCC) (a cavitary lesion) in the proximal bronchus, and large cell carcinoma (LCC) are the three histological subtypes of NSCLC. Their frequencies in proportion to all lung cancer diagnoses and the location at which each malignancy metastasizes in the lungs are displayed in Fig. 9.1. The source of malignant carcinoma to manifest from a healthy lung cell is genetic and epigenetic alterations such as DNA point mutations or chromosomal deletions that activate tumor suppressor genes (TSGs). Genetic susceptibility and exposure to carcinogens such as tobacco smoke and asbestos are probable causes for these alterations to occur since active smokers are currently responsible for 90% of NSCLC cases in the United States [11]. In the majority of NSCLC cases, the TSG loss of p53 (a transcription factor), which is used for various cell functions including cell progression, apoptosis, and DNA repair, is not properly functioning [12]. This mutated p53 exhibits reduced tumor suppressing ability, which results in apoptosis not occurring and DNA damage left unregulated.

A further factor in the progression of NSCLC is oncogene activation, which causes an upregulation of epidermal growth factor receptor (EGFR) [13]. This activates MAP and PI3-K-Akt pathways, which are important for resistance to apoptosis and cell proliferation. Moreover, in most NSCLC cases, telomerase activation occurs [14]. Telomerase is an enzyme used to lengthen telomeres, which are terminal repetitive DNA sequences, specifically hexameric TTAGGG repeats. The physiological function of a telomere is to "cap" the end of chromosomes, thus preventing them from unravelling. Telomeres shorten with each cell division that occurs, which consequently limits cell proliferation. The telomere activation expressed causes premalignant cells to progress to malignant cells by prolonged survival of cells via avoidance of programmed cell death. In order for an NSCLC malignancy to metastasize inside the lungs, the tumor requires an extensive blood supply, therefore consequently promoting angiogenesis.

9.3 Current FDA-Approved Therapies for Non-Small Cell Lung Cancer

Presently, there are a wide variety of FDA-approved therapies for NSCLC treatment that can be categorized into four main classes: chemotherapy, targeted therapy, immunotherapy, and radiotherapy. However, radiotherapy is becoming a decreasingly common method of treatment used due to new and emerging therapies being more efficacious [15]. The therapeutic regimen in the United Kingdom is determined via an extensive workup to evaluate the development of the malignancy and to observe different oncogenic driver alterations (biomarkers). In the United Kingdom, the four biomarkers tested are EGFR, anaplastic lymphoma kinase (ALK), c-ros oncogene 1 (ROS1), and programmed death-ligand 1 (PD-L1) [16–18]. However, currently in the European Union (EU), testing is more comprehensive with a variety of different patient biomarkers being utilized.

9.3.1 Chemotherapy

An aggressive systemic treatment method that affects the entire body and is considered a conventional method of treatment for NSCLC. Common chemotherapeutic agents used in the treatment of NSCLC include cisplatin, paclitaxel, and docetaxel. All of which are currently approved by the FDA and EMA for IV infusion delivery. A meta-analysis covering 16 randomized trials in patients with advanced stages of NSCLC displayed that 1 year survival in the chemotherapy arm was 29% versus 20% in patients who only received supportive care [19]. Additionally, another meta-analysis conducted by D'Addario, Pintilie [20] consisting of randomized control trials compared the use of platinum-based versus nonplatinum-based regimens in patients with advanced stages of NSCLC. The analysis showed that cisplatin and carboplatin provided an improved 1-year survival and an improved response rate compared to nonplatinum-based treatments but had higher gastrointestinal toxicity [20].

These studies confirm that both non-platinum and platinum-based therapies are effective in improving survival of patients. However, these cytotoxic and non-selective treatments often lead to vast adverse events due to their non-specific targeting. The body cannot tolerate higher dose-dependent toxicity and the dose of antineoplastic agents required to kill resistant tumors [21]. Consequently, this can result in a greatly reduced patient quality of life due to physical and biological factors, including serious hypertensive reactions, nephrotoxicity, neurotoxicity, therapeutic failure due to required reduced dose, and even patient mortality. Despite these drawbacks, the American Society for Clinical Oncology (ASCO) guidelines recommend a platinum combination therapy for first-line treatment of NSCLC [22]. Additionally, approximately 80% of patients diagnosed with NSCLC will undergo a chemotherapeutic treatment in combination or stand-alone over the period of their illness. Furthermore, in the United Kingdom, during 2013–14, 25.2% of people diagnosed with any stage of NSCLC were currently ongoing a chemotherapy treatment regimen [23].

Targeted and immunotherapies have become an attractive treatment option due to their reduced side effects leading to a better quality of life, improved overall survival rates, and greater efficacy in reducing tumor size. At present, if patients express the required biomarkers, it is usually preferential for them to undergo these treatments regimens.

9.3.2 Targeted Therapy

Under the circumstances of a patient with EGFR mutations or ALK translocation, targeted therapy is usually selected due to its high efficacy in reducing tumor size in comparison to immunotherapy and chemotherapy. Targeted therapies include monoclonal antibodies delivered via IV infusion and small molecule inhibitors delivered in an oral tablet dosage form. A key example of a small molecule inhibitor is gefitinib branded Iressa®. This tyrosine kinase inhibitor was developed by AstraZeneca and was approved in 2015 by the FDA and EMA for first-line targeted therapy for locally advanced or metastatic NSCLC, who show EGFR mutations either exon-19 deletions or exon-21 substitutions [24]. Gefitinib is now one of the leading targeted therapies, which is approved in 91 countries worldwide. However, the route of approval was not direct. Phase III clinical trials INTACT-1 and INTACT-2 help approve the drug back in 2003 [25, 26]. The FDA followed with a request for a confirmatory study. Consequently, this confirmatory study failed to meet its primary objective, thus it was withdrawn from the market. Gefitinib was later FDA approved in 2015 based on the results of phase IV trials named IPASS2 and IFUM1. IFUM1 was a single arm study of gefitinib as a first-line treatment in 106 Caucasian patients who are EGFR mutation positive with advanced NSCLC. The overall response rate was 70% as determined by the investigators and 50% by a Blinded Independent Centre Review [27].

Crizotinib marketed as Xalkori® developed by Pfizer is an ALK and ROS1 inhibitor delivered in an oral tablet dosage form. Currently, crizotinib is the only FDA-approved biomarker therapy that has two molecular targets being ALK and ROS1. A randomized Phase III study named PROFILE-1014 was performed on 343 patients with advanced ALK positive NSCLC who had no prior treatment [28]. This study resulted in a median progression free survival of 10.8 months versus 7.0 months in the pemetrexed-plus-platinum chemotherapy group. A further study, named PROFILE-1001, was a phase I single arm clinical trial, which led to the FDA approval for patients with a ROS1 rearrangement in 2016 [29]. The single arm trial involved

50 patients and demonstrated that 66% of patients witnessed a full or partial response to treatment. Considering that only 1–2% patients exhibited ROS1 rearrangement, this crizotinib opens up possibilities to patients who were previously unable to undergo a selected treatment regimen.

9.3.3 Immunotherapy

In contrast to chemotherapy and targeted therapy, immunotherapy is a modern technique that does not utilize the drug for cancer cell death. Instead, immunotherapies stimulate T-cell response in order to treat and may cure the malignancy. Their mechanism of action includes checkpoint inhibition of the PD-L1 pathway. Cancer cells that produce PD-L1 bind to PD-1 T-cell receptor proteins, which would typically inactivate the T-cells, thus allowing the body to recognize and kill these cancerous cells. However, there is a possibility of T-cells destroying healthy cells due these therapies removing an immune system checkpoint.

Pembrolizumab, branded as Keytruda® and developed by Merck & Co, was FDA approved in 2014 and is currently one of the leading immunotherapies used for both NSCLC and melanoma in patients who express PD-L1. Pembrolizumab is an FDA- and EMA-approved monoclonal antibody (mAb) checkpoint inhibitor for patients with stage III or metastatic NSCLC. The dose regimen is currently an IV infusion for 30 min every 3–6 weeks. A Phase III trial named KEYNOTE-024 concluded that the 5-year OS (overall survival) in patients with advanced NSCLC treated with Pembrolizumab was 31.9% versus 16.3% (95% CI) in the platinum-based chemotherapy group [30]. Thus, nearly doubling 5-year OS—a massive improvement, especially considering that prior to 2014, the 5-year survival rate for advanced stages of NSCLC was under 5%. Furthermore, patients treated with Pembrolizumab had fewer incidence-related grade 3–5 adverse events (31.2% versus 53.3% with platinum-based chemotherapy).

Nivolumab, branded Opdivo® and developed by Bristol Myers Squibb (BMS), was also approved in 2014 for the treatment of NSCLC. Similar to that of Pembrolizumab, Nivolumab also is a monoclonal antibody checkpoint inhibitor, which can be used to treat patients with advanced stages of NSCLC or melanoma who express PD-L1. Additionally, Nivolumab has recently been approved in October 2020 in combination therapy with ipilimumab for mesothelioma [31]. A Phase III trial named CHECKMATE-227 resulted in median duration of overall survival of 17.1 months with nivolumab plus ipilimumab versus 13.9 months with chemotherapy. Additionally, the percentage of patients who witnessed grade 3–5 adverse events was also lower in the nivolumab group (32.8% versus 36.0%) [32].

9.4 Oral Lipid-Based Formulations and Their Importance

Lipid-based formulations offer a new potential outlook on current anticancer treatments. As discussed in Section 9.3.2, targeted therapies showed improved response rates compared to that of chemotherapeutic regimens. However, these therapies are labeled to be taken on an empty stomach; thus, consequently administration in the post-prandial state can cause an increase in area under the plasma-time curve (AUC) and Cmax resulting in potential life-threatening risks such as QTc prolongation [9]. In addition, most patients undergoing NSCLC or other cancer treatment will be on multiple medications; therefore, the ability to take these treatments on an empty stomach can be challenging. Lipid nanocarriers, to a certain extent, mimic the fat content of a meal and, therefore, can reduce the effect of food on anticancer drugs, which can further help patient compliance and increase drug solubilization and absorption [33].

Additionally, as discussed in Section 9.3.1, the majority of patients diagnosed with NSCLC will undergo a chemotherapeutic regimen during their illness and consequently a range of adverse events can take place. Lipid nanocarriers can be utilized to entrap lipophilic anticancer chemotherapeutic drugs such as docetaxel and paclitaxel in order to improve oral bioavailability via uptake into the intestinal lymphatics [4, 5]. This route of administration could lead to an increase in patient compliance and improved quality of life in patients.

9.4.1 Mechanism of Intestinal Lymphatic Absorption

The lymphatic system is composed of vessels, capillaries, ducts, and lymphoid tissues. The system is a one-way street with the main physiological function being to maintain fluid balance in the body by returning excess fluid, proteins, and waste products from the interstitial tissue to the blood circulation [34]. Furthermore, the lymphatic system is an important part of the immunological defense system for the body since B and T lymphocytes are primarily transported around the body in the lymphatics. There are three potential pathways to target drugs to the intestinal lymphatic system. First, the paracellular route with the help of absorption enhancers [35]; the paracellular transport is drug movement across the epithelium via the intercellular spaces between the cells. Second, access via the microfold and gut-associated lymphoid tissue (GALT), which has largely been used for the delivery of particulates [36]. Last, the transcellular route in association with the chylomicron triglyceride core. Therefore, drugs mimicking the composition of chylomicrons such as lipid-based nanocarriers exhibit the same absorption route as that of chylomicrons [37]. The latter (transcellular) route is the predominant method for lipophilic compounds delivered in lipid-based carriers to access the lymphatic system, which will be reviewed in detail.

After the drug and its carrier are absorbed into the enterocyte in the intestinal milieu, it is usually diffused directly across the enterocytes and enters the portal vein leading to systemic circulation by the liver. Consequently, the first-pass metabolism of the drug transpires. This method of absorption has a higher affinity than lymphatic absorption due to 500:1 v/v portal vein versus mesenteric lymph flow [34]. However, lipophilic drugs with a log P > 5 and solubility > 50 mg in long-chain triglyceride lipids [34] have a greater affinity for lymphatic uptake compared to the portal vein. In order for drug absorption into the lymphatic system to occur, a series of steps are involved. These steps involve the smooth endoplasmic reticulum (SER) and rough endoplasmic reticulum (RER). The biological role of the SER is carbohydrate and lipid synthesis, including the synthesis of phospholipids and cholesterol; the SER also serves as a transitional area for transport vehicles. The RER contains attached ribosomes, which synthesize proteins via translation; additionally, the RER manufactures membranes. The SER and RER are connected to each other, so the production of membranes and proteins from the RER can quickly move to the SER to allow transport across the cells.

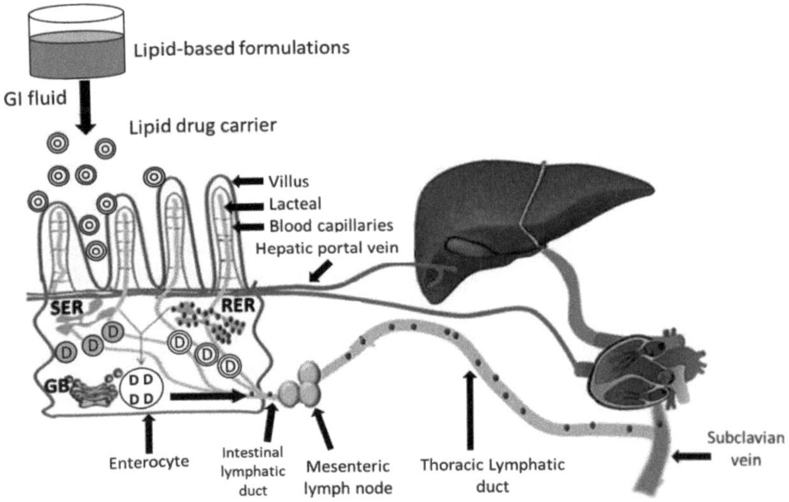

Figure 9.2 Mechanism of action for lipid-based formulations to be absorbed into the lymphatic system [38].

As displayed in Fig. 9.2 by Ahmad et al. [38], the lymphatic uptake of lipid-based drugs via the transcellular route through the enterocyte process mimics the same absorption as that of chylomicrons and because of this, the absorption method of chylomicrons will be stated as follows: First, the formation of primordial lipoprotein particles developing in the RER. Simultaneously in the SER, large triglyceride droplet synthesis occurs. Second, the core expansion phenomenon occurs, which is the fusion of the primordial lipoprotein particle with the large triglyceride-rich droplets [37]. The final step involves the chylomicrons, which are transported to the Golgi body (GB) in order to be processed. After processing, they are exocytosed from the enterocyte. Ensuing secretion by the enterocyte, these colloidal particles are too large for the blood capillaries and are consequently selectively taken up into the mesenteric lymphatic capillaries [39]. After the lymphatic circulation occurs, the chylomicrons or drug molecules enter the systemic circulation via drainage from the thoracic lymph duct into the subclavian vein, thus allowing for therapeutic action to occur [38].

9.4.2 Phospholipid Nanocarriers

Lipid nanocarriers are usually comprised of biologically accepted lipids such as phospholipids [40, 41], which are well accepted by the body due to being the main component of the cellular and subcellular membranes. Phospholipids provide many mandatory physiological functions such as lipoprotein formation to transport cholesterol and triglycerides through the blood. In addition, phospholipids are used as an emulsifier in the gall bladder to increase the absorption of lipophilic compounds, and as surface wetting agents in joints and alveoli of the lungs [42].

Phospholipids can be either derived from natural sources like soybean phosphatidylcholine or synthetically synthesized from glycerophosphocholine (GPC) [43, 44]. Phospholipids are amphiphilic molecules with a hydrophilic head and hydrophobic acyl chains linked to the alcohol [45–47]. When placed into aqueous solution, phospholipids form different structures or carriers based on their chain length, surface charge, and size. Differences in the composition of the hydrophilic head or the hydrophobic acyl chains lead to a large range of different phospholipids.

A variety of different carriers are formed from phospholipids, including self-emulsifying drug-delivery systems (SEDDS) [48], liposomes [49, 50], solid lipid nanoparticles (SLNs) [51], and nanostructured lipid carriers (NLCs) [52]. These lipid carriers can be used as a vehicle to transport and target these lipophilic anticancer drugs to the intestinal lymphatics following administration of an oral dosage form. Some examples of these carriers used to deliver different drugs can be found in Table 9.1. Phospholipid carriers can utilize a wide range of other excipients, which can provide versatility to a variety of drug molecules. Whereby relative percentage and selection of water-insoluble surfactants (HLB <12), water-soluble surfactants (HLB >12), and hydrophilic cosolvents can be used to improve the formulation properties, solubility, and dispersibility [9, 43, 45].

Table 9.1 Various lipophilic drugs and their respective lipid-based carrier systems, formulation components, and bioavailability

Drug	Carrier	Formulation Components	Bioavailability Effect	References
Methotrexate	Solid lipid nanoparticles	Compritol 888 ATO, soya lecithin	24.42%	[54]
Halofantrine	Long-chain triglyceride microemulsion	Soybean oil, Maisine 35-1, Cremophor EL and Ethanol	28.3%	[55]
Cefotaxime	Liposomes	Soybean phosphatidylcholine, glycerol and ethanol	2.7-fold higher versus standard suspension	[56]
Lopinavir	Solid lipid nanoparticles	Compritol 888 ATO, Pluronic F 127	2.13-fold higher versus standard suspension	[57]
Raloxifene	Solid lipid nanoparticles	Glyceryl tribehenate (GB) ($C_{69}H_{134}O_6$, molecular weight 1059.8), Kolliphor P407 (poloxamer 407)	3.24-fold increase in oral bioavailability	[58]
Doxorubicin	Nanostructured lipid carriers	Soybean oil, glyceryl monostearate, soya lecithin, Distearoyl phosphatidylethanolamine (DSPE-PEG) and Polyethylene glycol 40 stearate	2–3 times higher relative bioavailability	[59]
Sulforaphane	Nanostructured lipid carriers	Compritol 888 ATO, Vitamin E, Precirol ATO 5, Labrasol, Capryol90, Stearic acid, glyceryl monostearate, poloxamer 188, Tween20, Tween-80 and canola oil	5.04-fold increase in relative oral bioavailability	[60]
Paclitaxel	Liposomes Proliposomes	Soya phosphatidylcholine, Hydrogenated soya phosphatidylcholine, Dimyristoly phosphatidylcholine	—	[4]

Lipid-based formulations can be used to overcome many of the drawbacks of conventional dosage forms when targeting the intestinal lymphatic system. A study showed that following uptake into the intestinal lymphatic system, lipid-based formulations provided increased drug accumulation at target sites, thus lowering toxicity for the healthy cells. Also, if desired, specific tissues or organs such as the brain and liver can be targeted due to altered and improved pharmacokinetics and biodistribution via either active or passive targeting [53]. Furthermore, these carriers have enhanced cellular penetration and the ability to protect the drug from gastric and enzyme degradation, which occur in the stomach and the small intestine, resulting in a lower required dose of drug for therapeutic action to occur. Thus, reducing the potential of dose-dependent toxicity from arising in the patient.

A study conducted by Khoo et al. [55] has shown an increase in the bioavailability of lipophilic drugs when co-administered with lipids. The study in dogs showed 1.3% lymphatic transport of the administered dose of halofantrine in conventional form versus 27% when administered with a single capsule consisting of long-chain lipids [55]. In addition, another study showed that there is enhanced drug absorption of halofantrine in the post-prandial state where 53.9% of the administered dose reached lymphatic transport [61]. Whether the lipids are derived from food or the formulation itself, co-administration with lipids enhances drug solubilization and dissolution, and increases lipoprotein formation, resulting in enhanced lymphatic pathway targeting. Since drug access to the intestinal lymphatics is dependent on the association with developing lipoproteins in the enterocyte, careful consideration of the selected carrier to target the lymphatics must be undertaken.

9.4.2.1 Self-emulsifying drug-delivery systems

SEDDS consist of oil, water, surfactant, and cosurfactant, which rapidly self-emulsify in the stomach to form a fine oil-in-water emulsion (o/w). SEDDS can be further subcategorized into self-micro emulsifying systems (SMEDDS) and self-nano emulsifying systems (SNEDDS). This categorization is determined via the formed particle size, with SMEDDS having an oil particle size ranging from 100 to 200 nm and SNEDDS having a particle size <100 nm [62].

SEDDS can dissolve a large number of lipophilic drugs and P-gp inhibitors, which assists in intestinal lymphatic uptake [63]. Due to this, research has been conducted utilizing these vehicles for several lipophilic anticancer drugs such as paclitaxel [64]. A study done by Lo et al. [64] investigated nonionic self-emulsifying oil-in-water (o/w) formulations as drug-delivery systems for paclitaxel [64]. The study revealed that the optimal formulation contained oils, phosphatidylcholine, and Tween 80, which led to 500 ppm of paclitaxel being solubilized. Furthermore, the cytotoxicity effects were negligible, thus showing a strong biocompatibility profile and consequently promising applications in anticancer therapies.

Factors such as lipid selection need to be considered when administering a drug as a SEDDS. The fatty acid chain length plays a vital role in the efficacy of lymphatic uptake for SEDDS. Literature suggests that long-chain fatty acids with a hydrocarbon chain length greater than 14 have higher affinity to lymphatic absorption in contrast to short chain fatty acids [65], which have a greater water solubility, thus diffusing across the enterocyte and enter the portal vein. Furthermore, mono- and poly-unsaturated lipids produce larger lipoproteins in the enterocyte, resulting in greater uptake into the lymphatic system in comparison with saturated fats. Though not for anticancer treatment, the anti-psychotic and anti-emetic drug chlorpromazine exhibits poor aqueous solubility and high first-pass metabolism. However, a study found that when delivered as a long-chain triglyceride with a hydrocarbon chain length greater than 14 SNEDDS, chlorpromazine showed a sixfold increase in bioavailability compared to that of standard drug suspension [66].

9.4.2.2 Solid lipid nanoparticles (SLNs)

Solid lipid nanoparticles (SLNs) are 50–1000 nm in size, dependent on the manufacturing method, and are composed of biologically accepted lipids that remain solid in the body and at room temperature. SLNs can be composed of lipid acids, waxes, mono-, di-, or triglycerides, which can be further stabilized with biocompatible ionic or nonionic surfactants such as Pluronic F68 [38, 67]. Most commonly, hydrogenated cottonseed oil or hydrogenated soybean oil is the selected lipid for SLNs. These carriers stimulate the formation of chylomicrons in the enterocyte, thus increasing the absorption of the drug into the lymphatic system. Furthermore, similar to that of liposomes [68, 69], SLNs allow controlled drug release, increased drug stability, and protection of drug molecules labile to gastric degradation following oral administration of lipophilic drugs. The main advantage of SLNs is owed to their good storage stability, and some of the drawbacks of these carriers are due to poor relative drug-loading capacity and potential for expulsion of the encapsulated drug during storage [52, 67].

Investigations have been conducted using SLNs in order to improve the oral absorption of various anticancer drugs using surfaced-modified agents. A study conducted by Cho et al. [70] showed SLNs surface modification with D-alpha-tocopheryl polyethylene glycol (PEG) succinate (TPGS 1000), which led to an improved intestinal absorption and relative oral bioavailability in rats of docetaxel compared to that of Taxotere®—a standard IV docetaxel formulation. Another study in rats by Baek et al. [71] demonstrated an increase in the solubility of the surface-modified paclitaxel via strengthening of the carrier with hydroxypropyl-β-cyclodextrin. This led to a 15- to 17-fold higher solubility when compared to standard paclitaxel suspension. Moreover, the study also found that the surface-modified SLNs resulted in a greater drug concentration in the lymph nodes of 11.12 ng/mg versus 0.89 ng/mg in standard solution. This surface modification shows auspicious potential for greater affinity to intestinal lymphatics, which is promising for anticancer drugs. However, despite the large increase in drug concentration in the lymph, it consequently led to an increase in cytotoxicity. The relation between surface functionalization, drug concentration, and cytotoxicity needs to be investigated further to find an optimal dosage for an ideal therapeutic window with adequate therapeutic effect while minimizing cytotoxicity.

Cationic SLNs have been explored and shown to be suitable non-viral vectors for gene/drug delivery of anticancer drugs, since they can bind DNA via ionic interactions, which can facilitate gene transfection. The ability for adequate transfection has been improved through the use of surface functionalization moieties such as poly(styrene-4 sodium sulfonate) and poly(L-lysine hydrochloride) [72]. A study by Choi et al. [73] discovered that cationic SLNs can enhance p53 gene transfer to lung cancer cells, which could have auspicious prospects in a clinical application due to the effective induction of apoptosis and tumor growth inhibition.

Some examples of SLNs utilized for the delivery of low orally bioavailable drugs include the antiviral drug Lopinavir used for the treatment of HIV. It has been delivered as SLNs, which has helped to overcome obstacles associated with this drug previously. The problems include poor bioavailability, poor solubility, and extensive first-pass metabolism. A study found that the cumulative percentage dose of Lopinavir secreted into the lymph was 4.91-fold higher when compared with a conventional drug solution, resulting in a 2.13-fold higher bioavailability [57]. Additionally, all-trans retinoic acid (ATRA) has been used as a model poorly water-soluble drug for SLNs. It was determined that relative oral bioavailability in male rats was significantly improved compared with standard ATRA suspension. The study also indicated that surfactants and emulsifiers had a significant positive effect on oral absorption. Furthermore, ATRA is sensitive to light, heat, and oxidants and is easily degraded. However, when incorporated into SLNs after 3 months of storage at 4°C, more than 90% of the ATRA drug molecules in SLN remained chemically intact versus 50% degradation when stored in methanol solution [74].

9.4.2.3 Nanostructured lipid carriers (NLCs)

Recently, scientists have developed a second generation of SLNs named nanostructured lipid carriers (NLCs). The difference is due to the lipidic phase containing both solid lipid and liquid lipid at room temperature [45, 52]. These second-generation nanocarriers provide increased loading capacity and improved controlled drug release compared to that of standard SLNs. NLCs differ from SLNs in that the NLCs have been used to improve the bioavailability of various anticancer drugs, including docetaxel [75] and etoposide [76], both of which are chemotherapeutic medicines. Docetaxel exhibits very low drug-loading capacity into NLCs, which hinders its clinical application. A study by Sun et al. [77] employed the use of a novel lipophilic oleate prodrug of docetaxel encapsulated in an NLC [77]. The study showed a marked improvement in the drug-loading capacity of docetaxel in NLCs with an approximate fivefold increase (23% w/w versus 5.5% w/w). Moreover, colloidal stability and slower release properties were exhibited by this NLCs owing to a 4.04-fold increase in oral bioavailability compared to standard docetaxel solution. This research proposes that prodrug-NLCs are viable carriers to enable oral delivery of docetaxel.

Research surrounding NLCs and their ability to improve the oral bioavailability of poorly bioavailable drugs has been undertaken. A study where the incorporation of baicalin in NLCs showed enhanced drug-loading capacity as well as improved sustained release profiles [78]. Baicalin NLCs exhibited prolonged mean residence time and increased AUC, which led to a considerable increase in oral bioavailability in contrast to pure baicalin. NLCs have also shown promising signs for site-specific targeting via conjugating the receptors' ligands on the surface of the NLCs. A study employed by Tian et al. [79] investigated the effect of taurocholic acid as a ligand to improve the oral bioavailability of curcumin in rats, which discovered a 5- to 15-fold increase in AUC depending on the degree of modification, which is mediated by a bile-acid transporter.

9.4.2.4 Liposomes

Liposomes are composed of a phospholipid bilayer that has a hydrophilic core and a lipophilic bilayer region [40, 80]. So the drugs which are lipophilic in nature will come and interact/entrapped in the bilayers (e.g., paclitaxel) and the drug which are hydrophilic in nature will tend to encapsulate in the hydrophilic core (e.g., cefotaxime) of liposomes, and can be delivered for interstitial lymphatic uptake [38]. A study found that the oral bioavailability of cefotaxime administered in a liposomal formulation was 2.7-fold higher than the drug administered in aqueous solution [56].

Liposomes also offer enhanced permeability across the enterocyte, increased drug stability, decreased drug toxicity, and allow controlled release to occur. Additionally, the incorporation of cholesterol into liposomes can help reduce the permeability of water-soluble molecules through the liposomal membrane, decrease fluidity, and increase rigidity, which can improve stability in the blood or plasma [69, 81]. Dependent on the production technique and the phospholipid properties, liposomes can be either multilamellar or unilamellar. Multilamellar liposomes are advantageous when encapsulating lipophilic drugs. This is due to the increased surface area of lipid bilayers where the lipophilic drug is logged in.

The use of liposomes as a delivery system for chemotherapeutic agents has been investigated for some time with the first nano-scaled delivery system for cancer therapy being liposomal doxorubicin hydrochloride (marketed as Caelyx® in Europe), an I/V infusion, which received approval in 1995 for Kaposi's sarcoma [82]. Currently there are a variety of I/V liposomal-based chemotherapeutic drugs approved by the FDA and EMA, including Vincristine sulfate marketed as Marquibo® for acute lymphoblastic leukemia [83] and cytarabine branded DepoCyte® for lymphomatous meningitis [84]. Additionally, some liposomal chemotherapeutic formulations are still in different stages of development for NSCLC, including Lipoplatin, a cisplatin alternative currently in Phase III. A study showed that in 202 patients with advanced stages of NSCLC, a partial response was achieved by 59.22% of Arm A patients who received liposomal cisplatin combined with paclitaxel, versus 42.42% in comparison to Arm B patients with cisplatin in combination with paclitaxel. Furthermore, the median survival time was 10 months for Arm A versus 8 months for Arm B with the number of patient survivals in Arm A doubling after 18 months [85].

9.4.2.5 Lipid–polymer hybrid systems (LPHS)

Lipid-based nanocarriers, though advantageous for improved entrapment efficiency and cost-effective manufacturing, also have limitations such as limited surface functionalization, burst release of the drug, and high polydispersity. Polymer-based systems, including polymeric nanoparticles, polymeric micelles, and polymer-drug conjugates, offer improved stability, tailored drug release profile and better control over surface functionalization [86]. However, polymer-based systems have limited drug-loading capacities, require the use of organic solvents to manufacture, and the potential for toxicity due to polymer degradation.

More recently, systematic hybridization of these systems has resulted in a lipid–polymer hybrid system (LPHS). This drug-delivery system has improved loading capacity, which is well accepted by the body from the lipid-nanocarrier portion but also structural benefits that polymer-based systems provide, including improved rigidity and tailored release profile. LPHS consists of an inner polymer layer such as sodium alginate or a chitosan derivative, which entraps the desired drug; this layer is surrounded by phospholipids namely 1,2-dioleoyl-*sn*-glycero-3-phosphoethanolamine (DOPE) and 1,2-dioleoyl-*sn*-glycero-3-phosphocholine (DOPC) or cationic lipids such as 1,2-dioleoyl-3-trimethylammonium propane (DOTMA) and 1,2-dioleoyl-3-trimethylammonium propane (DOTAP). The outer layer of lipid is coated with PEG and is stabilized with the use of surfactants such as sodium cholate or cetyltrimthylammonium bromide [86–88].

For the use of anticancer therapy and more specifically, NSCLC therapy, LPHS provides a new outlook, which should be investigated further. These vesicles not only have the ability to entrap these lipophilic drugs, but they can also attach various ligands onto the surface of the vesicle. For example, vascular endothelial growth factor (VEGF) used in the promotion of angiogenesis in the lungs and tyrosine kinase inhibitors, which are the therapeutic targets of current targeted therapies for NSCLC, as discussed in Section 9.3.2 [89]. Furthermore, due to stability and biocompatibility properties, LPHS also displays promising signs to be viable non-viral vector agents for gene delivery of DNA, mRNA, siRNA, miRNA, and diagnostic agent delivery, which offer diminished toxicity and immunogenicity, deprivation from viral recombination, with the capability and feasibility of large-scale production [86, 90, 91]. Some of the examples of the application of gene delivery via LPHS are given in Table 9.2.

Table 9.2 Different genes delivered via LPH systems for specific cell types and their formulation constituents

Gene	Polymer	Lipid	Cell Type	References
DNA	Poly-ethleneimine	PEG-DSPE, triolein, phosphatidylcholine (PC)	Breast cancer cells	[92]
DNA	Poly(lactic-*co*-gylcolic acid)	DOTAP/[*N*-(*N*0,*N*0-dimethylaminoethane)-carbamoyl] cholesterol (Dc-Chol)	Prostate cancer cells	[93]
mRNA	Poly(β-amino ester) (PBAE)	PEG-DSPE, DOTAP	Dendritic cells	[94]

9.4.3 Future of Oral Dosage Forms for Anticancer Therapy and Lipid-Based Drug-Delivery Systems

The pursuit to find oral dosage forms of chemotherapeutic agents is currently underway. Companies such as Athenex, in February 2021, sought FDA approval for oral paclitaxel with Encequidar (a P-glycoprotein pump inhibitor for metastatic breast cancer). This would be the first oral taxane dosage form regimen. However, it received a complete response letter from the FDA with the request for the trial to be conducted with a sample, which is representative of the population in the United States. Despite this setback, the Phase III trial of 402 patients showed promising results with patients receiving oral paclitaxel exhibiting an overall response rate of 40.4% versus 25.6% in I/V paclitaxel group [95]. Moreover, the overall survival was a median of 27.9 months versus 16.9 months in oral paclitaxel regimen and I/V paclitaxel, respectively. The oral paclitaxel group also showed much lower neuropathic adverse events of 17% versus 57% in the I/V paclitaxel group. This company also has oral dosage drug candidates for different cancers, which are in various stages of clinical development such as oral docetaxel for solid tumor treatment and oral paclitaxel for angiosarcoma.

Over the past 10 years, lipid-based drug-delivery systems have progressed at an exponential rate with just recently the COVID-19 vaccine being delivered via lipid nanoparticles, which was a major breakthrough during the global pandemic. In January 2021, Sarepta therapeutics, a leader in precision genetic treatments for rare diseases announced a collaboration with Genevant, a leading nucleic acid delivery company, and has the most expansive lipid nanoparticle patents. The collaboration will focus on lipid nanoparticle delivery to four different neuromuscular indications, including Duchenne muscular dystrophy.

9.5 Conclusion

There are many challenges associated with current anticancer therapies, specifically chemotherapy. A range of potential life-threatening adverse events can transpire during treatment. Despite this, a majority of patients with NSCLC will undergo a chemotherapeutic regimen. Lipid-based drug-delivery systems have been investigated and shown to increase the oral bioavailability of lipophilic compounds such as many chemotherapeutic drugs, including paclitaxel and docetaxel. This is achieved via uptake into the intestinal lymphatics following drug entrapment into various lipid nanocarriers such as SNEDDS, SLNs, and NLCs. This route of administration could greatly increase patient compliance and reduce hospitalizations. Furthermore, LPHS showed promising anticancer clinical applications considering their ability to attach various ligands. Nonetheless, further research is required to find an optimal balance between the relation of oral bioavailability, drug concentration, cytotoxicity, and surface functionalization.

References

1. Ferlay, J., Soerjomataram, I., Dikshit, R., Eser, S., Mathers, C., Rebelo, M., Parkin, D. M., Forman, D., and Bray, F. (2015). Cancer incidence and mortality worldwide: Sources, methods and major patterns in GLOBOCAN 2012, *Int J Cancer*, **5**, pp. E359–E386.
2. Jemal, A., Bray, F., Center, M. M., Ferlay, J., Ward, E., and Forman, D. (2011). Global cancer statistics, *CA Cancer J Clin*, **2**, pp. 69–90.
3. Panchagnula, R. (1998). Pharmaceutical aspects of paclitaxel, *Int J Pharm*, **1–2**, pp. 1–15.
4. Khan, I., Lau, K., Bnyan, R., Houacine, C., Roberts, M., Isreb, A., Elhissi, A., and Yousaf, S. (2020). A facile and novel approach to manufacture paclitaxel-loaded proliposome tablet formulations of micro or nano vesicles for nebulization, *Pharm Res*, **6**, p. 116.
5. Khan, I., Apostolou, M., Bnyan, R., Houacine, C., Elhissi, A., and Yousaf, S. S. (2020). Paclitaxel-loaded micro or nano transfersome formulation into novel tablets for pulmonary drug delivery via nebulization, *Int J Pharm*, p. 118919.
6. Lipinski, C. A., Lombardo, F., Dominy, B. W., and Feeney, P. J. (2001). Experimental and computational approaches to estimate solubility and permeability in drug discovery and development settings, *Adv Drug Deliv Rev*, **1–3**, pp. 3–26.
7. Crew, M. (2014). Bioavailability enhancement: Analysis of the historical use of solubilization technologies, *Drug Dev Deliv*, pp. 22–31.
8. Wishart, D. S., Knox, C., Guo, A. C., Shrivastava, S., Hassanali, M., Stothard, P., Chang, Z., and Woolsey, J. (2006). DrugBank: A comprehensive

resource for in silico drug discovery and exploration, *Nucleic Acids Res*, Database issue, pp. D668–D672.

9. Savla, R., Browne, J., Plassat, V., Wasan, K. M., and Wasan, E. K. (2017). Review and analysis of FDA approved drugs using lipid-based formulations, *Drug Dev Ind Pharm*, **11**, pp. 1743–1758.
10. Dahan, A. and Hoffman, A. (2008). Rationalizing the selection of oral lipid based drug delivery systems by an in vitro dynamic lipolysis model for improved oral bioavailability of poorly water soluble drugs, *J Control Release*, **1**, pp. 1–10.
11. Khazaei, S., Mohammadian-Hafshejani, A., and Salehiniya, H. (2016). Role of smoking in lung cancer in United States, *Iran J Public Health*, **9**, pp. 1245–1246.
12. Ahrendt, S. A., Hu, Y., Buta, M., McDermott, M. P., Benoit, N., Yang, S. C., Wu, L., and Sidransky, D. (2003). p53 mutations and survival in stage I non-small-cell lung cancer: Results of a prospective study, *J Natl Cancer Inst*, **13**, pp. 961–970.
13. Midha, A., Dearden, S., and McCormack, R. (2015). EGFR mutation incidence in non-small-cell lung cancer of adenocarcinoma histology: A systematic review and global map by ethnicity (mutMapII), *Am J Cancer Res*, **9**, pp. 2892–2911.
14. Fernández-Marcelo, T., Gómez, A., Pascua, I., de Juan, C., Head, J., Hernando, F., Jarabo, J. R., Calatayud, J., Torres-García, A. J., and Iniesta, P. (2015). Telomere length and telomerase activity in non-small cell lung cancer prognosis: Clinical usefulness of a specific telomere status, *J Exp Clin Cancer Res*, **1**, p. 78.
15. Men, Y., Wang, L., Zhang, Y., Gao, S., Li, J., Wu, N., Yang, B., Liu, S., Ren, J., Huang, Y., Wang, D., Liao, X., Xing, X., Du, L., Yang, L., Liu, Y., Zhang, Y., Wei, D., Liu, Y., Zhang, K., Qiao, Y., Shi, J., Chen, W., Dai, M., and Hui, Z. (2019). Trends of postoperative radiotherapy for completely resected non-small cell lung cancer in China: A hospital-based multicenter 10-year (2005-2014) Retrospective Clinical Epidemiological Study, *Front Oncol*, p. 786.
16. Yates, L. R., Seoane, J., Le Tourneau, C., Siu, L.L., Marais, R., Michiels, S., Soria, J. C., Campbell, P., Normanno, N., Scarpa, A., Reis-Filho, J. S., Rodon, J., Swanton, C., and Andre, F. (2018). The European Society for Medical Oncology (ESMO) Precision Medicine Glossary, *Ann Oncol*, **1**, pp. 30–35.
17. Smolle, E., Leithner, K., and Olschewski, H. (2020). Oncogene addiction and tumor mutational burden in non-small-cell lung cancer: Clinical significance and limitations, *Thorac Cancer*, **2**, pp. 205–215.

18. Bodor, J. N., Boumber, Y., and Borghaei, H. (2020). Biomarkers for immune checkpoint inhibition in non-small cell lung cancer (NSCLC), *Cancer*, **2**, pp. 260–270.

19. Marino, P., Pampallona, S., Preatoni, A., Cantoni, A., and Invernizzi, F. (1994). Chemotherapy vs supportive care in advanced non-small-cell lung cancer. Results of a meta-analysis of the literature, *Chest*, **3**, pp. 861–865.

20. D'Addario, G., Pintilie, M., Leighl, N. B., Feld, R., Cerny, T., and Shepherd, F. A. (2005). Platinum-based versus non-platinum-based chemotherapy in advanced non-small-cell lung cancer: A meta-analysis of the published literature, *J Clin Oncol*, **13**, pp. 2926–2936.

21. Devlin, E. J., Denson, L. A., and Whitford, H. S. (2017). Cancer treatment side effects: A meta-analysis of the relationship between response expectancies and experience, *J Pain Symptom Manage*, **2**, pp. 245–258. e2.

22. Leighl, N. B., Rekhtman, N., Biermann, W. A., Huang, J., Mino-Kenudson, M., Ramalingam, S. S., West, H., Whitlock, S., and Somerfield, M. R. (2014). Molecular testing for selection of patients with lung cancer for epidermal growth factor receptor and anaplastic lymphoma kinase tyrosine kinase inhibitors: American Society of Clinical Oncology endorsement of the College of American Pathologists/International Association for the study of lung cancer/association for molecular pathology guideline, *J Clin Oncol*, **32**, pp. 3673–3679.

23. NCRAS. (2017). Chemotherapy, Radiotherapy and Tumour Resections in England: 2013-2014. National Cancer Registration & Analysis Service and Cancer Research UK: workbook, London.

24. Kazandjian, D., Blumenthal, G. M., Yuan, W., He, K., Keegan, P., and Pazdur, R. (2016). FDA approval of Gefitinib for the treatment of patients with metastatic EGFR mutation-positive non-small cell lung cancer, *Clin Cancer Res*, **6**, pp. 1307–1312.

25. Giaccone, G., Herbst, R. S., Manegold, C., Scagliotti, G., Rosell, R., Miller, V., Natale, R. B., Schiller, J. H., Von Pawel, J., Pluzanska, A., Gatzemeier, U., Grous, J., Ochs, J. S., Averbuch, S. D., Wolf, M. K., Rennie, P., Fandi, A., and Johnson, D. H. (2004). Gefitinib in combination with gemcitabine and cisplatin in advanced non-small-cell lung cancer: A phase III trial--INTACT 1, *J Clin Oncol*, **5**, pp. 777–784.

26. Herbst, R. S., Giaccone, G., Schiller, J. H., Natale, R. B., Miller, V., Manegold, C., Scagliotti, G., Rosell, R., Oliff, I., Reeves, J. A., Wolf, M. K., Krebs, A. D., Averbuch, S. D., Ochs, J. S., Grous, J., Fandi, A., and Johnson, D. H. (2004). Gefitinib in combination with paclitaxel and carboplatin in advanced

non-small-cell lung cancer: A phase III trial--INTACT 2, *J Clin Oncol*, **5**, pp. 785–794.

27. Douillard, J. Y., Ostoros, G., Cobo, M., Ciuleanu, T., McCormack, R., Webster, A., and Milenkova, T. (2014). First-line gefitinib in Caucasian EGFR mutation-positive NSCLC patients: A phase-IV, open-label, single-arm study, *Br J Cancer*, **1**, pp. 55–62.

28. Solomon, B. J., Mok, T., Kim, D. W., Wu, Y. L., Nakagawa, K., Mekhail, T., Felip, E., Cappuzzo, F., Paolini, J., Usari, T., Iyer, S., Reisman, A., Wilner, K. D., Tursi, J., and Blackhall, F. (2014). First-line crizotinib versus chemotherapy in ALK-positive lung cancer, *N Engl J Med*, **23**, pp. 2167–2177.

29. Shaw, A. T., Riely, G. J., Bang, Y. J., Kim, D. W., Camidge, D. R., Solomon, B. J., Varella-Garcia, M., Iafrate, A. J., Shapiro, G. I., Usari, T., Wang, S. C., Wilner, K. D., Clark, J. W., and Ou, S. I. (2019). Crizotinib in ROS1-rearranged advanced non-small-cell lung cancer (NSCLC): Updated results, including overall survival, from PROFILE 1001, *Ann Oncol*, **7**, pp. 1121–1126.

30. Reck, M., Rodríguez-Abreu, D., Robinson, A. G., Hui, R., Csőszi, T., Fülöp, A., Gottfried, M., Peled, N., Tafreshi, A., Cuffe, S., O'Brien, M., Rao, S., Hotta, K., Leiby, M. A., Lubiniecki, G. M., Shentu, Y., Rangwala, R., and Brahmer, J. R. (2016). Pembrolizumab versus chemotherapy for PD-L1-positive non-small-cell lung cancer, *N Engl J Med*, **19**, pp. 1823–1833.

31. Baas, P., Scherpereel, A., Nowak, A. K., Fujimoto, N., Peters, S., Tsao, A. S., Mansfield, A. S., Popat, S., Jahan, T., Antonia, S., Oulkhouir, Y., Bautista, Y., Cornelissen, R., Greillier, L., Grossi, F., Kowalski, D., Rodríguez-Cid, J., Aanur, P., Oukessou, A., Baudelet, C., and Zalcman, G. (2021). First-line nivolumab plus ipilimumab in unresectable malignant pleural mesothelioma (CheckMate 743): A multicentre, randomised, open-label, phase 3 trial, *Lancet*, **10272**, pp. 375–386.

32. Hellmann, M. D., Paz-Ares, L., Bernabe Caro, R., Zurawski, B., Kim, S. W., Carcereny Costa, E., Park, K., Alexandru, A., Lupinacci, L., de la Mora Jimenez, E., Sakai, H., Albert, I., Vergnenegre, A., Peters, S., Syrigos, K., Barlesi, F., Reck, M., Borghaei, H., Brahmer, J. R., O'Byrne, K. J., Geese, W. J., Bhagavatheeswaran, P., Rabindran, S. K., Kasinathan, R. S., Nathan, F. E., and Ramalingam, S. S. (2019). Nivolumab plus Ipilimumab in advanced non-small-cell lung cancer, *N Engl J Med*, **21**, pp. 2020–2031.

33. Cerpnjak, K., Zvonar, A., Gašperlin, M., and Vrečer, F. (2013). Lipid-based systems as a promising approach for enhancing the bioavailability of poorly water-soluble drugs, *Acta Pharm*, **4**, pp. 427–445.

34. Porter, C. J., Trevaskis, N. L., and Charman, W. N. (2007). Lipids and lipid-based formulations: Optimizing the oral delivery of lipophilic drugs, *Nat Rev Drug Discov*, **3**, pp. 231–248.

35. Hiroshi, Y., Shozo, M., Chiharu, K., and Hitoshi, S. (1981). Bifunctional delivery system for selective transfer of bleomycin into lymphatics via enteral route, *Int J Pharm*, **4**, pp. 291–302.

36. Hawley, A. E., Davis, S. S., and Illum, L. (1995). Targeting of colloids to lymph nodes: Influence of lymphatic physiology and colloidal characteristics, *Adv Drug Deliv Rev*, **1**, pp. 129–148.

37. Hussain, M. M. (2000). A proposed model for the assembly of chylomicrons, *Atherosclerosis*, **1**, pp. 1–15.

38. Ahmad, J., Kohli, K., Mir, S., and Amin, S. (2013). Lipid based nanocarriers for oral delivery of cancer chemotherapeutics: an insight in the intestinal lymphatic transport, *Drug Del Lett*, **1**, pp. 38–46.

39. O'Driscoll, C. M. (2002). Lipid-based formulations for intestinal lymphatic delivery, *Eur J Pharm Sci*, **5**, pp. 405–415.

40. Khan, I., Elhissi, A., Shah, M., Alhnan, M. A., and Ahmed, W. (20132013). Liposome-based carrier systems and devices used for pulmonary drug delivery, in: *Biomaterials and Medical Tribology*, J. P. Davim (Ed.), Woodhead Publishing, pp. 395–443.

41. Khan, I., Yousaf, S., Subramanian, S., Alhnan, M. A., Ahmed, W., and Elhissi, A. (2018). Proliposome powders for the generation of liposomes: The influence of carbohydrate carrier and separation conditions on crystallinity and entrapment of a model antiasthma steroid, *AAPS PharmSciTech*, **1**, pp. 262–274.

42. Nagao, K. and Yanagita, T. (2002). Physiological functions of phospholipids, *Oleoscience*, **3**, pp. 129–135.

43. Bnyan, R., Khan, I., Ehtezazi, T., Saleem, I., Gordon, S., O'Neill, F., and Roberts, M. (2018). Surfactant effects on lipid-based vesicles properties, *J Pharm Sci*, **5**, pp. 1237–1246.

44. Bnyan, R., Khan, I., Ehtezazi, T., Saleem, I., Gordon, S., O'Neill, F., and Roberts, M. (2019). Formulation and optimisation of novel transfersomes for sustained release of local anaesthetic, *J Pharm Pharmacol*, **10**, pp. 1508–1519.

45. Apostolou, M., Assi, S., Fatokun, A. A., and Khan, I. (2021). The effects of solid and liquid lipids on the physicochemical properties of nanostructured lipid carriers, *J Pharm Sci*, **110**, pp. 2859–2872.

46. Bnyan, R., Cesarini, L., Khan, I., Roberts, M., and Ehtezazi, T. (2020). The effect of ethanol evaporation on the properties of inkjet produced liposomes, *DARU J Pharm Sci*, **1**, pp. 271–280.

47. Subramanian, S., Khan, I., Korale, O., Alhnan, M. A., Ahmed, W., Najlah, M., Taylor, K. M. G., and Elhissi, A. (2016). A simple approach to predict the stability of phospholipid vesicles to nebulization without performing aerosolization studies, *Int J Pharm*, **1–2**, pp. 18–27.

48. Rani, S., Rana, R., Saraogi, G. K., Kumar, V., and Gupta, U. (2019). Self-emulsifying oral lipid drug delivery systems: Advances and challenges, *AAPS PharmSciTech*, **3**, pp. 129.

49. Gala, R. P., Khan, I., Elhissi, A. M., and Alhnan, M. A. (2015). A comprehensive production method of self-cryoprotected nano-liposome powders, *Int J Pharm*, **1–2**, pp. 153–158.

50. Khan, I., Yousaf, S., Najlah, M., Ahmed, W., and Elhissi, A. (2021). Proliposome powder or tablets for generating inhalable liposomes using a medical nebulizer, *J Pharm Investig*, **1**, pp. 61–73.

51. Scioli Montoto, S., Muraca, G., and Ruiz, M. E. (2020). Solid lipid nanoparticles for drug delivery: Pharmacological and biopharmaceutical aspects, *Frontiers in Molecular Biosciences*, **319**.

52. Khan, I., Hussein, S., Houacine, C., Khan Sadozai, S., Islam, Y., Bnyan, R., Elhissi, A., and Yousaf, S. (2021). Fabrication, characterization and optimization of nanostructured lipid carrier formulations using Beclomethasone dipropionate for pulmonary drug delivery via medical nebulizers, *Int J Pharm*, p. 120376.

53. Kuentz, M. (2012). Lipid-based formulations for oral delivery of lipophilic drugs, *Drug Discov Today Technol*, **2**, pp. e71–e174.

54. Paliwal, R., Rai, S., Vaidya, B., Khatri, K., Goyal, A. K., Mishra, N., Mehta, A., and Vyas, S. P. (2009). Effect of lipid core material on characteristics of solid lipid nanoparticles designed for oral lymphatic delivery, *Nanomedicine*, **2**, pp. 184–191.

55. Khoo, S. M., Shackleford, D. M., Porter, C. J., Edwards, G. A., and Charman, W. N. (2003). Intestinal lymphatic transport of halofantrine occurs after oral administration of a unit-dose lipid-based formulation to fasted dogs, *Pharm Res*, **9**, pp. 1460–1465.

56. Ling, S. S., Magosso, E., Khan, N. A., Yuen, K. H., and Barker, S. A. (2006). Enhanced oral bioavailability and intestinal lymphatic transport of a hydrophilic drug using liposomes, *Drug Dev Ind Pharm*, **3**, pp. 335–345.

57. Aji Alex, M. R., Chacko, A. J., Jose, S., and Souto, E. B. (2011). Lopinavir loaded solid lipid nanoparticles (SLN) for intestinal lymphatic targeting, *Eur J Pharm Sci*, **1–2**, pp. 11–18.

58. Ravi, P. R., Aditya, N., Kathuria, H., Malekar, S., and Vats, R. (2014). Lipid nanoparticles for oral delivery of raloxifene: optimization, stability,

in vivo evaluation and uptake mechanism, *Eur J Pharm Biopharm*, **1**, pp. 114–124.

59. Zhang, H. W., Dang, Q., Zhang, Z. W., and Wu, F. S. (2017). Development, characterization and evaluation of doxorubicin nanostructured lipid carriers for prostate cancer, *J buon*, **1**, pp. 102–111.

60. Soni, K., Rizwanullah, M., and Kohli, K. (2018). Development and optimization of sulforaphane-loaded nanostructured lipid carriers by the Box-Behnken design for improved oral efficacy against cancer: In vitro, ex vivo and in vivo assessments, *Artif Cells Nanomed Biotechnol*, **sup1**, pp. 15–31.

61. Khoo, S. M., Edwards, G. A., Porter, C. J., and Charman, W. N. (2001). A conscious dog model for assessing the absorption, enterocyte-based metabolism, and intestinal lymphatic transport of halofantrine, *J Pharm Sci*, **10**, pp. 1599–1607.

62. Zanchetta, B., Chaud, M., and Santana, M. (2015). Self-emulsifying drug delivery systems (SEDDS) in pharmaceutical development, *J Adv Chem Eng*, **3**, p. 1000130.

63. Gursoy, R. N. and Benita, S. (2004). Self-emulsifying drug delivery systems (SEDDS) for improved oral delivery of lipophilic drugs, *Biomed Pharmacother*, **3**, pp. 173–182.

64. Lo, J. T., Chen, B. H., Lee, T. M., Han, J., and Li, J. L. (2010). Self-emulsifying O/W formulations of paclitaxel prepared from mixed nonionic surfactants, *J Pharm Sci*, **5**, pp. 2320–2332.

65. Caliph, S. M., Charman, W. N., and Porter, C. J. (2000). Effect of short-, medium-, and long-chain fatty acid-based vehicles on the absolute oral bioavailability and intestinal lymphatic transport of halofantrine and assessment of mass balance in lymph-cannulated and non-cannulated rats, *J Pharm Sci*, **8**, pp. 1073–1084.

66. Baloch, J., Sohail, M. F., Sarwar, H. S., Kiani, M. H., Khan, G. M., Jahan, S., Rafay, M., Chaudhry, M. T., Yasinzai, M., and Shahnaz, G. (2019). Self-nanoemulsifying drug delivery system (SNEDDS) for improved oral bioavailability of chlorpromazine: In vitro and in vivo evaluation, *Medicina (Kaunas)*, **5**.

67. Nasirizadeh, S. and Malaekeh-Nikouei, B. (2020). Solid lipid nanoparticles and nanostructured lipid carriers in oral cancer drug delivery, *J Drug Deliv Sci Technol*, p. 101458.

68. Khan, I., Yousaf, S., Subramanian, S., Albed Alhnan, M., Ahmed, W., and Elhissi, A. (2018). Proliposome tablets manufactured using a slurry-driven lipid-enriched powders: Development, characterization and stability evaluation, *Int J Pharm*, **1**, pp. 250–262.

69. Khan, I., Yousaf, S., Subramanian, S., Korale, O., Alhnan, M. A., Ahmed, W., Taylor, K. M. G., and Elhissi, A. (2015). Proliposome powders prepared using a slurry method for the generation of beclometasone dipropionate liposomes, *Int J Pharm*, **2**, pp. 342–350.

70. Cho, H. J., Park, J. W., Yoon, I. S., and Kim, D. D. (2014). Surface-modified solid lipid nanoparticles for oral delivery of docetaxel: Enhanced intestinal absorption and lymphatic uptake, *Int J Nanomedicine*, pp. 495–504.

71. Baek, J. S., So, J. W., Shin, S. C., and Cho, C. W. (2012). Solid lipid nanoparticles of paclitaxel strengthened by hydroxypropyl-β-cyclodextrin as an oral delivery system, *Int J Mol Med*, **4**, pp. 953–959.

72. Bondi, M. L., Azzolina, A., Craparo, E. F., Lampiasi, N., Capuano, G., Giammona, G., and Cervello, M. (2007). Novel cationic solid-lipid nanoparticles as non-viral vectors for gene delivery, *J Drug Target*, **4**, pp. 295–301.

73. Choi, S. H., Jin, S. E., Lee, M. K., Lim, S. J., Park, J. S., Kim, B. G., Ahn, W. S., and Kim, C. K. (2008). Novel cationic solid lipid nanoparticles enhanced p53 gene transfer to lung cancer cells, *Eur J Pharm Biopharm*, **3**, pp. 545–554.

74. Hu, L., Tang, X., and Cui, F. (2004). Solid lipid nanoparticles (SLNs) to improve oral bioavailability of poorly soluble drugs, *J Pharm Pharmacol*, **12**, pp. 1527–1535.

75. Liu, D., Liu, Z., Wang, L., Zhang, C., and Zhang, N. (2011). Nanostructured lipid carriers as novel carrier for parenteral delivery of docetaxel, *Colloids Surf B Biointerfaces*, **2**, pp. 262–269.

76. Zhang, T., Chen, J., Zhang, Y., Shen, Q., and Pan, W. (2011). Characterization and evaluation of nanostructured lipid carrier as a vehicle for oral delivery of etoposide, *Eur J Pharm Sci*, **3**, pp. 174–179.

77. Sun, B., Luo, C., Li, L., Wang, M., Du, Y., Di, D., Zhang, D., Ren, G., Pan, X., Fu, Q., Sun, J., and He, Z. (2016). Core-matched encapsulation of an oleate prodrug into nanostructured lipid carriers with high drug loading capability to facilitate the oral delivery of docetaxel, *Colloids Surf B Biointerfaces*, pp. 47–55.

78. Luan, J., Zheng, F., Yang, X., Yu, A., and Zhai, G. (2015). Nanostructured lipid carriers for oral delivery of baicalin: In vitro and in vivo evaluation, *Colloid Surf A: Physicochem Eng Aspects*, pp. 154–159.

79. Tian, C., Asghar, S., Wu, Y., Chen, Z., Jin, X., Yin, L., Huang, L., Ping, Q., and Xiao, Y. (2017). Improving intestinal absorption and oral bioavailability of curcumin via taurocholic acid-modified nanostructured lipid carriers, *Int J Nanomedicine*, pp. 7897–7911.

80. Khan, I., Yousaf, S., Alhnan, M. A., Ahmed, W., Elhissi, A., and Jackson, M. J. (2016). Design characteristics of inhaler devices used for pulmonary delivery of medical aerosols, in: *Surgical Tools and Medical Devices*, W. Ahmed and M. J. Jackson (Eds), Springer International Publishing, pp: 573–591: Cham. pp. 573–591.

81. Allen, T. M. and Cullis, P. R. (2013). Liposomal drug delivery systems: From concept to clinical applications, *Adv Drug Deliv Rev*, **1**, pp. 36–48.

82. Bogner, J. R., Kronawitter, U., Rolinski, B., Truebenbach, K., and Goebel, F. D. (1994). Liposomal doxorubicin in the treatment of advanced AIDS-related Kaposi sarcoma, *J Acquir Immune Defic Syndr (1988)*, **5**, pp. 463–468.

83. Raj, T. A., Smith, A. M., and Moore, A. S. (2013). Vincristine sulfate liposomal injection for acute lymphoblastic leukemia, *Int J Nanomedicine*, pp. 4361–4369.

84. Rueda Domínguez, A., Olmos Hidalgo, D., Viciana Garrido, R., and Torres Sánchez, E. (2005). Liposomal cytarabine (DepoCyte) for the treatment of neoplastic meningitis, *Clin Transl Oncol*, **6**, pp. 232–238.

85. Stathopoulos, G. P., Antoniou, D., Dimitroulis, J., Stathopoulos, J., Marosis, K., and Michalopoulou, P. (2011). Comparison of liposomal cisplatin versus cisplatin in non-squamous cell non-small-cell lung cancer, *Cancer Chemother Pharmacol*, **4**, pp. 945–950.

86. Date, T., Nimbalkar, V., Kamat, J., Mittal, A., Mahato, R. I., and Chitkara, D. (2018). Lipid-polymer hybrid nanocarriers for delivering cancer therapeutics, *J Control Release*, pp. 60–73.

87. Clawson, C., Ton, L., Aryal, S., Fu, V., Esener, S., and Zhang, L. (2011). Synthesis and characterization of lipid-polymer hybrid nanoparticles with pH-triggered poly(ethylene glycol) shedding, *Langmuir*, **17**, pp. 10556–10561.

88. Trotta, M., Debernardi, F., and Caputo, O. (2003). Preparation of solid lipid nanoparticles by a solvent emulsification-diffusion technique, *Int J Pharm*, **1-2**, pp. 153–160.

89. Chaudhary, Z., Ahmed, N., Rehman, A., and Khan, G. M. (2018). Lipid polymer hybrid carrier systems for cancer targeting: A review, *Int J Polym Mater Polym Biomater*, **2**, pp. 86–100.

90. Parker, A. L., Newman, C., Briggs, S., Seymour, L., and Sheridan, P. J. (2003). Nonviral gene delivery: Techniques and implications for molecular medicine, *Expert Rev Mol Med*, **22**, pp. 1–15.

91. Hadinoto, K., Sundaresan, A., and Cheow, W. S. (2013). Lipid-polymer hybrid nanoparticles as a new generation therapeutic delivery platform: A review, *Eur J Pharm Biopharm*, **3 Pt A**, pp. 427–443.

92. Li, J. (2010). Multi-tasking of somatic embryogenesis receptor-like protein kinases, *Curr Opin Plant Biol*, **5**, pp. 509–514.

93. Zhong, Q., Chinta, D. M., Pamujula, S., Wang, H., Yao, X., Mandal, T. K., and Luftig, R. B. (2010). Optimization of DNA delivery by three classes of hybrid nanoparticle/DNA complexes, *J Nanobiotechnology*, p. 6.

94. Su, X., Fricke, J., Kavanagh, D. G., and Irvine, D. J. (2011). In vitro and in vivo mRNA delivery using lipid-enveloped pH-responsive polymer nanoparticles, *Mol Pharm*, **3**, pp. 774–787.

95. Umanzor, G., Cutler, D. L., Barrios, F. J., Vassallo, R. H., Chivalan, M. A., Bejarano, S. A., Ramirez, J. R., Fein, L., Kramer, E. D., Kowalyszyn, R. D., Wang, H., Goldfinch, J., Moore, T., and Kwan, R. M. F. (2020). Abstract GS6-01: Oral paclitaxel with encequidar: The first orally administered paclitaxel shown to be superior to IV paclitaxel on confirmed response and survival with less neuropathy: A phase III clinical study in metastatic breast cancer, *Cancer Res*, **4** Supplement, pp. GS6–01.

Chapter 10

Skin Aging and Vesicular Delivery Systems

Eleni Axioti,[a] Maria Apostolou,[a] Sakib Yousaf,[b] and Iftikhar Khan[a]
[a]*School of Pharmacy and Biomolecular Sciences, Liverpool John Moores University, Liverpool L3 3AF, United Kingdom*
[b]*School of Pharmacy and Biomedical Sciences, University of Central Lancashire, Preston PR1 2HE, United Kingdom*
i.khan@ljmu.ac.uk, iftikharkhans@yahoo.com

Skin aging is a structural and physiological alteration in skin appearance accompanied by loss of homeostasis. It leads to thin, fragile, and aged skin. This aging process is linked with phenotypic changes in cutaneous cells. Moreover, extracellular matrix components such as collagen and elastin are subject to changes. In older skin, the content of collagen decreases and that results in weak bonds between the dermis and the epidermis, and therefore wrinkles appear. Also, free radicals can damage DNA, lipids, and proteins and hence speed up aging. Unfortunately, skin aging is an unavoidable

Science and Applications of Nanoparticles
Edited by Waqar Ahmed and Ehsan Nourafkan

ISBN 978-981-4267-34-2 (Hardcover), 978-1-003-28029-3 (eBook)
www.jennystanford.com

stage in everyone's life and is in direct line with lifestyle, genetics, and environmental factors. Skin aging is one of the most crucial and urgent issues linked to skin appearance. Formulating cosmetics with antiaging properties has always been a substantial field of research. However, when delivering active ingredients across the skin, the barrier function of the skin renders the transport of molecules difficult. Therefore, vesicular delivery systems have been developed in order to obtain the maximum benefit from cosmetic products and to overcome the problems associated with their skin penetration. It has been identified that vesicular carriers improve the topical delivery of active ingredients. The first phospholipid nanocarriers that have been used were conventional liposomes, but their inability to penetrate deeper layers of the skin resulted in the introduction of modified liposomes (i.e., transferosomes, ethosomes, and niosomes). Throughout the years, even more active antiaging ingredients have been incorporated and investigated into those delivery systems.

10.1 Introduction

10.1.1 Skin

Skin is a vital, complex organ. Being the body's largest organ, it covers the entire outside of the body, forming a protective barrier against pathogens and injuries from the environment [1–3]. It is a multilayered organ composed of "heterogeneous cell types" [3], keratinocytes, melanocytes, Merkel cells, Langerhans cells, and extracellular substances, such as water and collagen. The main function of the skin is to provide a protective barrier against radiation, variations in conditions such as pressure and temperature, different chemicals or microorganisms in addition to preventing excessive water loss. In addition to this, it also preserves the body's homeostasis and body conditions like body's temperature, water content, and carbon dioxide levels, by limiting sweat evaporation, heat loss, and body temperature with its blood supply. Moreover, it is also composed of an extensive network of nerve cells that detect and relay changes in the environment [3, 4]. Skin consists of three main layers: the epidermis (the outer layer), the dermis (the middle layer), and the hypodermis (subcutaneous layer or superficial fascia) (Fig. 10.1).

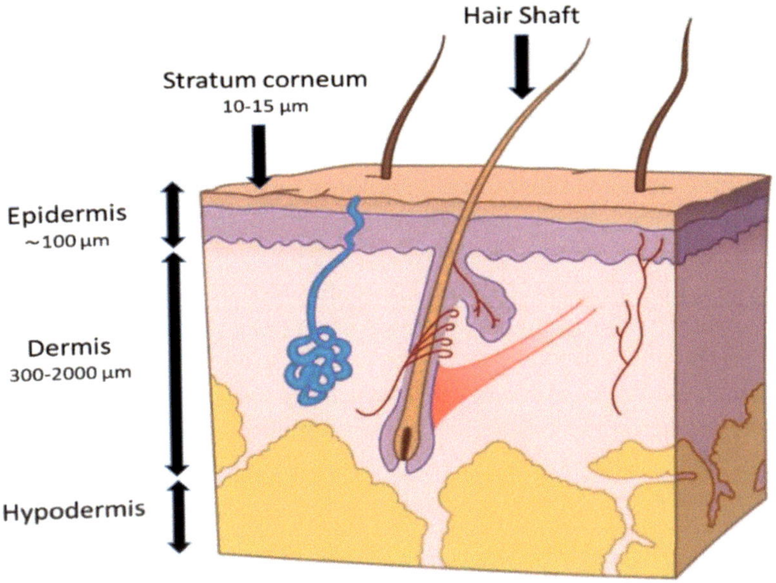

Figure 10.1 Layers of the skin: (a) full-thickness skin and (b) expansion of the upper ~200 μm thickness [5].

The epidermis layer provides a barrier to infection from environmental pathogens and balances the amount of water released from the body into the atmosphere through transepidermal water loss. The dermis lies between two other layers and is the thickest layer of the skin, composed of fibrous and elastic tissue. Therefore, it provides strength and flexibility to the skin. The hypodermis is present as the interface between the skin and the underlying fascia (fibrous tissue) of the bones and muscles [2, 3, 5].

10.1.1.1 Epidermis

The epidermis consists of the stratum corneum, which is the outermost layer exposed to the environment and is further divided into stratum granulosum (granular layer), stratum spinosum (prickle cell layer), and stratum basale (basal layer) (Fig. 10.2).

The outermost layer of the skin, the "stratum corneum," is circa 10–15 μm in thickness and is composed of a multilayered matrix of flattened, dehydrated, and nonviable keratolytic cells (corneocytes), which are surrounded by an extracellular lipid mortar, which is established as the main barrier to topical or transdermal delivery [7, 8]. Stratum corneum mainly consists of proteins and lipids, which are structurally organized as "bricks and mortar" [1, 3] creating the outer water barrier. The corneocytes, which are terminally differentiated keratinocytes stacked up to 18–20 layers, are featured as the bricks and the intercellular lipids and esters form the lamellar membranes, which enclose the corneocytes and hold them together [9]. This structure provides functional complexity, metabolic adaptations, and the potential for autopoiesis. In other words, due to the stratum corneum, the skin layer will prevent dehydration and the entry of toxins, bacteria, as well as some topically applied drugs/actives, apart from those which are lipophilic and with a low molecular weight (<500 Da). At the same time, it is also protecting the skin layers underneath [1, 9]. Keratinocytes are produced in the lower layers of the epidermis, and upon entering the stratum corneum, they convert into corneocytes with a more stable cell envelope. One layer of corneocytes is discarded each day, and the corneocytes are replaced with new keratinocytes from the lower layer of the epidermis called the stratum granulosum [9]. Apart from corneocytes and keratinocytes, other structural components are corneodesmosomes, epidermal lamellar bodies, and lipolytic and proteolytic enzymes. All these units are fundamental to the stratum corneum barrier, which is crucial for transdermal drug delivery.

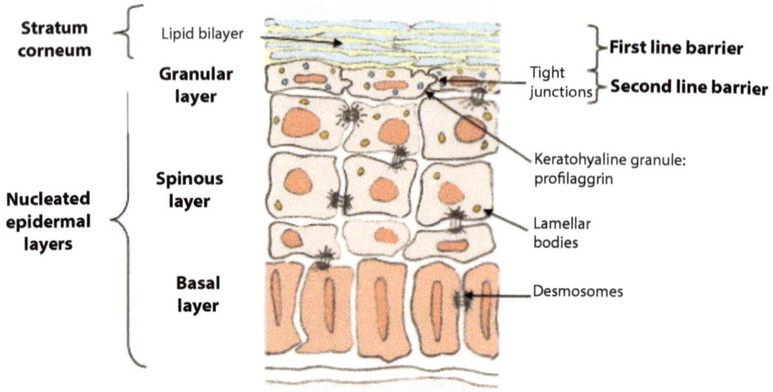

Figure 10.2 Structure of the skin epidermis. The stratum corneum is rendered as first line barrier. The second-line barrier starts in the granular layer, one of the sub-layers in the nucleated epidermal layers [6].

The next layer, the stratum granulosum (1–3 cell layers), is composed of enzymes that have the ability to diminish vital cell organelles such as nuclei. In this layer, keratin is also synthesized. The stratum spinosum, the next viable epidermal layer, consists of two to six layers of columnar keratinocytes, which convert themselves into polygonal shapes. The keratin in this layer aggregates to form filaments called tonofilaments. The combination of the stratum spinosum and the lower stratum basal layer is known as the malpighian layer. The only layer that is capable of cell division is the stratum basal, which possesses all the typical cell organelles. Keratinocytes originate in this layer and move up to the final barrier layer of the skin, the stratum corneum. Keratinocytes represent the major cell type of the epidermis (the outermost of the layer of the skin), making up about 90% of the cells there. Along with keratinocytes, other specialized cells existent in the basal layer are melanocytes, Langerhans cells, and Merkel cells. Melanocytes produce specific organelles, melanosomes, which contain melanin (eumelanin or phaeomelanin). Melanin protects the skin from ultraviolet radiation and free radicals. Langerhans cells are a part of the immune system, functioning as antigen-presenting cells of the skin. Merkel cells together with nerve endings are present in the dermis and are responsible for cutaneous sensation [9].

10.1.1.2 Dermis

The dermis (or corium) is 3–5 mm thick and consists of an extracellular matrix of connective tissue, namely elastic and fibrous, which is composed of predominantly collagen and elastin (Fig. 10.3). It aims to provide support, strength, flexibility, and firmness [10]. It is composed of a reticulate network of blood vessels, lymphatic vessels, nerve endings, and numerous appendages, such as hair follicles, sebaceous glands, and eccrine glands. Hair follicles cover the entire body surface except for soles of feet, palms of hands, and lips (Fig. 10.3). Sebaceous glands produce sebum, which consists of free fatty acids, triglycerides, and waxes, and have the potential to lubricate the skin surface and maintain a surface pH (approximately 5). Another liquid, sweat, is made by the eccrine glands (or sweat glands). It is a dilute salt solution of pH 5, a response to physical and emotional stress, and is responsible for absorbing the heat from the body upon evaporation. Another type of gland is specialized (or apocrine) glands, which are located in the dermo-epidermal layer in selective regions. The role of skin appendages is to act as barriers for all "intruders" that can enter the lower layers of the skin and serve a particular function, including sensation, contractility, lubrication, and heat loss [9].

The major component of the dermis, collagen, comprises 70–80% of the dry weight of the dermis and is capable of preventing overstretching and tearing due to its high tensile strength. The second component, elastin, is an elastic protein that maintains skin tension, flexibility, and resilience [12]. The ground substance of the dermal extracellular matrix is composed of glycosaminoglycans, glycoproteins, and water. The glycosaminoglycan/proteoglycan macromolecules constitute around 0.1–0.3% of the dry weight. Glycosaminoglycans are polysaccharides termed proteoglycans when linked to a protein core. Hyaluronic acid and dermatan sulfate are the key constituents of glycosaminoglycans in the skin. Chondroitin 4- and 6-sulfate are also found in smaller amounts. These compounds are called natural moisturizing factors and have the potential to provide hydration to the skin, binding a volume of water in the dermis up to 10,000 times the size of the molecule itself, mainly through the water-binding capacity of hyaluronic acid. These properties allow the rapid diffusion of water and water-soluble molecules and the transport of nutritional material in the matrix [13]. The major cell type in the dermis is fibroblasts, which act as the mother cell of all components of the matrix. Some other important ones are collagen, elastin, and ground substances that are responsible for the remodeling of the collagen fibrils. The dermis also contains macrophages and mast cells [14].

The dermis is divided into two main layers: the papillary dermis, which directly underlies the epidermis, and the reticular layer, which underlies the papillary dermis and is superficial to the hypodermis [12]. The papillary dermis is the uppermost layer of the dermis and is supplied with thin bundles of collagen fibers that are interwoven with elastic fibers. It interlaces the rete ridges of the epidermis and is composed of fine and loosely arranged connective tissues that are highly vascular. The reticular dermis consists of thick undulating bundles of collagen fibrils that are almost parallel to the plane of the epidermis. The collagen bundles are linked to each other through the interlacing of their fibers and are surrounded by a network of large, fully mature elastic fibers. When moving toward the hypodermis, the size of the elastic fibers and the collagen bundles increases.

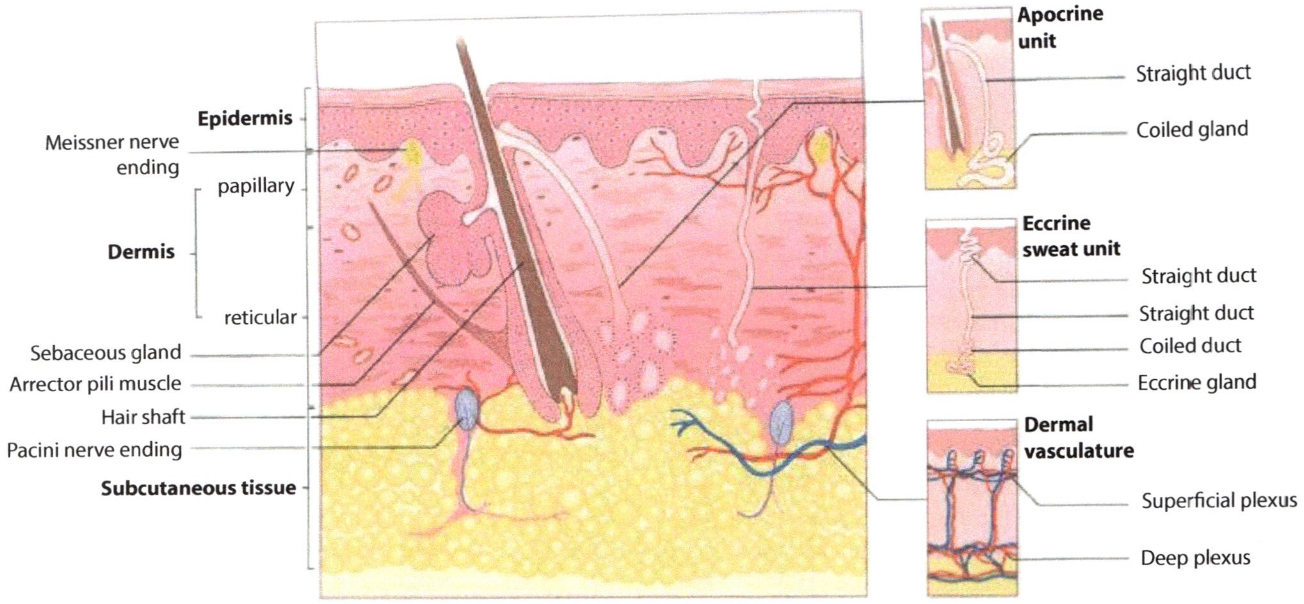

Figure 10.3 Layers of the skin and their composition [11].

10.1.1.3 Hypodermis

As demonstrated in Fig. 10.3, the hypodermis (or subcutaneous fat layer), being the deepest layer of the skin, provides the connection between the dermis with the underlying organs. Adipose lobules along with some skin appendages like the hair follicles, sensory neurons, and blood vessels lay in the hypodermis [10, 15]. Regarding the cellular composition, fibroblasts, adipose cells, and macrophages are able to influence the adipocyte homeostasis [16], remodel tissue [17], and arouse thermogenesis of fats during cold exposure and exercise [18, 19]. It provides insulation to the body and protects it from mechanical shock.

The hypodermis is mainly composed of loose connective tissue, which is supplied with hyaluronan, proteoglycan, and glycosaminoglycans and enables its fluid captivation when the capillary osmotic pressures are exceeded, for example during inflammation, providing it with mucous-like properties [20]. This fluid can alter the structural stiffness of tissue. The fibrous components of this tissue prevent their swelling and influence the interstitial compartment pressure through cytoskeletal tension, which is exerted on the collagen fibrils [21].

10.2 Skin Aging

Skin aging is a structural and physiological time-dependent alteration in skin appearance. The aging process is triggered from the moment someone is born and can be recognized throughout the years. This occurs due to the fact that the skin is the most visible organ. Unfortunately, the signs of aging are typically more visible with time, with people desiring to look younger, particularly focusing on eliminating wrinkles [22]. Over the past decades, the scientific interest in reducing the appearance of skin aging has been significantly increased. Skin aging indicates reduced biological activities of cells, diminished regenerative processes, and weak defenses toward the environmental factors [22–24].

10.2.1 Categories of Aging

Skin aging is an intricate biological mechanism affected by a combination of endogenous (genetics, cellular metabolism, and hormones) and exogenous factors (lifestyle, pollution, toxins, ultraviolet radiation). Simultaneously, hormonal disregulations, poor nutrition, or vitamin deficiencies can also affect the functioning of the skin [25–27]. Therefore, aging can be divided into two different categories based on the changes that occur in the skin. Intrinsic/chronological aging is the result of genetic factors and inherent changes during the normal aging process. It refers to alterations in the skin due to the passage of time and is less frequent than extrinsic aging. According to studies, only 3% of all aging factors have a genetic background [28, 29]. Extrinsic aging refers to the exogenous factors affecting the skin aging process. For example, ultraviolet radiation, environmental pollution, and smoking are capable of altering skin conditions. Photoaging, a significant subcategory of extrinsic aging, is driven by the changes from chronic sun exposure [30].

10.2.1.1 Chronological aging

It has been proven that chronologically aged skin results in xerosis, laxity, wrinkles, slackness, dryness, dullness, and the appearance of a variety of benign neoplasms such as seborrheic keratosis and cherry angioma [22]. The flattening of the dermal–epidermal junction is the main result of aging [31]. Also, an impact upon the thickness of the extracellular matrix is observed, and therefore, the number of fibroblasts decreases, as well as the levels of collagen and elastin [32, 33]. In addition, the skin gets dryer and gradually loses function to serve as the first barrier against the environment [33]. One major manifestation of innate (intrinsic) aging is the appearance of porous elastic fibers in the connective tissues [30]. Another common setting in the innate aging of the skin is the “deterioration” of the elastic tissue network. As a result, the skin becomes loose and incapable of returning to its former state. It has been elucidated that this destruction of the elastic fiber architecture starts at the age of 30 and is further noticeable after the age of 70 years [34].

Three primary structural components of the dermis—collagen, elastin, and glycosaminoglycans—have gained interest in antiaging research. In older skin, collagen degradation occurs due to various matrix metalloproteinases, such as serine, and other proteases, which are not able to produce the same collagen type. Therefore, the content of collagen type VII (Col-7) is decreased, resulting in weak bonds between the dermis and epidermis, and as a result wrinkles.

Hyaluronic acid, which is constructed in both the epidermis and dermis, is only demolished in the epidermis. Glycosaminoglycans, which are responsible for binding water, are not able to function properly in aged skin. By the thinning of the epidermis, the levels of elastin are reduced, resulting in the loss of resilience and elasticity of the skin tissues [35].

As the skin ages, the transdermal absorption of molecules into it is significantly impacted. As the skin is getting older, dehydration occurs. Hydration is crucial for transdermal absorption. Through aging, both the enzymatic activity and the blood flow decrease. A lower blood flow results in a slower drug absorption in an older stratum corneum, which further reduces the concentration of intracellular lipids and hence affect the lipid structure [36].

Regarding the hair, depigmentation and loss of hair can occur. Lastly, alterations in the nail plate can be observed. There are fewer glands in aged skin [8]. Throughout the years, there have been several theories regarding the mechanisms that occur in an organism during aging.

- **Genetic theory:** The first theory highlights that external factors have a limited impact on aging compared to genes. It illustrates that a group of nuclear DNA includes genes, known as gerontogenes, which are responsible for aging and are also able to determine the duration of life. According to this theory, gerontogenes are key components that influence the speed of aging [37].
- **Theory of a limited number of cell divisions (Hayflick theory):** According to this theory, there is a limit in the division potency of each cell. This limit is referred to as cellular aging or as the Hayflick phenomenon. This division potency is reduced with age. Division potency simply means that each cell endures a specific number of divisions. If a cell exceeds this number, then it enters the apoptosis stage, when the telomeres are shortened, and therefore cells are leading to cellular death [38].
- **Stochastic theory of skin aging:** This theory demonstrates that not only different biochemical abnormalities but also disfunctions in the human body are to be blamed for the aging of cells [37]. Both affect the functioning of tissues

and organs, including the skin, leading to an abnormal response to external stimuli, epidermal barrier disfunction, or pathological immune reactions. The excessive formation of reactive oxygen radicals is rendered as one of the most significant biochemical abnormalities. Reactive oxygen species (ROS) damage proteins and DNA, causing a loss of the ability of cellular self-repair and tissue regeneration. Another biochemical disturbance is amino acid "racemization," which produces inaccurate D-isomers from natural L-amino acids. The third is the non-enzymatic glycosylation of proteins, which generates improper crosslinking of different proteins, including collagen fibers [37, 38].

10.2.1.2 Extrinsic aging

Extrinsic aging can be accumulated by ultraviolet radiation, which has a negative impact on exposure to sunlight. Three exogenous factors are considered to be the main accelerators of skin aging: ultraviolet radiation, cigarette smoking, and environmental pollution. These may stimulate the formation of ROS. ROS can result in mutagenesis and the activation of the aging process, known as photoaging [39–41]. ROS production is the outcome of a combination of UVA and UVB radiation. UVA, in the wavelength range of 320–400 nm, leads to alterations in the middle layers (dermis), while the UVB light can interact only with the outer layer of the skin (epidermis) as it cannot penetrate the deeper layers of the skin [42]. UVB, with a wavelength of 280–315 nm, leads to damage in keratinocytes, melanocytes, and Langerhans cells. UVA affects the functioning of cells in the dermis (fibroblasts, dendritic cells, and endothelial cells of vessels). This can result in the production of aggressive inflammatory cells (lymphocytes, monocytes, granulocytes) or inducing synthesis of free oxygen radicals that cause lipid peroxidation, activate transcription factors destroying the extracellular matrix. Additionally, due to these free radicals, UVA damages both nuclear and mitochondrial DNA, while UVB wavelengths are also capable of generating free oxygen radicals and, therefore, affect cellular DNA by producing mutations. Even small daily doses of UVA can cause the thinning of both the stratum corneum and the whole epidermis [39, 42].

Over-exposed skin to ultraviolet radiation is characterized by a dry, earthen appearance, full of small wrinkles and thick furrows.

Ultraviolet radiation can also cause several horny cysts and open blackheads in the skin, a condition known as Favre–Racouchot disease [43]. It can also be linked to uneven pigmentations that resemble freckles and lentigines, which demonstrate a spotty pattern called hypomelanosis guttata. From a histopathological point of view, keratinocytes in the epidermis are observed to be abnormally matured, melanocytes are accumulated and distributed throughout the epidermal basal layer, the number of Langerhans cells is diminished, and lymphocytes can be infiltrated into the basal membrane, leading to an atrophic and thickened epidermis. The dermis is also affected by ultraviolet radiation, namely UVA. The connective tissue in the dermis breaks down due to the induction of metalloproteinases. This results in matrix proteolysis and, therefore, a degradation of collagen bundles. Moreover, the hypertrophication of some elastin fibers leads to the formation of a condition called senile elastosis. In the meantime, blood vessels become dilated and the sebaceous glands are overgrown. Thus, the innermost epithelial layer thickens [43, 44]. Additionally, due to the long-term exposure to sunlight, the immunological and regenerative abilities of the skin are undermined. Conditions, such as seborrheic keratosis or a precancerous condition called actinic keratosis, and skin cancers like basal cell carcinoma and squamous-cell carcinoma, are also linked to chronic exposure to ultraviolet radiation [45, 46].

Despite their differences, photoaging and intrinsic aging can demonstrate similar molecular mechanisms. First, the amount of matrix metalloproteinases (MMPs) is increased in aged skin. This increase is much higher in photodamaged skin. The number of fibroblasts is decreased, while abnormalities in the connective tissue, disorganization of the collagen, and reduced synthesis of new collagen are observed [17]. As interactions between fibroblasts and the collagen peptides affect the synthesis of procollagen by fibroblasts, collagen degradation disturbs these interactions and reduces the procollagen synthesis. This phenomenon is more significant in photodamaged skin due to the "ultraviolet-induced upregulation" in MMPs and results in collagen degradation [46, 47]. Neutrophils, which are present in the sunburned skin, are potent producers of a wide array of proteolytic substances, including the pluripotent neutrophil elastase, MMP-8, and MMP-9 [48].

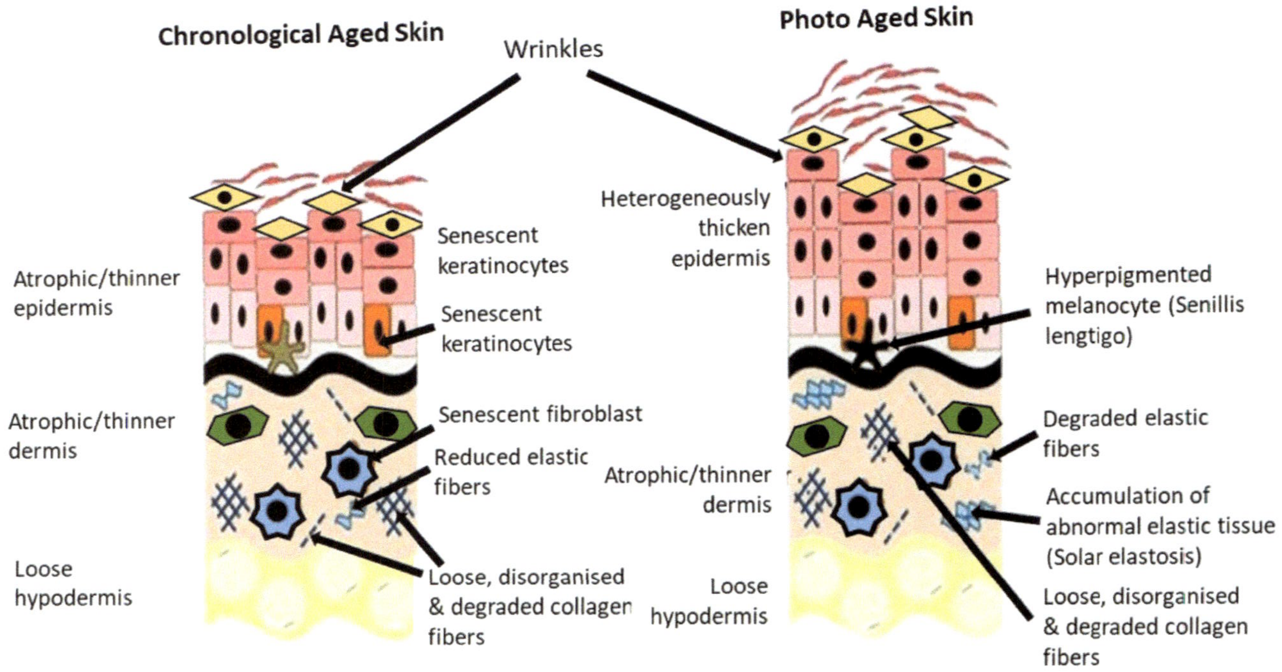

Figure 10.4 Morphological features of young, chronologically and photo-aged skin. Schematic organization of the chronologically aged and photo-aged skin tissue [52].

As illustrated in Fig. 10.4, both chronologically aged and ultraviolet-irradiated skin share important molecular features among their differences. For example, they share alterations in signal transduction pathways that promote MMP expression [49, 50], decreased procollagen synthesis, and connective tissue damage [44]. It has been demonstrated that photoaging includes dryness of the skin, irregular pigmentation-freckles, lentigines, hyperpigmentation, wrinkling, and inelasticity. From a histological point of view, the stratum corneum is being condensed, the granular cell layer is thicker, while the epidermis is becoming thinner, epidermal rete ridges are elongated, and the amount of hypertrophic dopa-positive melanocytes is expanding [51].

10.2.2 ROS and Skin Aging

During aging, damaged mitochondria produce less adenosine triphosphate and ROS accumulate. The current consensus is that ROS cause oxidative stress, damaging mitochondria and resulting in an energetic crisis that triggers neurodegenerative diseases, such as Alzheimer's disease, Parkinson's disease, and accelerate aging [53]. ROS are produced during mitochondrial oxidative metabolism and are involved in free radical production and skin aging. In the skin, approximately ~1.5–5% of oxygen consumption is transformed to ROS. The free radicals are products either of normal cell metabolisms or from external sources (pollution, cigarette smoke, radiation, medication) [54]. A free radical can be described as a chemical species, atom or molecule, containing one or more unpaired electrons in the shells around the atomic nucleus, existing independently. The odd number of electrons of a free radical renders it unstable, short-lived, and highly reactive, and therefore can attract electrons from other compounds to attain stability. In addition, the attacked molecule can develop into a free radical itself by losing its electron, provoking chain reactions, which result in damaging the living cell [55].

The most significant free radicals are radicals derived from oxygen, ROS. ROS can be classified into two groups of compounds, namely radicals and non-radicals. Some examples of radicals include superoxide, oxygen radical, hydroxyl, alkoxyradical, peroxyl radical, nitric oxide (nitrogen monoxide), and nitrogen dioxide; while non-radicals include hydrogen peroxide, hypochlorous acid, hypobromous acid, and ozone [56]. At low or moderate concentrations, ROS are necessary for the maturation process of cellular structures and can act as weapons for the host defense system. However, at higher concentrations, ROS are capable of provoking oxidative stress, causing potential damage to the biomolecules. The oxidative stress is developed when there is an excess production of ROS, which cannot be destroyed. While this excess occurs, ROS can interact with various biomolecules and damage them. This process includes lipids [57], proteins [58], and DNA [59], leading to the development of chronic and degenerative illnesses such as cancer, and demonstrates a major part in skin aging.

10.2.3 Aging within the Epidermis

With aging, the epidermis become thin, due to changes in the collagen structure, elastosis, and water accumulation. Besides this, the number of melanocytes also decreases. However, despite the thinning of the epidermis, the number of cell layers remains stable [44, 60]. While the activity of keratinocytes within the epidermis is affected, the number of both keratinocytes and pigment-containing cells (melanocytes) is decreased (alongside loss of appropriate function). The decrease in melanocyte density is combined with the formation of lentigines from the remaining melanocytes, resulting in their disappearance from other sites, causing cutaneous discoloration. This generalized pigmentation is, however, uneven and not uniform [30, 43]. Simultaneously, the synthesis of vitamin D3 declines, resulting in impaired cornification in the epidermis. Additionally, the epidermis tends to lose attachment to the dermis due to the ability of the dermal–epidermal junction to depress [3]. The projections between the epidermis and the dermis flatten with age, leading to the weakening of the epidermis. This effect eventually leads to the characteristic dry aged skin [9].

Additionally, at a cellular level, alterations in the differentiation and morphology of the basal layer cells also occur. Langerhans cell density is also reduced. Langerhans cells are responsible for the recognition of foreign antigens and subsequent immune response. As a result, these changes conclude to a reduced immunological response in older skin [61]. Moreover, telomere length reduces with age in the epidermis. Telomeres are small stretches of DNA at each end of every chromosome. Their role is to protect the chromosome against degradation and guarantee chromosomal integrity. The telomeres, however, shorten with each replication until the chromosome is unable to replicate further [48]. Last, epidermal cell turnover decreases with age [62]. Along with aging, sarcopenia, an age-related loss in skeletal muscle mass and function, can also be observed. Sarcopenia is a condition that affects all layers of the epidermis except the stratum corneum, which has a tendency to increase its thickness. Its results can be highlighted as a decrease in mobility and independence [8]. All the alterations that may occur within the skeletal muscle during aging are illustrated in Fig. 10.5.

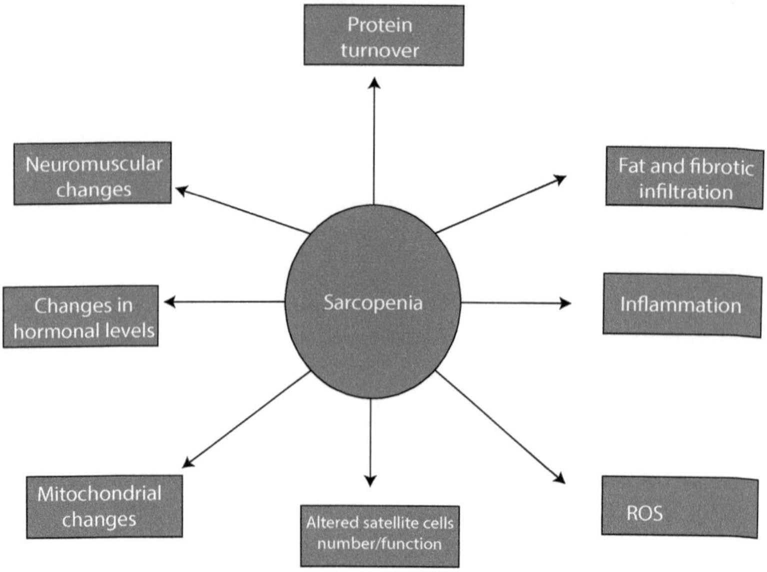

Figure 10.5 Alterations that occur within the skeletal muscle. Increased ROS levels, alterations in function and morphology of mitochondria, neuromuscular changes, reduce and proliferation of satellite cells, fat and fibrotic infiltration, and inflammation can be observed [7].

10.2.4 Aging within the Dermis

Regarding the dermis, reduction in the amount and organization of dermal connective tissue in aged skin leads to its thinning. Due to these alterations, wrinkles form, flexibility decreases, and sensitivity of the skin increases [3]. Modifications occur in the structure and number of components of the dermis. Collagen synthesis decreases over time, while dermal elastin fibers vanish mainly in the papillary layer with some of the remaining fibers becoming hypertrophied in the reticular layer, which results in a condition known as senile elastosis. The elastic network endures conversions and develops disorganization. The amount of collagen and elastin is reduced leading to a skin with impaired strength and resiliency. Therefore, the skin loses its youthful appearance and develops a propensity to bruise easily. Moreover, the number and size of fibroblasts significantly reduces [63], and andoxytalan fibers also disappear. The collapsing of elaunic and elastic fibers connected with the formation of lacunae or cysts leads to increasing dystrophy, and the appearance of the skin becomes porous [34]. Old skin loses the vertical fine elastic fibers in the subepidermal regions and, therefore, contributes to the appearance of superficial laxity and the finely wrinkled surface. The terminal elastic fibers are thicker leading to shrunken and saggy elastic fiber structure in the papillary dermis [64, 65]. Additionally, with age, changes in the levels of elastin messenger RNA and as a result elastin gene expression are inevitable, activating changes in the elastin fiber network [66].

Significant reductions in the content of extracellular components such as hyaluronic acid, dermatan sulfate, or proteoglycan gel have been observed with aging. This decrease has the potential to lead to a reduction in water content and, therefore, changes in skin thickness. The water-binding capacity of the dermis is decreased due to the loss of hyaluronic acid and other glycosaminoglycans. This loss of water-retaining capacity could result in a more compact appearance and account for the dried and wrinkled appearance of aged skin [67]. Oxygen demonstrates a lesser ability to be metabolized in the cells, leading to reduced synthesis of intracellular adenosine triphosphate and extracellular proteins. The blood supply to the skin is reduced, resulting in an abnormal vascular growth that causes visible cutaneous telangiectasia [3]. A disintegration, derangement, and finally a loss of superficial blood vessels is observed [40].

Regarding proteins, two major processes occur in the dermis with aging, which are enzyme-controlled or non-enzymatic processes. The "immature and reducible divalent crosslinks" are converted into mature and stable trivalent crosslinks of histidinohydroxylysinonorleucine, pyridinoline, and deoxypyridinoline by the enzyme-controlled process. Histidinohydroxylysinonorleucine is the major part of mature cross-linking in the skin. The non-enzymatic glycosylation process, known as the Maillard reaction, results in the generation of advanced glycosylation end products. Advanced glycosylation end products activate molecular damage by forming crosslinks in collagen. An additional cutaneous process that is produced with aging is angiogenesis. The term angiogenesis refers to the physiological process in which new blood vessels are formed [68].

10.2.5 Aging of the Skin Appendages and Nerves

Aging also affects the nails through an increase in the size of nail plates. Calcium and strontium levels are increased, and the nails become more brittle, full of longitudinal furrows, and undulate due to reduced hydration. As a result, their recovery process slows down [9]. Lastly, skin aging also influences hair growth. A decrease in hair follicle diameter in both frontal and parietal scalp is observed [3]. The cutaneous nerve network also develops alterations, at first gradually and then rapidly, with age. The density of nerve structures is decreased, while their complexity is expanded during old age [69]. The number of Meissner corpuscles, which are composed of a cutaneous nerve ending and, therefore, act as transmitters for the sensations, also decreases [70]. All the above-mentioned changes are summarized in Fig. 10.6.

10.2.6 Changes on Facial Skin throughout the Years

In accordance with age brackets, a number of changes may be observed [3]:

- **30+ years:** Excess skin on the upper eyelids leads to the presence of ptosis in the upper eyelids. Moreover, laxation of the skin around the corners of the eyes occurs.
- **40+ years:** Nasolabial folds, which are the lines that start from the mouth and extend to the edge of the nose, can be observed. Also, forehead creases and frown lines (vertical lines between the eyebrows) are crucial in this stage.

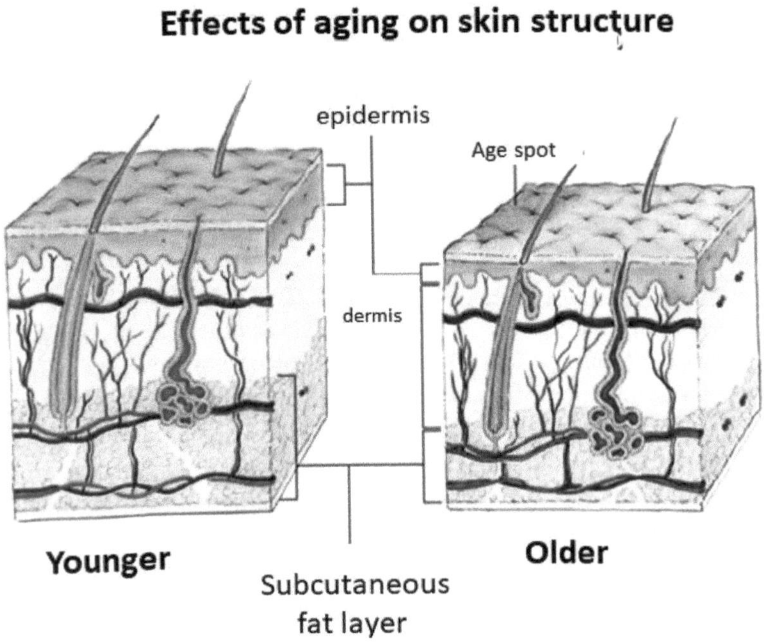

Figure 10.6 Structural differences in young and older skin. In older skin, age spots and decrease in the diameter of hair follicles can be observed. Also, the epidermis and subcutaneous fat layer are thinner [71].

- **50+ years:** Display of horizontal bands of platysma in the lateral sides of the cheeks.
- **60–80+ years:** Atrophy leads to the thinning of the skin, especially the epidermis and the subcutaneous layer. Simultaneously, wrinkles are more pronounced, and the skin becomes saggy.

10.2.7 Antiaging Strategies

There are numerous strategies in order to slow down skin aging, but at present they cannot halt the process of skin aging altogether. One of the identified factors with the potential to influence antiaging is lifestyle. A balanced diet, in combination with physical exercise, stress reduction, and an appropriate skincare regime are necessary factors to fight skin aging. Regarding skincare, at present, a range of cosmetic (i.e., cosmeceutical) products with biologically active substances that can inhibit skin aging are under manufacture. These products need to be applied systematically in order to gain better penetration into the deeper parts of the epidermis. All these products can enhance epidermal regeneration, protect the skin, and improve skin-nourishing and hydration [3].

The two main groups of ingredients that can be employed as antiaging cream components are the antioxidants and cell regulators. Antioxidants such as vitamins, polyphenols, and flavonoids can reduce collagen degradation. Among the antioxidants, the most effective are vitamin C, krill oil, zinc, vitamin D, sea buckthorn berry oil, cocoa, red isoflavones, and hormone-substances (e.g., the soybean known as phytoestrogen). Cell regulators such as retinols, peptides, and growth factors can affect collagen metabolism and produce collagen. It has been demonstrated that components of the extracellular matrix in the dermis can significantly prevent skin aging. The most significant being hyaluronic acid. Hyaluronic acid is a polymer of disaccharides and is responsible for the water-binding capacity of the dermis. It is distributed in spaces between collagen bundles and allows for the maintenance of skin elasticity. Other components with an equivalent effect against skin aging are mucopolysaccharidies and chondroitin sulfate. Both of these ingredients are extracellular matrix components that regulate the moisture of the skin. Other substances that effectively prevent skin aging include Vitamin A and its derivatives, UVA and UVB protective filters, and hydroxyl acids (especially smaller molecules as glycolic acid) [3, 72, 73].

Antiaging ingredients should provide:

- Moisturizing, plumping outcome
- Cell renewal/repair
- Antioxidant effect
- MMPs inhibition—to address dermis network degradation
- Collagen synthesis
- Photoaging prevention

Many of the ingredients may prevent aging through a number of mechanisms. For instance, hyaluronic acid, based on the molecular weight in which it will be incorporated in a formulation, can either achieve a moisturizing effect or cell regeneration [74]. Over the years, a range of ingredients were studied and investigated for their antiaging effects. Table 10.1 below shows the most used antiaging ingredients, their purpose of use, and the source that each ingredient can be isolated from.

Table 10.1 List of the most commonly used ingredients in antiaging formulations, their action, and their source [75]

Ingredient	Action	Source
Allantoin skin	Soothing, moisturizing	Comfrey
Aloe vera	Softens skin, moisturizing	*Aloe barbadensis*
Ascorbic acid	Antioxidant, photoprotectant	Vitamin C (citrus fruits)
Arnica	Astringent and soothing	*Arnica montane*
Arjunolic extract	Antioxidant and anti-inflammatory	*Terminalia arjuna*
Beta-Carotene	Minimizes lipid peroxidation and cellular antioxidant	Carrots and tomatoes
Boswellia	Anti-inflammatory and antiaging	*Boswellia serrata*
Calendula	Soothes softens skin, promotes cell formation	*Calendula officinalis*
Centella	Skin conditioning agent, increases collagen production, improves texture and integrity of skin, and reduces the appearance of stretch marks	*Centella asciatica*
Coriander seed oil	Anti-inflammatory and anti-irritant, skin- lightening properties	*Coriandrum sativa*

(*Continued*)

Table 10.1 (*Continued*)

Ingredient	Action	Source
CoenzymeQ10 (Ubiquinone)	Cellular antioxidant	Naturally occurring in the skin
Cucumber	Refreshes and tightens pores	*Cucumis sativus*
Essential fatty acids	Smoothens, moisturizes, and protects	Linolenic and arachidonic acid
Furfuryladenine	Improves hydration and texture of the skin	Plant growth hormone
Lupeol	Antioxidant and skin conditioning agent	*Cratacva nurvula*
Ginkgo	An antioxidant that smooths, rejuvenates, promotes a youthful appearance	*Ginkgo biloba* extract
Green tea	Antioxidant	Green teas (*Camellia sinensis*)
Kinetin	Antioxidant	Plants and yeast
Licorice extract	Skin whitening properties	*Glycyrrhiza glabra*
Oleanolic extract	Antioxidant, improves texture and integrity of the skin	Olive leaf
Panthenol	Builds moisture and soothes irritation	Provitamin B5 (broccoli, calf's liver, turnip greens)
Pycnogenol	Antiaging protects collagen	Grape seed extract

Ingredient	Action	Source
Retinoic acid	Promotes cell renewal and improves circulation to the skin	Vitamin A (green leafy vegetables)
Sodium hyaluronate	Lubricant between skin tissues and maintains natural moisture	Natural protein
Rosemary extract	Antioxidant, antimicrobial, and anti-inflammatory	Rosemarinus officinalis
Tetrahydrocurcuminoides, turmeric oil	Antioxidant and antiaging	*Curcuma longa*
Tocopherol	Antioxidant improves hydration and cellular regeneration of skin	Vitamin E (vegetable oils)
Ursolic acid	Anti-inflammatory, collagen buildup	*Rosemarinus officinalis*
Witch hazel	Tightens pores, tones	*Hamamelis virginiana*

Multiple strategies that can be applied in order to reverse the dermal and epidermal signs of skin aging are summarized below [73]:

- **Cosmetological care**: Daily skincare, appropriate sun protection, aesthetic noninvasive procedures, chemical peelings, and visible light devices.
- **Topical medical agents**: Antioxidants and cell regulators.
- **Invasive procedures**: Chemical peelings, visible light devices, intense pulsed light, ablative and non-ablative laser photo-rejuvenation, radiofrequency, injectable skin biostimulation and rejuvenation, prevention of dynamic wrinkles, correction of static, anatomical wrinkles, restoration (redistribution) of fat and volume loss, skin augmentation and contouring.

- **Systemic agents**: Hormone replacement therapy and antioxidants.
- **Avoiding exogenous factors of aging**: Smoking, environmental pollution, solar ultraviolet irradiation, and stress/lifestyle.
- **Correction of lifestyle and habits**: Nutrition, diet restrictions, supplements, physical activity, and control of general health.

Careful, daily skincare can improve skin regeneration, resulting in a healthy and functioning skin barrier, which provides protection against dehydration, penetration of microorganisms, ROS, and radiation [72, 73]. Lately, more noninvasive (non-surgical) techniques have been applied. However, a strategy that is becoming even more popular is based mainly on superficial chemical peelings, using Alpha-hydroxy acids, Beta-hydroxy acids, and retinoids. These can influence the function and morphology of the skin due to the turnover and maturation of keratinocytes. They stimulate proliferation, prevent abnormal cornification of the epidermis, and accelerate the growth and differentiation of follicular orifices preventing their occlusion [75]. Moreover, vitamins can be used for antiaging treatments. For example, retinoids (derivatives of vitamin A) are marked for their antioxidant effect and their ability for cell renewal. Additionally, ascorbic acid (vitamin C) stimulates collagen production and vitamin E (tocopherol) demonstrates strong antioxidant activity [76].

The skin is exposed daily to stressful environmental factors, including pollutants and ultraviolet radiation, which potentially generate a number of aggressive oxidants that can damage the cell membrane of biological skin. A large number of plants and plant extracts have been studied for their anti-oxidative action, including *Alpinia zerumbet* (Zingiberaceae), *Libidibia ferrea, Aconitum carmichaelii, Cuscuta chinensis, Emblica officinalis, Sonchus oleraceus, Asphodelus microcarpus, Nephelium lappaceum, Camellia sinensis, Curcuma longa, Labisia pumila,* and *Terminalia chebula* [77]. Moreover, Epigallocatechin-3-gallate, which can be found in green tea, also demonstrates a powerful antioxidant effect [76]. All these natural product extracts can scavenge free radicals and defend the skin matrix over the inhibition of enzymatic degradation. The use of plant extracts in cosmetic industry is very positive (as they are naturally derived) and constructive; however, further and detailed investigation is still required in this field to explore new avenues [77].

10.3 Delivery through the Skin

Skin, despite its strong barrier properties, demonstrates a potential route of delivery. Dermal drug delivery via topical application can deposit high concentrations of active ingredients for local effect in the upper skin layers, while minimizing their systemic exposure in the body. Successful drug/actives delivery depends on the physicochemical characteristics of the actives/drugs as well as their carriers in the formulation. Active delivery refers to the process by which the active ingredient penetrates and reaches the right site, at the right concentration through the skin [74].

10.3.1 Factors Affecting Permeation with Age

The dermal and transdermal permeation/penetration of ingredients is age dependent. The structural and biochemical changes in skin can lead to changes in the permeability of actives. A variety of physiological changes in the skin occur with age. These changes include decrease in the moisture content in the skin, lipid composition, lipid content, and density of skin appendages. The resistance in permeability caused by the epidermal stratum corneum is called "percutaneous penetration" and the rate of clearance by the microvasculature in the dermis is known as "percutaneous absorption." Some of the factors that affect permeation are as follows:

- **Stratum corneum barrier:** The stratum corneum is the limiting layer for skin permeation; therefore, the permeation in the stratum corneum is a major factor. Permeability alterations occur in both epidermal lipids and water. One example includes ceramides, which are a major lipid constituent of lamellar sheets present in the intercellular spaces of the stratum corneum. Ceramides play a key role in the retention and permeability in the stratum corneum, and a decrease in these would imply improved water permeation. It was observed that the total ceramide content significantly decreased with increasing age. It was also found that with aging, secretions of sebaceous glands decrease, which may increase water permeation, accompanied by increased volumetric porosity [78, 79]. Additionally, in aged skin, the skin barrier function affects the epidermis. As a result, a damaged epidermis can be expected to produce a defective stratum corneum. Hence, this process allows irritants to penetrate the stratum corneum [80]. Moreover, the thinning of the stratum corneum decreases skin permeability. The permeability coefficients through intact skin also decrease with increasing age, and the extent of this reduction is higher for lipophilic permeants than that for hydrophilic permeants [81].

- **Cutaneous blood flow:** Another factor that is accountable for changes in the absorption of compounds through the skin with age is the difference in blood flow. Blood flow is decreased with age in areas such as the lips, the cushion of the third finger, nasal tip, and forehead [82]. Moreover, the dermal papillae is eliminated due to the disappearance of the capillary loops that supply them [83]. The extinction of the blood vessels around the appendages leads to a decrease in glandular function, such as a reduction in sweating and hair growth. Changes in vessel walls result in a damaged, thickened forearm skin [80]. In some cases, the permeation of ingredients, such as sodium fluorescein and tetrachlorosalicylanilide, can be increased with age due to their absorption through cutaneous blood vessels [84].
- **Enzyme activity:** An important influencing factor that has only recently gained attention in age-related permeation studies is the role of skin enzymes in the permeation of actives. First, metabolic saturation of enzymes during the transdermal process can occur in all ages. However, the transdermal permeation of active ingredients is affected by age. The reason behind this is the difference in the development of the enzymes in the growth process [85].
- **Electrophysiological properties:** The resistance of the stratum corneum reflects the properties of pathways for hydrophilic permeants or pore pathways in skin permeation

and is markedly higher than the resistance generated by the viable epidermis and dermis. Factors affecting the resistance of stratum corneum include the thickness of the stratum corneum, the volume of pore pathways, and water content. Also, the volume of pore pathways (which include pores through hair follicles and sweat glands) or porosity of the skin decreases with increasing age. A decrease in water content, which defines the aqueous pathways and reduces the fluidity of stratum corneum lipids, has been observed. A decreased water content and increased stratum corneum thickness are responsible for the reduced *in vitro* permeation of hydrophilic permeants [86]. The resistance by the epidermis and dermis also increases with aging. This occurs due to the increased thickness of the viable epidermis and dermis. Furthermore, capacitance, which represents insulating properties of the lipid-cellular matrix, also decreases with age and hence additionally reduces the *in vitro* permeation of lipophilic compounds [86].

10.4 Delivery Systems

Embracing the structure and function of the skin, transdermal delivery systems offer a viable route for the delivery of key active ingredients into the skin. Topical application of cosmetic products and therapeutic substances allows penetration of skin tissues with desired effects obtained [87]. It is clear that there are many barriers to developing a successful therapeutic treatment. The complications occur mainly due to the existence of first-pass metabolism in addition to crucial issues, such as severe aftereffects, the exclusion of invasive operations, and complications with patients' acquiescence. A range of studies demonstrated that drug-delivery systems such as lipid- or polymer-based nanoparticles can be employed to overcome some of the aforementioned issues and can positively improve pharmacological and therapeutic outcomes [88–91].

The need of delivery systems via the dermal route is partially due to the administered drug via the oral route facing intervention by digestive enzymes [92]. Dermal delivery systems are potentially more successful since skin absorption is a safe noninvasive route with a large and accessible surface area [93], eliminating a number of factors associated with the oral route, including pH, food, enzymes, and gastrointestinal motility [94]. It also eliminates the pain and side effects of drugs/actives when compared to the traditional transvenous injection route [95]. For these reasons, dermal delivery systems gained popularity, not only for drug administration in pharmaceuticals, but also in cosmetics.

Skin demonstrates a complex structure, with the three layers (epidermis, dermis, and hypodermis) acting as a barrier against external elements [96, 97]. As the stratum corneum is a lipophilic membrane, hydrophilic drugs/actives cannot penetrate the skin, while a high lipophilic drug tends to remain in the layers of the stratum corneum. Therefore, the transport and delivery of drugs/active ingredients through the skin are a complex process. When a cosmetic or pharmaceutical product is applied to the skin surface, the molecules in it can penetrate the epidermis through three different pathways: through sweat ducts, across the continuous stratum corneum (transepidermal route), or through the hair follicles (Fig. 10.7) [98, 99].

The stratum corneum has been shown to be the tissue that mainly affects drug permeation into the skin. The routes by which molecules can penetrate the stratum corneum have been extensively studied and investigated [101, 102]. The transepidermal route (or the route through stratum corneum) is divided into the intracellular (passing through the cells) and the intercellular (passing in the spaces between the cells) routes. The intracellular route is the passage through which a molecule (mainly highly hydrophilic) permeates through the stratum corneum moving across and through the corneocytes (located in the stratum corneum). The intercellular route is adopted by lipophilic molecules. They can travel around the corneocytes through the extracellular region, which is lipid rich. Hydrophilic compounds that lay in the lipid bilayers support the transport of water-soluble active molecules. Recent studies demonstrated that the intercellular lipid route is the primary permeation pathway, especially for small, uncharged molecules. Due to the stability and strength of the barrier of the stratum corneum, the range of molecules that can accomplish topical therapeutic activity is firmly narrow [103, 104]. However, the low transappendageal route is preferred by ions and large polar molecules [98].

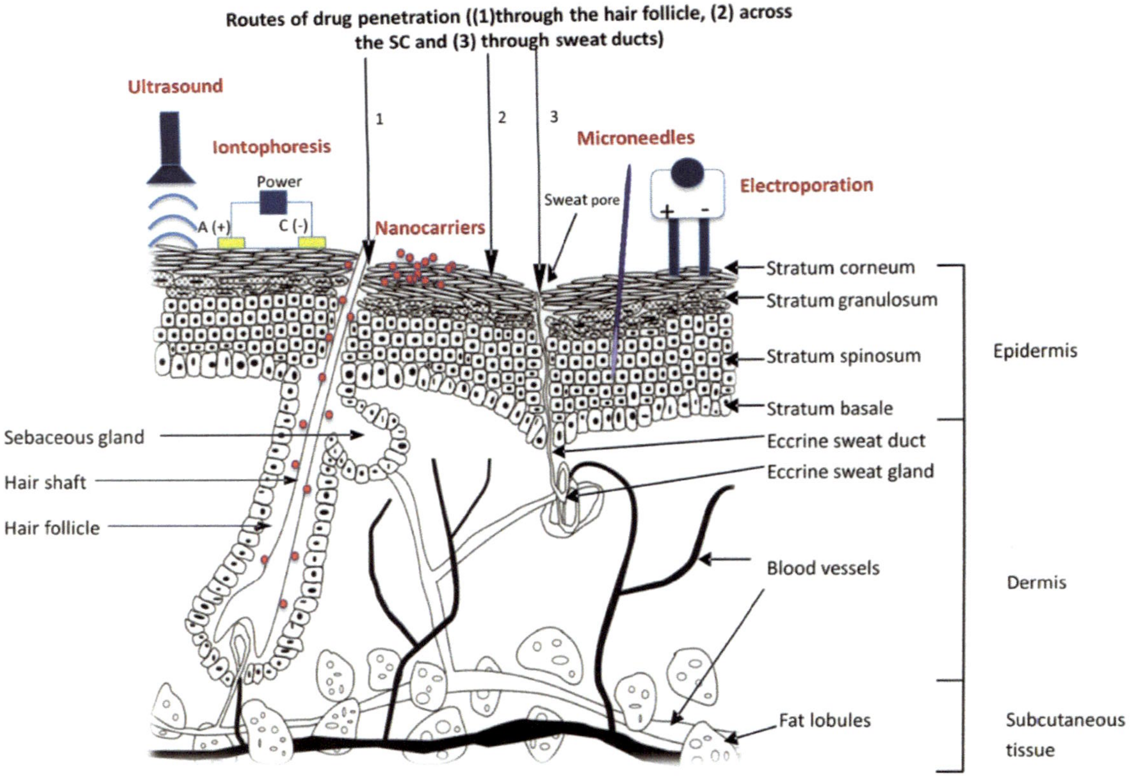

Figure 10.7 Simplified representation of skin showing routes of penetration: (1) through the hair follicles, (2) directly across the stratum corneum, and (3) via the sweat ducts [100].

In order to achieve a profitable delivery through a wider range of components, several delivery systems have been studied and developed. Nanoencapsulation using a lipid-based vesicular system such as liposomes has been used widely. Nano-delivery systems have been advantageous due to the protection they offer and controlled delivery at low doses that they provide [105–110]. Therefore, ingredients can pass through the skin by contact with the stratum corneum and skin appendages. Nanocarriers are capable of increasing skin contact time and penetration of drugs/actives through the skin, protecting against physical and chemical instability and enhancing their effectiveness [111, 112]. The encapsulation of actives and their delivery significantly improve efficiency as well as tolerance level offered by the skin.

10.4.1 Topical Delivery Systems

The use of topical delivery systems provides efficacy, stability and consumer compliance. The topical delivery route is not only safe but non-immunogenic [113] and provides optimal penetration and sustained release of actives to the skin [88]. While designing topical delivery systems, some important factors that need to be considered include the properties of the active compound (i.e., chemical and physical stability, solubility, lipophilicity or hydrophilicity, chemical and physical form (liquid/solid or acid/base/salt), the properties of active molecule carriers (i.e., the ability to transport the active ingredients successfully into cell membranes), as well as particle characteristics (i.e., shape, size, surface charge/zeta potential, and structure can also dramatically influence the release properties). Moreover, skin properties, including moisture content, age, race, and the site of application, also affect delivery patterns and effectiveness [114].

10.4.2 Nanocarriers as Cosmetic Delivery System

Nanocarriers essentially encapsulate active ingredients and transport them to the site of action. There are numerous types of nanocarriers; however, some key carriers are liposomes, nanocapsules using polymeric membranes, solid lipid nanoparticles, nanocrystals, and niosomes [115, 116]. Nanocarriers offer targeted delivery of active substances to deeper tissues and enhance local topical effects. They provide a range of advantages over conventional passive delivery, such as increased surface area, higher solubility, improved stability, controlled release of active ingredients, reduced skin irritancy, protection from degradation, increased drug loading, and improved permeation of actives into the skin [117, 118].

Different properties of nanoparticles such as size, shape, coating, and purity may affect the skin permeation of nanoparticles. Active ingredients that are encapsulated in nanocarriers can be delivered into the skin through the intercellular, transcellular, or transappendageal routes. These nanoparticles are capable of passing through epidermis, dermis, hypodermis, and finally reach the blood vessels, as is demonstrated in Fig. 10.8.

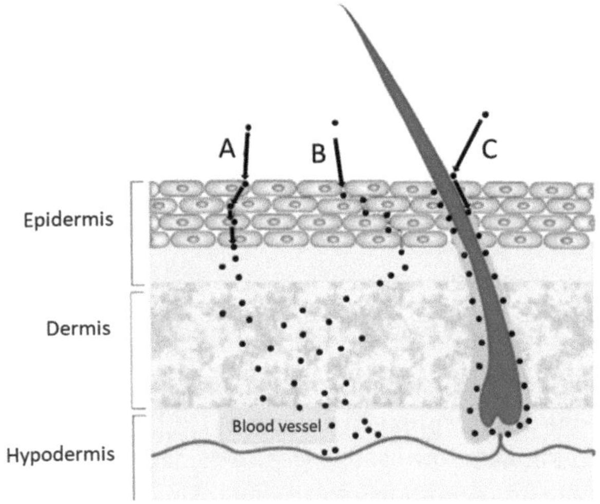

Figure 10.8 Possible ways of penetration of nanocarriers: (A) transcellular, (B) intercellular, and (C) transappendageal [119].

10.4.3 Factors Influencing Nano-Based Delivery Systems

There are a number of factors that can influence the delivery of nanoparticles to their site of action.

10.4.3.1 Zeta potential

Zeta potential plays a significant role in determining the stability of colloidal suspensions of nanoparticles. This measures the electrokinetic potential at the shear plane, where the compact layer of charged particles surrounding the nanoparticle meets the diffuse layer [120, 121].

Nanoparticles are capable of interacting with cell membranes. However, this is affected by surface charge. While cell membranes repel negatively charged nanoparticles, the incorporation of positively charged molecules enhances cellular uptake. By changing their surface charge, nanoparticles can transform into particular intracellular targets (mitochondria, lysosomes, cytoplasm, etc.) to affect their uptake and intracellular dissemination [120, 122]. The surface charge can alter the physical stability of nano-vesicles. Zeta potential, as an indicator of the surface charge, is increased when physical stability is achieved due to the electrostatic repulsion against the natural tendency to aggregate [74].

10.4.3.2 Particle size and shape

Particle size and shape also affect drug release rate, bio-distribution, mucoadhesion, cellular uptake of water, and buffer exchange to the interior of the nanoparticles, as well as protein diffusion, by controlling the adhesion of the particles and their interaction with cells [120, 123–125].

10.4.3.3 Saturation

If a compound or phospholipid elucidates saturation, it is usually more prone or susceptible to oxidation [126]. When a formulation is oxidized, it breaks down and leads to a damaged product, with color and consistency alterations, and a short shelf life [74]. Oxidation can also lead to degradation of the product, which may result in toxicity post administration.

10.4.3.4 Temperature and viscosity

Depending on the temperature and the viscosity of the organic and aqueous phases (formulation steps), the stability of the formulation can alter [127]. For instance, viscosity may also affect the surface tension and may influence the rate at which the droplets of an emulsion/suspension tend to form. This can lead to the destabilization of the formulation [128, 129].

10.4.4 Use of Vehicles in Delivery Systems

Many cosmeceutical active molecules with different polarities and stabilities can be incorporated into vesicular delivery systems. Through this incorporation, delivery systems maintain their unique properties and their skin absorption can be increased. The interest to use vesicular systems for cosmetic applications is based on the following points [130]:

- Vesicles carry and deliver the entrapped active molecules into the skin.
- Vesicles have the ability to entrap both hydrophilic and lipophilic active components within the vesicles.
- Vesicles provide sufficient penetration due to their lipid composition.
- Vesicles offer sustained release of active compounds.
- Vesicles are biodegradable and biocompatible, possess low toxicity, and are relatively non-immunogenic.

Examples of vesicular delivery systems that have been used in the cosmetics industry are liposomes, niosomes, transferosomes, solid lipid nanoparticles, nanostructured lipid carriers, and ethosomes [131]. Vesicles are "hollow colloidal particles" containing amphiphilic molecules. As a result, both water-soluble and water-insoluble compounds can be incorporated into the vesicle [126, 132–134]. In order to understand properly the components of these phospholipid nanocarriers, a comparison of their liposomal structure is demonstrated in Fig. 10.9.

10.4.4.1 Liposomes

The first established vesicle nanocarriers were liposomes, introduced by Bingham in 1965. The word liposome has origins in the Greek language: "Lipos" meaning fat and "Soma" meaning body [135]. In pharmaceutical and cosmetic products, phospholipids are used as the main component of vesicular formulations and can be obtained from different sources. Liposome morphology is similar to that of cell membranes as both contain phospholipids in their composition [136]. Liposomes, as part of the drug-delivery systems, form a suspension, where phospholipids (forming liposomes via self-assembly) are suspended in the aqueous phase [110, 126, 137].

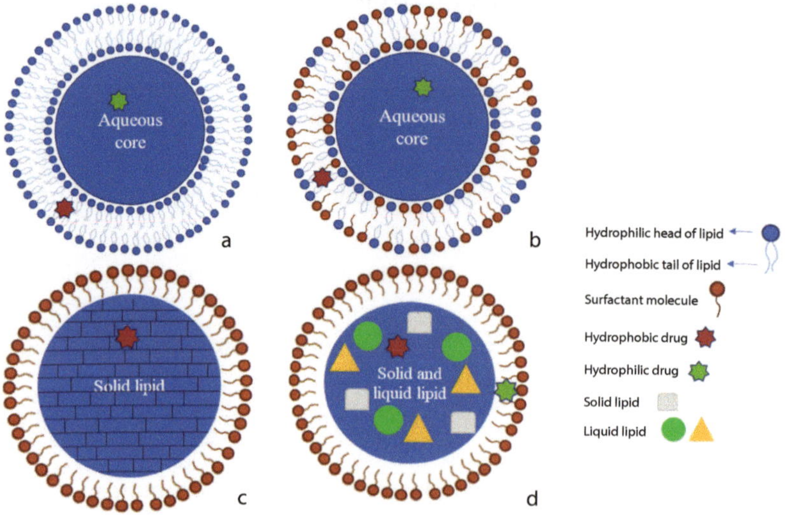

Figure 10.9 Schematic structure of nanocarriers and their components, including (a) liposome, (b) transferosome, (c) solid lipid nanoparticle, and (d) nanostructured lipid carrier [91].

Composition of liposomes: Liposomes are observed to be spherical self-closed vesicles of colloidal dimensions composed of lipid bilayers [138]. Natural phospholipids such as soya phosphatidylecholine and egg phosphatidylecholine possess low toxicity. Regarding synthetic phospholipids, dipalmitoyl phosphatidylcholine, dimyristoyl phosphatidylcholine, and dilauroyl phosphatidylcholine are the most commonly used [123, 139]. Cholesterol can also be incorporated into the liposomal structure improving the bilayer characteristics of the liposomes, such as the membrane fluidity and bilayer stability, reducing the leakage of drugs/actives from the vesicles [126, 140]. Due to their vesicle structure, they can encapsulate both hydrophilic and hydrophobic active ingredients (hydrophilic in the central core and hydrophobic in the concentric bilayers) [90, 109].

Mechanism of liposomal formulation: Phospholipids consist of a hydrophobic tail that has two fatty acid chains and a hydrophilic head group (Fig. 10.10). As a result, phospholipids are considered amphiphilic and they can arrange themselves in the form of a sheet, where the polar component of the phospholipid interacts with the polar environment of an aqueous medium with bonding occurring with water-soluble molecules. A sufficient amount of energy is needed for phospholipids to form layers of lipids in close alignment (i.e., sonication, homogenization, hand shaking, or heating) to generate liposome vesicles [134, 138, 141].

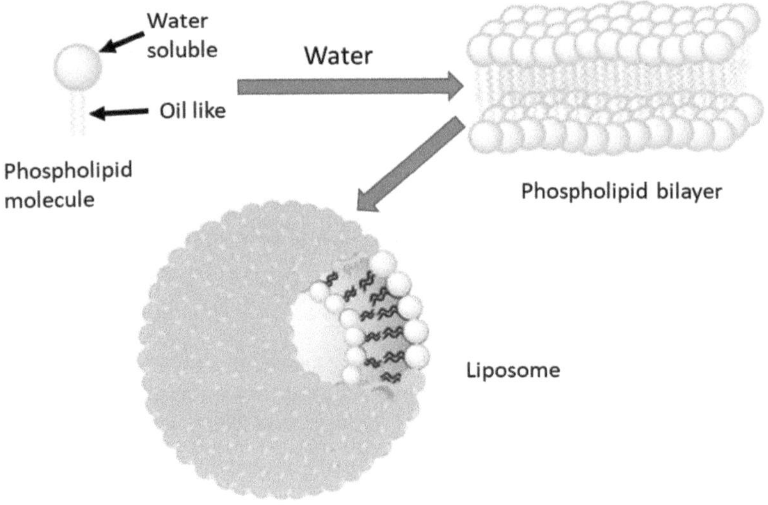

Figure 10.10 Mechanism of liposomal formation. The phospholipids form bilayers. These bilayers are packed converting to liposomes [138].

Mechanism of action in cosmetic formulations: Due to their lipid composition, liposomes achieve skin moisturization by depositing lipids onto the stratum corneum [142]. Lipids are well hydrated and can reduce the dryness of the skin, which is a primary cause for skin aging [143]. Therefore, liposomes produce an occlusive effect via topical delivery. Occlusive compounds promote the penetration of active compounds in the skin [82] (Fig. 10.11).

Phospholipids are capable of increasing the barrier function of the skin while decreasing water loss, quickly after application (due to the high content of esterified essential fatty acids). This function renders them as an appropriate delivery system for topical applications in cosmetics [145]. Topical liposomes reduce skin roughness because of their interaction with the corneocytes and intercellular lipids, resulting in skin softening and smoothening [83].

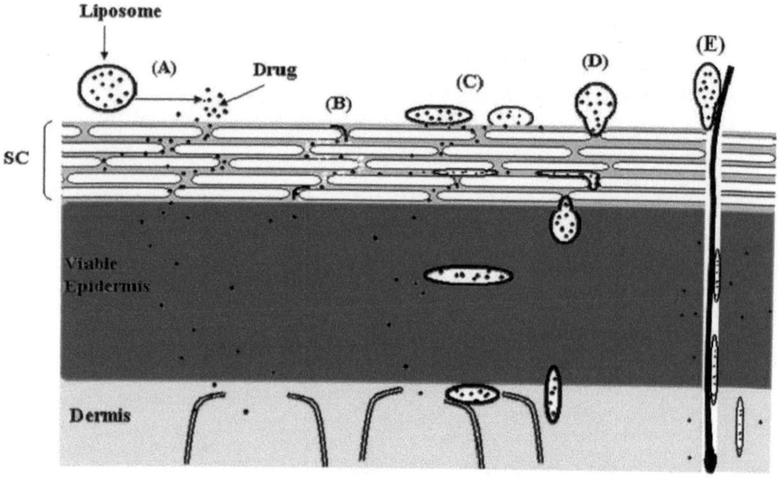

Figure 10.11 Skin penetration behavior of liposomes showing (a) diffusion of drug in the epidermis (stratum corneum; SC), (b) vesicle absorption, (c) intact penetration, (d) vesicle fusion, and (e) deformation of liposomes [144].

Advantages and limitations of using liposomes: Liposomes are highly organized structures [146] and protect entrapped molecules from degradation, while providing sustained and targeted release [123, 137, 147]. They differ from other release systems, and the drugs/actives release occurs either in plasma or at the site of administration, by offering selective and sufficiently high localization of active drugs/actives at the infected sites. Liposomes, due to their lipid composition, do not inhibit the ability to reach deeper layers of the skin [148]. When applied on the skin, most of the time they can only permeate through the upper layers of the stratum corneum and, therefore, can reach the epidermal layer, but are unable to penetrate the lower layers of the skin, such as the dermis or blood circulation. Moreover, they are nontoxic and do not lead to systemic toxicity, when employed as drug nanocarriers. Also, liposomes are easily re-absorbed from the epidermis into the deepest layers, minimizing unwanted systemic side effects [82]. They induce increased moisturization, restoring action, biodegradability, biocompatibility, and extended slow dermal release [82, 149]. Although liposomes have been widely used as drug carriers, they lack stability [150]. The vesicles possess a tendency to fuse and aggregate, leading to possible drug leakage [109, 126].

10.4.4.2 Ultradeformable liposomes (transferosomes)

The word transferosome originates from the Latin word "transferre" meaning "to carry across" and the Greek word "soma" meaning "body." They were first introduced in 1991 by Çevc and Blume, and since then, these vesicles have been studied and investigated for use in many different areas [87]. Transferosomes are also referred to as ultradeformable liposomes, due to their elasticity or flexibility, owing to component surfactants in combination with phospholipids [124, 132, 134]. Surfactants are also amphiphilic molecules composed of a single lipophilic alkyl chain and a hydrophilic head group. In general, nonionic surfactants are preferred due to their ability to act as solubilizers, stabilizers, emulsifiers, and glycoprotein inhibitors, which increases drug absorption [151]. They are capable of carrying active compounds such as liposomes to the site of action. Transferosomal formulations are efficient in "peripheral drug targeting," "transdermal immunization," and "transdermal delivery" employing a range of therapeutic agents [152]. It has been demonstrated that bioactive molecules of different molecular weights ($200 \leq MW \leq 10^6$) can be transported by transferosomes. Generally, transferosomes can deliver molecules regardless of their size, structure, molecular weight, or polarity [103]. They are also capable of carrying both hydrophilic and hydrophobic molecules through membrane, establishing them as a drug-delivery system that can carry active ingredients with a wide range of solubilities [103, 132] (Fig. 10.12). Transferosomes and surfactants (also referred to as edge activators) can be employed in different ratios based on phospholipid and surfactant type. They can be prepared in different size ranges depending on the method of preparation and size reduction process. Surfactants are generally used in concentrations ranging from 10 to 25%, which are assembled in the bilayers, increasing the flexibility and permeability of the transferosome [132]. Polysorbate-20 (Tween 20), Polysorbate-80 (Tween 80), Span 80, and sodium cholate are commonly used as surfactants due to the membrane-softening and ultra-deformable properties they provide. Moreover, the other two components are alcohol (in percentage approximately 10%) as a solvent and water or saline phosphate buffer (pH 6.5–7) as a vehicle.

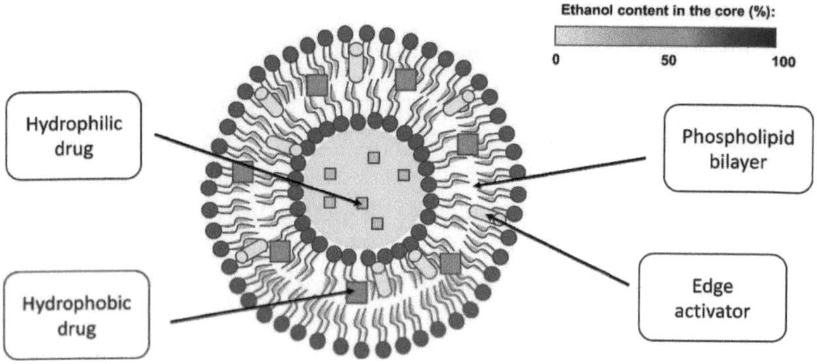

Figure 10.12 Structure of transferosomes. Transferosomes consist of an aqueous core that can incorporate a hydrophilic drug and is surrounded by a phospholipid bilayer, where the edge activator lays and where the hydrophobic drug can be capsuled [89].

As illustrated in Fig. 10.12, transferosomes consist of an aqueous core, where the hydrophilic active ingredients are captured in the core, and the hydrophobic active ingredients in the bilayer. This structure provides transferosomes their main ability, which renders them as successful delivery systems for the release of desired therapeutic agents. Transferosomes possess the efficiency in altering their membrane flexibility and passing through the skin pores, regardless of their narrow size (i.e., even up to 5 or 10 times narrower than their size) [103]. In other words, they are capable of self-optimizing deformability [153], and therefore can penetrate the skin, even through the stratum corneum (being able to penetrate more efficiently to the deep layers of the skin and then resume their initial form) [154].

Transferosomes generally fall into three generations. The first generation of transferosomes consists of phospholipid vesicle nanocarriers, which are only composed of phospholipids and an edge activator. In the second generation, a combination of a bilayer component (e.g., phosphatidylcholine) and more than one amphipathic membrane-destabilizing element is present in the transferosome structure. In the third generation, transferosomes are composed of amphipathic non-phospholipidic bilayers. What distinguishes the third type of transferosomes from the first and second generations is the replacement of surfactant with water-soluble modulator compounds (e.g., organic ions) that possess the same efficacy [127].

10.4.4.3 Flexible vesicular nanocarriers containing ethanol (ethosome)

To overcome the limitations of conventional liposomes, ethosomes were also introduced. Ethosomes are vesicular nanocarrier systems where apart from the phospholipid, ethanol is also used to form the membranous or vesicular structure [155]. Ethosomes can be introduced as the second generation of liposomes. The presence of ethanol in these nanocarriers contributes to the stability of vesicles and makes them more efficient at transporting active molecules [156]. Ethosomes, as opposed to conventional liposomes, which demonstrate little skin absorption and remain in the upper layers of the stratum corneum, are able to reach the deeper layers of the skin due to their greater flexibility. Ethosomes enhance the permeation of the drugs/actives through the stratum corneum and subsequent retention in the skin layers [157]. As an alternative form of flexible liposomes, ethosomes incorporate alcohol (ethanol) into the lipid bilayer to give the structure its flexibility. These flexible vesicular nanocarriers illustrate significant advantages in the dermal and transdermal delivery of active molecules that are limited in conventional liposomes. They also provide controlled drug release [86]. Ethosomes can encapsulate hydrophilic, hydrophobic, or amphiphilic drugs/actives and are able to reach the deeper skin layers and systemic circulation.

10.4.4.4 Niosomal nanocarriers (Niosomes)

Niosomes are microscopic lamellar structures, ranging in size from 10 to 1000 nm [158]. While they offer similarities to liposomes, they are more flexible. Their flexibility and stability are driven by their bilayer composition containing nonionic surfactants. In comparison to liposomes, niosomes do not contain phospholipids; this provides niosomes with a better stability profile because of improved oxidative stability. By using nonionic surfactants, niosomes provide high entrapment efficiency, improved chemical stability, and enhanced penetration. Both hydrophilic and hydrophobic active ingredients can be incorporated into niosomes. Nonionic surfactants have demonstrated low toxicity, which further improves their use in niosome preparations for cosmetic formulations as a transport vesicle [159]. Sustained drug release, better penetration, and higher skin retention distinguish niosomes from other lipid-based nanocarriers. Various oils can be incorporated into niosome vesicles to improve flexibility and influence transdermal delivery [160].

10.4.5 Use of Delivery Systems in Cosmetics

The purpose of each cosmetic product is to improve skin appearance and treat various skin conditions. In order to do so, cosmetic formulations must promote skin permeability, and that can be achieved using various drugs/actives delivery systems, which are able to deliver active ingredients to the skin. Usually, in cosmetic products, active ingredients such as deodorants and sun filters are supposed to remain on the external skin surface or within stratum corneum, while other active ingredients, such as anti-acne, antiaging, skin lightening, and hair regrowth agents, should penetrate through stratum corneum or diffuse through the skin appendages; for example through hair follicles [114]. When applying delivery systems on cosmetic formulations, the main task of the delivery system is the delivery of active ingredients to different layers of the skin, while minimizing systemic involvement [161]. Presently, biological molecules such as peptides and proteins, vitamins, and components of the skin matrix are among the ingredients that are extensively used in cosmetic products in order to fight skin aging [162]. With the use of liposomes, transferosomes, ethosomes, or niosomes as carriers, these ingredients are effectively delivered into the skin [163]. Over the years, a wide range of novel delivery systems such as microneedles and patches, as well as polymeric and emulsion-based delivery systems, have been discovered to deliver cosmetic actives to the skin [164]. Literature suggests that a range of drugs and actives have been studied and investigated using vesicular delivery systems in skin-related areas (Table 10.2).

Table 10.2 Commercial cosmetic formulations incorporating vesicular delivery systems

Product	Key Ingredient	Technology Used	Intended Use
Capture™	Thymus extract, collagen and elastin peptides, hyaluronic acid	100 nm liposomes_ made from soya lecithin	Antiaging
Revitalift® creams and lotions	Pro-Retinol A	Nanosome (liposome)	Antiaging and antiwrinkle
Liposomal serum (sesderma)	Azelaic acid, 4-butylresorcinol	Liposomes	Skin whitening
Niosome+	Glyceropolyether	Niosomes	Antiaging
Deep action lightening serum	Vitamin C	Niosomes	Lightening
Effect du Soleil	Tanning agents in liposomes	Niosomes	Self-tanning
Eye perfector	Peptides	Liposomes	Smoothing eye cream
Formule liposome gel	Thymoxin, hyaluronic acid	Liposomes	Hydration
Future perfect skin gel	TMF, vitamins E, A palmitate, cerebroside ceramide	Liposomes	Antiaging, antiwrinkle
Flawless finish Liquid make up	Carob tree extract, wild mushroom extract	Hydro-pigment technology, soy lipid technology	Foundation-covering
Royal Jelly Lift Concentrate	Royal jelly	Liposomes	Antiwrinkle

10.4.6 Antiaging Formulations Using Vesicular Delivery Systems

Apart from their ability to maintain the stability of active ingredients, vesicular systems are able to deliver active antiaging ingredients through the layers of skin right down to a cellular level, enhancing their effectiveness. With age, the cell membrane induces rigidness due to the function of cholesterol. It leads to decreased cell activity. Finally, the cells lose the ability to renew themselves, and therefore, essential collagen and elastin cannot be produced. Lipid-based vesicular systems can penetrate beneath the stratum corneum and deliver ingredients into living cells resulting in restored fluidity of the membrane, resulting in younger cells [119]. Moreover, a number of studies have been conducted to examine and investigate the effect of various actives using vesicular delivery systems (Table 10.3).

Table 10.3 List of researches referring to vesicular delivery systems and their antiaging effects

Vesicular Delivery System	Active Ingredients	Characteristics or Properties	Ref.
Liposomes	Licorice extract	Antioxidant activity, proliferation of fibroblasts, re-epitelization process	[165]
	Sodium ascorbyl phosphate	Increased penetration, antioxidant activity	[166]
	Vitamin E	Drug deposition	[167]
	Catechin	Antioxidant activity, anionic surfactants increase skin permeation	[168]
	C-Phycocyanin	Higher drug accumulation, anti-inflammatory activity	[169]
	Ursolic acid	Increased ceramide and collagen content	[170]

Vesicular Delivery System	Active Ingredients	Characteristics or Properties	Ref.
	Aloe vera	Higher keratinocytes proliferation and increased collagen synthesis	[171]
	Resveratrol	Increased release, antioxidant activity	[172]
	Resveratrol	Cell proliferation and photoprotection. Decreased cytotoxicity of resveratrol	[173]
	Palmitoyl hexapeptide and Vitamin E acetate	Controlled drug release	[174]
	Linoleic acid	Whitening effect	[175]
	Arbutin	Whitening activity and increased deposition amount in the epidermis/dermis layers	[176]
	Quercetin	Enhanced skin permeation, antioxidant activity	[177]
	Quercetin and curcumin	Enhanced skin penetration, antioxidant activity	[178]
	Octyl methoxycinnamate (OMC)	Multilammelar liposomes are good vehicle for OMC. Increased skin accumulation of OMC Small unilamellar liposome deeper penetration, increased SPF of OMC	[179]

(*Continued*)

Table 10.3 (*Continued*)

Vesicular Delivery System	Active Ingredients	Characteristics or Properties	Ref.
	Glycolic acid	Liposomes with chitosan illustrate better results in 5:1 glycolic acid/lipid molar ratio	[180]
	Resveratrol	Chitosan-coated liposomes had increased liposome stability and enhanced resveratrol skin permeation	[181]
	Rosmarinic acid	Antioxidant activity and inhibitory effects. Ethosomal formulation had better entrapment results than liposomal	[182]
	Curcumin	Increased skin permeation	[183]
	Apigenin	Antioxidant and anti-inflammatory activity with superior skin targeting	[184]
Transferosomes	Resveratrol	Antioxidant activity	[185]
	Epigallocatechin-3-gallate (EGCG) and hyaluronic acid	Increased cell viability and reduced lipid peroxidation	[186]
	Curcumin	Increased skin permeation	[183]
	Apigenin	Antioxidant and anti-inflammatory activity with superior skin targeting	[184]

Vesicular Delivery System	Active Ingredients	Characteristics or Properties	Ref.
Ethosomes	Curcumin	Improvement of skin viscoelasticity, biological elasticity, and sagginess	[187]
	Gammaoryzanol	Antioxidant effect	[188]
	Apigenin	Antioxidant and anti-inflammatory activity with superior skin targeting	[184]
	Rosmarinic acid	Antioxidant activity and inhibitory effects. Ethosomal formulation had better entrapment results than liposomal	[182]
Niosomes	Gallic acid (GA)	Improved GA skin permeation, antioxidant activity	[189]
	Zingiber Cassumunar	Anti-inflammatory effect	[190]
	Ellagic acid (EA)	Increased EA delivery	[191]
	Coenzyme Q10 (CoQ_{10})	protection of skin from UV	[192]
Phytosomes	(Quercevita®)	Reduced redness, itching, and inflammation of UV-damaged skin as well as the water loss of the skin, restoration of the skin barrier function and increased skin hydration	[193]

References

1. Binic, I., Lazarevic, V., Ljubenovic, M., Mojsa, J., and Sokolovic, D. (2013). Skin ageing: Natural weapons and strategies, *J Evid Based Complementary Altern Med*, p. 827248.

2. Rinnerthaler, M., Bischof, J., Streubel, M. K., Trost, A., and Richter, K. (2015). Oxidative stress in aging human skin, *Biomolecules*, **2**, pp. 545–89.
3. Bergler-Czop, B. and Miziołek, B. (2017). Aging: What do we know? *Acta Dermatovenerologica Croatica*, pp. 50–56.
4. Cevc, G., Blume, G., Schätzlein, A., Gebauer, D., and Paul, A. (1996). The skin: A pathway for systemic treatment with patches and lipid-based agent carriers, *Adv Drug Deliv Rev*, **3**, pp. 349–378.
5. Michniak-Kohn, B., Wertz, P., Al-Khalili, M., and Meidan, V. (2005). Skin: Physiology and penetration pathways. In *Delivery System Handbook for Personal Care and Cosmetic Products*, pp. 77–100. William Andrew.
6. Baroni, A., Buommino, E., De Gregorio, V., Ruocco, E., Ruocco, V., and Wolf, R. (2012). Structure and function of the epidermis related to barrier properties, *Clin Dermatol*, **3**, pp. 257–262.
7. Hu, F. (1979). Aging of melanocytes, *J Invest Dermatol*, **1**, pp. 70–79.
8. McCormick, R. and Vasilaki, A. (2018). Age-related changes in skeletal muscle: Changes to life-style as a therapy, *Biogerontology*, **6**, pp. 519–536.
9. Thakur, R., Batheja, P., Kaushik, D., and Michniak, B. (2009). Structural and biochemical changes in aging skin and their impact on skin permeability barrier. In *Skin Aging Handbook*, N. Dayan (Ed). William Andrew Publishing: Norwich, NY. pp. 55–90.
10. Yousef, H., Alhajj, M., and Sharma, S. (2019). Anatomy, Skin (Integument), Epidermis.
11. Wu, Q. (2019). Andrews' diseases of the skin: Clinical dermatology, 13th edition, *J Am Acad Dermatol*, **6**, p. e187.
12. Boer, M., Duchnik, E., Maleszka, R., and Marchlewicz, M. (2016). Structural and biophysical characteristics of human skin in maintaining proper epidermal barrier function, *Postepy dermatologii i alergologii*, **1**, pp. 1–5.
13. Fenske, N. A. and Lober, C. W. (1986). Structural and functional changes of normal aging skin, *J Am Acad Dermatol*, **4 Pt 1**, pp. 571–585.
14. Brown, T. M. and Krishnamurthy, K. (2021). *Histology, Dermis*. StatPearls Publishing: Treasure Island (FL).
15. Wong, R., Geyer, S., Weninger, W., Guimberteau, J. C., and Wong, J. K. (2016). The dynamic anatomy and patterning of skin, *Exp Dermatol*, **2**, pp. 92–98.

16. Stienstra, R., Dijk, W., Beek, L., Jansen, H., Heemskerk, M., Houtkooper, R., Denis, S., Harmelen, V., Dijk, K., Tack, C., and Kersten, S. (2014). Mannose-binding lectin is required for the effective clearance of apoptotic cells by adipose tissue macrophages during obesity, *Diabetes*, **63**, pp. 4143–4153.

17. Fjeldborg, K., Pedersen, S. B., Møller, H. J., Christiansen, T., Bennetzen, M., and Richelsen, B. (2014). Human adipose tissue macrophages are enhanced but changed to an anti-inflammatory profile in obesity, *J Immunol Res*, p. 309548.

18. Nguyen, K. D., Qiu, Y., Cui, X., Goh, Y. P., Mwangi, J., David, T., Mukundan, L., Brombacher, F., Locksley, R. M., and Chawla, A. (2011). Alternatively activated macrophages produce catecholamines to sustain adaptive thermogenesis, *Nature*, **7375**, pp. 104–108.

19. Lee, S. D. and Tontonoz, P. (2014). Eosinophils in fat: Pink is the new brown, *Cell*, **6**, pp. 1249–1250.

20. Guimberteau, J. C., Delage, J. P., McGrouther, D. A., and Wong, J. K. (2010). The microvacuolar system: how connective tissue sliding works, *J Hand Surg Eur Vol*, **8**, pp. 614–622.

21. Reed, R. K., Lidén, A., and Rubin, K. (2010). Edema and fluid dynamics in connective tissue remodelling, *J Mol Cell Cardiol*, **3**, pp. 518–523.

22. Serri, R. and Iorizzo, M. (2008). Combating aging skin, *Clin Dermatol*, **2**, p. 105.

23. Shiffman M. A., Mirrafati, S., Lam, S. M., and Cueteaux, C. G. (2008). *Simplified Facial Rejuvenation*. Springer.

24. Hunt, K. J., Hung, S. K., and Ernst, E. (2010). Botanical extracts as anti-aging preparations for the skin: A systematic review, *Drugs Aging*, **12**, pp. 973–985.

25. Wysong, A., Kim, D., Joseph, T., MacFarlane, D. F., Tang, J. Y., and Gladstone, H. B. (2014). Quantifying soft tissue loss in the aging male face using magnetic resonance imaging, *Dermatol Surg*, **7**, pp. 786–793.

26. Ichibori, R., Fujiwara, T., Tanigawa, T., Kanazawa, S., Shingaki, K., Torii, K., Tomita, K., Yano, K., Sakai, Y., and Hosokawa, K. (2014). Objective assessment of facial skin aging and the associated environmental factors in Japanese monozygotic twins, *J Cosmet Dermatol*, **2**, pp. 158–163.

27. Simić-Krstić, J. B., Kalauzi, A. J., Ribar, S. N., Matija, L. R., and Misevic, G. N. (2014). Electrical properties of human skin as aging biomarkers, *Exp Gerontol*, pp. 163–167.

28. Farage, M.A., Miller, K. W., Elsner, P., and Maibach, H. I. (2008). Intrinsic and extrinsic factors in skin ageing: A review, *Int J Cosmet Sci*, **2**, pp. 87–95.

29. Poljšak, B., Dahmane, R. G., and Godić, A. (2012). Intrinsic skin aging: The role of oxidative stress, *Acta Dermatovenerol Alp Pannonica Adriat*, **2**, pp. 33–36.

30. El-Domyati, M., Attia, S., Saleh, F., Brown, D., Birk, D. E., Gasparro, F., Ahmad, H., and Uitto, J. (2002). Intrinsic aging vs. photoaging: A comparative histopathological, immunohistochemical, and ultrastructural study of skin, *Exp Dermatol*, **5**, pp. 398–405.

31. Südel, K. M., Venzke, K., Mielke, H., Breitenbach, U., Mundt, C., Jaspers, S., Koop, U., Sauermann, K., Knussman-Hartig, E., Moll, I., Gercken, G., Young, A. R., Stäb, F., Wenck, H., and Gallinat, S. (2005). Novel aspects of intrinsic and extrinsic aging of human skin: Beneficial effects of soy extract, *Photochem Photobiol*, **3**, pp. 581–587.

32. Quatresooz, P. and Piérard, G. E. (2009). Immunohistochemical clues at aging of the skin microvascular unit, *J Cutan Pathol*, **1**, pp. 39–43.

33. Zouboulis, C. C. and Makrantonaki, E. (2011). Clinical aspects and molecular diagnostics of skin aging, *Clin Dermatol*, **1**, pp. 3–14.

34. Braverman, I. M. and Fonferko, E. (1982). Studies in cutaneous aging: I. The elastic fiber network, *J Invest Dermatol*, **5**, pp. 434–443.

35. Lee, D. H., Oh, J. H., and Chung, J. H. (2016). Glycosaminoglycan and proteoglycan in skin aging, *J Dermatol Sci*, **3**, pp. 174–181.

36. Mangoni, A. A. and Jackson, S. H. D. (2004). Age-related changes in pharmacokinetics and pharmacodynamics: Basic principles and practical applications, *British J Clin Pharm*, **1**, pp. 6–14.

37. Chen, Y. and Lyga, J. (2014). Brain-skin connection: Stress, inflammation and skin aging, *Inflamm Allergy Drug Targets*, **3**, pp. 177–190.

38. Babizhayev, M. A., Vishnyakova, K. S., and Yegorov, Y. E. (2014). Oxidative damage impact on aging and age-related diseases: Drug targeting of telomere attrition and dynamic telomerase activity flirting with imidazole-containing dipeptides, *Recent Pat Drug Deliv Formul*, **3**, pp. 163–192.

39. Poljšak, B. and Dahmane, R. (2012). Free radicals and extrinsic skin aging, *Dermatol Res Pract*, p. 135206.

40. Xu, X., Wang, B., Ren, C., Hu, J., Greenberg, D. A., Chen, T., Xie, L., and Jin, K. (2017). Age-related impairment of vascular structure and functions, *Aging Disease*, **5**, pp. 590–610.

41. Hollmerus, S., Yousaf, S., Islam, Y., and Khan, I. (2018). Isoflavones-based liposome formulations as anti-aging for skincare, *Novel Approaches in Drug Designing & Development*, pp. 1–4.

42. Krutmann, J. and Schroeder, P. (2009). Role of mitochondria in photoaging of human skin: The defective powerhouse model, *J Investig Dermatol Symp Proc*, **1**, pp. 44–49.

43. Gilchrest, B. A., Blog, F. B., and Szabo, G. (1979). Effects of aging and chronic sun exposure on melanocytes in human skin, *J Invest Dermatol*, **2**, pp. 141–143.

44. Waller, J. M. and Maibach, H. I. (2005). Age and skin structure and function, a quantitative approach (I): Blood flow, pH, thickness, and ultrasound echogenicity, *Skin Res Technol*, **4**, pp. 221–235.

45. Uitto, J., Fazio, M. J., and Olsen, D. R. (1989). Molecular mechanisms of cutaneous aging. Age-associated connective tissue alterations in the dermis, *J Am Acad Dermatol*, **3 Pt 2**, pp. 614–622.

46. Schiro, J. A., Chan, B. M., Roswit, W. T., Kassner, P. D., Pentland, A. P., Hemler, M. E., Eisen, A. Z., and Kupper, T. S. (1991). Integrin alpha 2 beta 1 (VLA-2) mediates reorganization and contraction of collagen matrices by human cells, *Cell*, **2**, pp. 403–410.

47. Varani, J., Spearman, D., Perone, P., Fligiel, S. E., Datta, S. C., Wang, Z. Q., Shao, Y., Kang, S., Fisher, G. J., and Voorhees, J. J. (2001). Inhibition of type I procollagen synthesis by damaged collagen in photoaged skin and by collagenase-degraded collagen in vitro, *Am J Pathol*, **3**, pp. 931–942.

48. Sugimoto, M., Yamashita, R., and Ueda, M. (2006). Telomere length of the skin in association with chronological aging and photoaging, *J Dermatol Sci*, **1**, pp. 43–47.

49. Gilchrest, B. A., Murphy, G. F., and Soter, N. A. (1982). Effect of chronologic aging and ultraviolet irradiation on Langerhans cells in human epidermis, *J Invest Dermatol*, **2**, pp. 85–88.

50. Giacomoni, P. U. and Rein, G. (2001). Factors of skin ageing share common mechanisms, *Biogerontology*, **4**, pp. 219–229.

51. Gilchrest B. A. and Krutmann, J. (2006). *Skin Aging*, 1 ed.: Springer-Verlag Berlin Heidelberg, p. 198.

52. Orioli, D. and Dellambra, E. (2018). Epigenetic regulation of skin cells in natural aging and premature aging diseases, *Cells*, **12**, p. 268.

53. Stefanatos, R. and Sanz, A. (2018). The role of mitochondrial ROS in the aging brain, *FEBS Lett*, **5**, pp. 743–758.

54. Pham-Huy, L. A., He, H., and Pham-Huy, C. (2008). Free radicals, antioxidants in disease and health, *Int J Biomed Sci*, **2**, pp. 89–96.

55. Phaniendra, A., Jestadi, D. B., and Periyasamy, L. (2015). Free radicals: Properties, sources, targets, and their implication in various diseases, *Indian J Clin Biochem*, **1**, pp. 11–26.

56. Halliwell, B. Free Radicals and Other Reactive Species in Disease, in eLS.

57. Ylä-Herttuala, S. (1999). Oxidized LDL and atherogenesis, *Ann N Y Acad Sci*, pp. 134–137.

58. Stadtman, E. R. and Levine, R. L. (2000). Protein oxidation, *Ann N Y Acad Sci*, pp. 191–208.

59. Marnett, L. J. (2000). Oxyradicals and DNA damage, *Carcinogenesis*, **3**, pp. 361–370.

60. BLACK, M. M. (1969). A modified radiographic method for measuring skin thickness, *Brit J Dermatol*, **9**, pp. 661–666.

61. Sauder, D. N. (1989). Effect of age on epidermal immune function, *Clin Geriat Med*, **1**, pp. 149–160.

62. Baker, H. and Kligman, A. M. (1967). Technique for estimating turnover time of human stratum corneum, *Arch Dermatol*, **4**, pp. 408–411.

63. Verzár, F. (1964). Aging of the collagen fiber, in *International Review of Connective Tissue Research*, D. A. Hall (Ed.). Elsevier. pp. 243–300.

64. Tsuji, T. and Hamada, T. (1981). Age-related changes in human dermal elastic fibres, *Brit J Dermatol*, **1**, pp. 57–63.

65. Montagna, W. and Carlisle, K. (1979). Structural changes in aging human skin, *J Invest Dermatol*, **1**, pp. 47–53.

66. Fazio, M. J., Olsen, D. R., Kuivaniemi, H., Chu, M. L., Davidson, J. M., Rosenbloom, J., and Uitto, J. (1988). Isolation and characterization of human elastin cDNAs, and age-associated variation in elastin gene expression in cultured skin fibroblasts, *Laboratory Investigation*, **3**, pp. 270–277.

67. Scott, J. E. (1988). Proteoglycan-fibrillar collagen interactions, *Biochem J*, **2**, pp. 313–323.

68. Bailey, A. J., Paul, R. G., and Knott, L. (1998). Mechanisms of maturation and ageing of collagen, *Mech Ageing Dev*, **1-2**, pp. 1–56.

69. Winkelmann, R. K. (1956). The cutaneous innervation of human newborn prepuce, *J Invest Dermatol*, **1**, pp. 53–67.

70. Mathewson, R. C. and Nava, P. B. (1985). Effects of age on Meissner corpuscles: A study of silver-impregnated neurites in mouse digital pads, *J Comp Neurol*, **2**, pp. 250–259.

71. Farage, M., Miller, K., Elsner, P., and Maibach, H. (2007). Structural characteristics of the aging skin: A review, *Cutaneous Ocular Toxicol*, pp. 343–357.

72. Zhang, S. and Duan, E. (2018). Fighting against skin aging: The way from bench to bedside, *Cell Transplant*, **5**, pp. 729–738.

73. Ganceviciene, R., Liakou, A. I., Theodoridis, A., Makrantonaki, E., and Zouboulis, C. C. (2012). Skin anti-aging strategies, *Dermato-endocrinology*, **3**, pp. 308–319.

74. Cernasov, D. (2009). The design and development of anti-aging formulations, *Skin Aging Handbook*, pp. 291–325.

75. Ravikumar, P. (2014). Liposomes as carriers in skin ageing, *Int J Curr Pharm Res*, pp. 1–7.

76. Wilson, N. K. (2009). Market evolution of topical anti-aging treatments. In *Skin Aging Handbook*, N. Dayan (Ed.), William Andrew Publishing: Norwich, NY. pp. 15–34.

77. Sekar, M. (2020). Natural products in aging skin. In *Aging*, Second edition, V. R. Preedy and V. B. Patel (Eds.), Academic Press, pp. 267–273.

78. Squier, C. A. (1991). The permeability of oral mucosa, *Crit Rev Oral Biol Med*, **1**, pp. 13–32.

79. Akimoto, K., Yoshikawa, N., Higaki, Y., Kawashima, M., and Imokawa, G. (1993). Quantitative analysis of stratum corneum lipids in xerosis and asteatotic eczema, *J Dermatol*, **1**, pp. 1–6.

80. Ghadially, R. (1998). Aging and the epidermal permeability barrier: Implications for contact dermatitis, *Am J Contact Dermat*, **3**, pp. 162–169.

81. Ngawhirunpat, T., Yoshikawa, H., Hatanaka, T., Koizumi, T., and Adachi, I. (2001). Age-related changes in skin permeability of hydrophilic and lipophilic compounds in rats, *Pharmazie*, **3**, pp. 231–234.

82. Rahimpour, Y. and Hamishehkar, H. (2012). Liposomes in cosmeceutics, *Expert Opin Drug Deliv*, **4**, pp. 443–455.

83. Kaur, I. P. and Agrawal, R. (2007). Nanotechnology: A new paradigm in cosmeceuticals, *Recent Pat Drug Deliv Formul*, **2**, pp. 171–182.

84. Roskos, K. V. and Maibach, H. I. (1992). Percutaneous absorption and age, *Drugs Aging*, **5**, pp. 432–449.

85. Ngawhirunpat, T., Hatanaka, T., Kawakami, J., and Adachi, I. (2001). Age difference in simultaneous permeation and metabolism of ethyl nicotinate in rat skin, *Biol Pharm Bull*, **4**, pp. 414–417.

86. Peram, M. R., Jalalpure, S., Kumbar, V., Patil, S., Joshi, S., Bhat, K., and Diwan, P. (2019). Factorial design based curcumin ethosomal nanocarriers for the skin cancer delivery: In vitro evaluation, *J Liposome Res*, **3**, pp. 291–311.

87. Modi, C. D. and Bharadia, P. (2012). Transfersomes: New dominants for transdermal drug delivery, *Am J PharmTech Res*, pp. 71–91.

88. Kim, B., Cho, H.-E., Moon, S. H., Ahn, H.-J., Bae, S., Cho, H.-D., and An, S. (2020). Transdermal delivery systems in cosmetics, *Biomed Dermatol*, **1**, p. 10.

89. Fernández-García, R., Lalatsa, A., Statts, L., Bolás-Fernández, F., Ballesteros, M. P., and Serrano, D. R. (2020). Transferosomes as nanocarriers for drugs across the skin: Quality by design from lab to industrial scale, *Int J Pharm*, p. 118817.

90. Khan, I., Elhissi, A., Shah, M., Alhnan, M. A., and Ahmed, W. (2013). Liposome-based carrier systems and devices used for pulmonary drug delivery. In *Biomaterials and Medical Tribology*, J. P. Davim (Ed.), Woodhead Publishing. pp. 395–443.

91. Apostolou, M., Assi, S., Fatokun, A. A., and Khan, I. (2021). The effects of solid and liquid lipids on the physicochemical properties of nanostructured lipid carriers, *J Pharm Sci*, **110**, pp. 2859–2872.

92. Kalluri, H. and Banga, A. K. (2011). Transdermal delivery of proteins, *AAPS PharmSciTech*, **1**, pp. 431–441.

93. Cevc, G., Schätzlein, A., and Blume, G. (1995). Transdermal drug carriers: Basic properties, optimization and transfer efficiency in the case of epicutaneously applied peptides, *J Controlled Release*, **1**, pp. 3–16.

94. Elias, P. M. and Menon, G. K. (1991). Structural and lipid biochemical correlates of the epidermal permeability barrier. In *Advances in Lipid Research*, P. M. Elias (Ed.), Elsevier. pp. 1–26.

95. Cevc, G., Blume, G., and Schätzlein, A. (1997). Transfersomes-mediated transepidermal delivery improves the regio-specificity and biological activity of corticosteroids in vivo, *J Controlled Release*, **45**, pp. 211–226.

96. Lee, S. H., Jeong, S. K., and Ahn, S. K. (2006). An update of the defensive barrier function of skin, *Yonsei Med J*, **3**, pp. 293–306.

97. Lai, F., Caddeo, C., Manca, M. L., Manconi, M., Sinico, C., and Fadda, A. M. (2020). What's new in the field of phospholipid vesicular nanocarriers for skin drug delivery, *Int J Pharm*, p. 119398.

98. Bolzinger, M.-A., Briançon, S., Pelletier, J., and Chevalier, Y. (2012). Penetration of drugs through skin, a complex rate-controlling membrane, *Curr Opin Colloid Interface Sci*, pp. 156–165.

99. Kumar, A., Pathak, K., and Bali, V. (2012). Ultra-adaptable nanovesicular systems: A carrier for systemic delivery of therapeutic agents, *Drug Discovery Today*, **21–22**, pp. 1233–1241.

100. Dragicevic, N. and Maibach, H. (2018). Combined use of nanocarriers and physical methods for percutaneous penetration enhancement, *Adv Drug Deliv Rev*, pp. 58–84.

101. Mathur, V., Satrawala, Y., and Rajput, M. (2010). Physical and chemical penetration enhancers in transdermal drug delivery system, *Asian J Pharm*, **4**.

102. Haque, T. and Talukder, M. M. U. (2018). Chemical enhancer: A simplistic way to modulate barrier function of the stratum corneum, *Adv Pharm Bull*, **2**, pp. 169–179.

103. Opatha, S. A. T., Titapiwatanakun, V., and Chutoprapat, R. (2020). Transfersomes: A promising nanoencapsulation technique for transdermal drug delivery, *Pharmaceutics*, **9**, p. 855.

104. Rajan, R., Jose, S., Mukund, V. P. B., and Vasudevan, D. T. (2011). Transferosomes: A vesicular transdermal delivery system for enhanced drug permeation, *J Adv Pharm Technol Res*, **3**, pp. 138–143.

105. Hua, S. (2015). Lipid-based nano-delivery systems for skin delivery of drugs and bioactives, *Frontiers in Pharmacology*, **219**.

106. Bochicchio, S., Dalmoro, A., De Simone, V., Bertoncin, P., Lamberti, G., and Barba, A. A. (2020). Simil-microfluidic nanotechnology in manufacturing of liposomes as hydrophobic antioxidants skin release systems, *Cosmetics*, **2**, p. 22.

107. Khan, I., Yousaf, S., Najlah, M., Ahmed, W., and Elhissi, A. (2021). Proliposome powder or tablets for generating inhalable liposomes using a medical nebulizer, *J Pharm Investigation*, **1**, pp. 61–73.

108. Bnyan, R., Cesarini, L., Khan, I., Roberts, M., and Ehtezazi, T. (2020). The effect of ethanol evaporation on the properties of inkjet produced liposomes, *DARU J Pharm Sci*, **1**, pp. 271–280.

109. Khan, I., Yousaf, S., Subramanian, S., Alhnan, M. A., Ahmed, W., and Elhissi, A. (2018). Proliposome powders for the generation of liposomes: The influence of carbohydrate carrier and separation conditions on crystallinity and entrapment of a model antiasthma steroid, *AAPS PharmSciTech*, **1**, pp. 262–274.

110. Gala, R. P., Khan, I., Elhissi, A. M. A., and Alhnan, M. A. (2015). A comprehensive production method of self-cryoprotected nano-liposome powders, *Int J Pharm*, **1**, pp. 153–158.

111. Contri, R., Fiel, L., Guterres, Pohlmann, A., and Beck, R. (2011). Transport of substances and nanoparticles across the skin and in vitro models to

evaluate skin permeation and/or penetration. In *Nanocosmetics and Nanomedicines*. Springer Berlin Heidelberg, pp. 3–35.

112. Uchechi, O., Ogbonna, J., and Attama, A. (2014). Nanoparticles for dermal and transdermal drug delivery. In *Application of Nanotechnology in Drug Delivery*. IntechOpen.

113. Hsiao, P. F., Tsai, H.-C., Peng, S., Prasannan, A., Tang, T.-C., Chang, H.-M., Chen, Y.-S., Lin, S.-Y., and Hsiue, G.-H. (2019). Transdermal delivery of poly(ethylene glycol)-co-oleylamine modified gold nanoparticles: Effect of size and shape, *Mater Chem Phys*, pp. 22–28.

114. Rosen, H. and Abribat, T. (2005). The rise and rise of drug delivery, *Nat Rev Drug Discov*, **5**, pp. 381–385.

115. Gökçe, E. H., Yapar, E. A., Tanrıverdi, S. T., and Özer, Ö. (2016). Nanocarriers in cosmetology. In *Nanobiomaterials in Galenic Formulations and Cosmetics*, A. M. Grumezescu (Ed.), William Andrew Publishing. pp. 363–393.

116. Vinardell, M. P. and Mitjans, M. (2015). Nanocarriers for delivery of antioxidants on the skin, *Cosmetics*, **4**, pp. 342–354.

117. Ansam, M., Bnyan, R., Khan, I., and Yousaf, S. (2018). Anti-aging liposomal formulation: A mini review, *Novel Approaches Drug Designing Development*, **3**, pp. 66–68.

118. Abramovits, W., Granowski, P., and Arrazola, P. (2010). Applications of nanomedicine in dermatology: Use of nanoparticles in various therapies and imaging, *J Cosmet Dermatol*, **2**, pp. 154–159.

119. Khezri, K., Saeedi, M., and Maleki Dizaj, S. (2018). Application of nanoparticles in percutaneous delivery of active ingredients in cosmetic preparations, *Biomed Pharmacother*, pp. 1499–1505.

120. Honary, S. and Zahir, F. (2013). Effect of zeta potential on the properties of nano-drug delivery systems: A review (Part 1), *Tropical J Pharm Res*, **12**.

121. Soema, P. C., Willems, G.-J., Jiskoot, W., Amorij, J.-P., and Kersten, G. F. (2015). Predicting the influence of liposomal lipid composition on liposome size, zeta potential and liposome-induced dendritic cell maturation using a design of experiments approach, *Eur J Pharm Biopharm*, pp. 427–435.

122. Makino, K. and Shibata, A. (2006). Surface properties of liposomes depending on their composition. In *Advances in Planar Lipid Bilayers and Liposomes*, A. L. Liu (Ed.), Academic Press, pp. 49–77.

123. Khan, I., Lau, K., Bnyan, R., Houacine, C., Roberts, M., Isreb, A., Elhissi, A., and Yousaf, S. (2020). A facile and novel approach to manufacture

paclitaxel-loaded Proliposome tablet formulations of micro or nano vesicles for nebulization, *Pharm Res*, **6**, p. 116.

124. Khan, I., Apostolou, M., Bnyan, R., Houacine, C., Elhissi, A., and Yousaf, S.S. (2020). Paclitaxel-loaded micro or nano transfersome formulation into novel tablets for pulmonary drug delivery via nebulization, *Int J Pharm*, p. 118919.

125. Bnyan, R., Khan, I., Ehtezazi, T., Saleem, I., Gordon, S., O'Neill, F., and Roberts, M. (2018). Surfactant effects on lipid-based vesicles properties, *J Pharm Sci*, **5**, pp. 1237–1246.

126. Khan, I., Yousaf, S., Subramanian, S., Korale, O., Alhnan, M. A., Ahmed, W., Taylor, K. M. G., and Elhissi, A. (2015). Proliposome powders prepared using a slurry method for the generation of beclometasone dipropionate liposomes, *Int J Pharm*, **2**, pp. 342–350.

127. Khogta, S., Patel, J., Barve, K., and Londhe, V. (2020). Herbal nano-formulations for topical delivery, *J Herbal Med*, p. 100300.

128. Khan, I., Hussein, S., Houacine, C., Khan Sadozai, S., Islam, Y., Bnyan, R., Elhissi, A., and Yousaf, S. (2021). Fabrication, characterization and optimization of nanostructured lipid carrier formulations using Beclomethasone dipropionate for pulmonary drug delivery via medical nebulizers, *Int J Pharm*, p. 120376.

129. Khan, I., Yousaf, S., Alhnan, M. A., Ahmed, W., Elhissi, A., and Jackson, M. J. (2016). Design characteristics of inhaler devices used for pulmonary delivery of medical aerosols. In *Surgical Tools and Medical Devices*, W. Ahmed and M. J. Jackson (Eds.), Springer International Publishing: Cham. pp. 573–591.

130. Gupta, P., Singh, P., Mishra, V., Jain, S., Dubey, P., and Vyas, S. (2003). Topical immunization: Mechanistic insight and novel delivery systems, *Ind J Bio*, **3**.

131. Singh, D., Pradhan, M., Nag, M., and Singh, M. R. (2015). Vesicular system: Versatile carrier for transdermal delivery of bioactives, *Artif Cells Nanomed Biotechnol*, **4**, pp. 282–290.

132. Bnyan, R., Khan, I., Ehtezazi, T., Saleem, I., Gordon, S., O'Neill, F., and Roberts, M. (2019). Formulation and optimisation of novel transfersomes for sustained release of local anaesthetic, *J Pharm* , **10**, pp. 1508–1519.

133. Kaul, S., Gulati, N., Verma, D., Mukherjee, S., and Nagaich, U. (2018). Role of nanotechnology in cosmeceuticals: A review of recent advances, *J Pharm*, p. 3420204.

134. Subramanian, S., Khan, I., Korale, O., Alhnan, M. A., Ahmed, W., Najlah, M., Taylor, K. M. G., and Elhissi, A. (2016). A simple approach to

predict the stability of phospholipid vesicles to nebulization without performing aerosolization studies, *Int J Pharm*, **1–2**, pp. 18–27.

135. Daraee, H., Etemadi, A., Kouhi, M., Alimirzalu, S., and Akbarzadeh, A. (2016). Application of liposomes in medicine and drug delivery, *Artificial Cells, Nanomed, Biotechnol*, **1**, pp. 381–391.

136. Bozzuto, G. and Molinari, A. (2015). Liposomes as nanomedical devices, *Int J Nanomed*, pp. 975–999.

137. Khan, I., Yousaf, S., Subramanian, S., Albed Alhnan, M., Ahmed, W., and Elhissi, A. (2018). Proliposome tablets manufactured using a slurry-driven lipid-enriched powders: Development, characterization and stability evaluation, *Int J Pharm*, **1**, pp. 250–262.

138. Mirtaleb, M. S., Shahraky, M. K., Ekrami, E., and Mirtaleb, A. (2021). Advances in biological nano-phospholipid vesicles for transdermal delivery: A review on applications, *J Drug Deliv Sci Technol*, p. 102331.

139. Chouhan, N., Sharma, M., and Vaidya, S. (2020). Liposome: A comprehensive approach for researchers. In *Molecular Pharmacology*, IntechOpen.

140. Laouini, A., Jaafar-Maalej, C., Limayem-Blouza, I., Sfar, S., Charcosset, C., and Fessi, H. (2012). Preparation, characterization and applications of liposomes: State of the art, *J Colloid Sci Biotechnol*, **2**, pp. 147–168.

141. Dwivedi, C., Sahu, R., Tiwari, S., Satapathy, T., and Roy, A. (2014). Role of liposome in novel drug delivery system, *J Drug Deliv Therap*, **4**.

142. Pierre, M. B. R. and dos Santos Miranda Costa, I. (2011). Liposomal systems as drug delivery vehicles for dermal and transdermal applications, *Arch Dermatol Res*, **9**, p. 607.

143. Farris, P. K. (2018). Topical skin care and the cosmetic patient. In *Master Techniques in Facial Rejuvenation*, Second Edition, B. Azizzadeh, et al. (Ed.), Elsevier. pp. 68–72.e2.

144. El Maghraby, G. M. and Williams, A. C. (2009). Vesicular systems for delivering conventional small organic molecules and larger macromolecules to and through human skin, *Expert Opin Drug Deliv*, **2**, pp. 149–163.

145. Egbaria, K. and Weiner, N. (1990). Liposomes as a topical drug delivery system, *Adv Drug Deliv Rev*, **3**, pp. 287–300.

146. Ravikumar, P. (2014). Liposomes as carriers in skin ageing, *Int J Pharm Sci Res*, pp. 1–7.

147. Gang, W., Ping, Z., Ming, D., Ying, L., Lei, W., and Fang, Y. (2012). Liposomal quercetin: Evaluating drug delivery in vitro and biodistribution in vivo, *Expert Opin Drug Deliv*, pp. 599–613.

148. Brewer, J., Bloksgaard, M., Kubiak, J., Sørensen, J. A., and Bagatolli, L. A. (2013). Spatially resolved two-color diffusion measurements in human skin applied to transdermal liposome penetration, *J Invest Dermatol*, **5**, pp. 1260–1268.

149. Yadav, A. V., Murthy, M. S., Shete, A. S., and Sakhare, S. (2011). Stability aspects of liposome, *Indian J Pharm Educ*, pp. 402–413.

150. Arnardóttir, H. B., Sveinsson, S. J., and Kristmundsdóttir, T. (1996). The release of clindamycin phosphate from a suspension of different types of liposomes and selected topical dosages forms, *Int J Pharm*, **1**, pp. 71–77.

151. Ghanbarzadeh, S., Khorrami, A., and Arami, S. (2015). Nonionic surfactant-based vesicular system for transdermal drug delivery, *Drug Deliv*, **8**, pp. 1071–1077.

152. Kalyani, V. and Babu, M. (2014). Transferosomes: An emerging tool for effective transdermal drug delivery, *J Chem Pharm Sci*, pp. 236–241.

153. Rai, S., Pandey, V., and Rai, G. (2017). Transfersomes as versatile and flexible nano-vesicular carriers in skin cancer therapy: The state of the art, *Nano Rev Exp*, **1**, pp. 1325708–1325708.

154. Cevc, G., Schätzlein, A., and Richardsen, H. (2002). Ultradeformable lipid vesicles can penetrate the skin and other semi-permeable barriers unfragmented. Evidence from double label CLSM experiments and direct size measurements, *Biochimica et Biophysica Acta (BBA) - Biomembranes*, **1**, pp. 21–30.

155. Limsuwan, T., Boonme, P., Khongkow, P., and Amnuaikit, T. (2017). Ethosomes of phenylethyl resorcinol as vesicular delivery system for skin lightening applications, *BioMed Res Int*, p. 8310979.

156. Sguizzato, M., Mariani, P., Spinozzi, F., Benedusi, M., Cervellati, F., Cortesi, R., Drechsler, M., Prieux, R., Valacchi, G., and Esposito, E. (2020). Ethosomes for coenzyme Q10 cutaneous administration: From design to 3D skin tissue evaluation, *Antioxidants (Basel, Switzerland)*, **6**, p. 485.

157. Arora, D. and Nanda, S. (2019). Quality by design driven development of resveratrol loaded ethosomal hydrogel for improved dermatological benefits via enhanced skin permeation and retention, *Int J Pharm*, p. 118448.

158. Khan, R. and Irchhaiya, R. (2016). Niosomes: A potential tool for novel drug delivery, *J Pharm Investig*, **3**, pp. 195–204.

159. Lu, B., Huang, Y., Chen, Z., Ye, J., Xu, H., Chen, W., and Long, X. (2019). Niosomal nanocarriers for enhanced skin delivery of quercetin with functions of anti-tyrosinase and antioxidant, *Molecules*, **12**.

160. Mohamed, M. I., Abdelbary, A. A., Kandil, S. M., and Mahmoud, T. M. (2019). Preparation and evaluation of optimized zolmitriptan niosomal emulgel, *Drug Dev Ind Pharm*, **7**, pp. 1157–1167.

161. Montenegro, L. (2014). Nanocarriers for skin delivery of cosmetic antioxidants, *J Pharm Res*, **4**, pp. 73–92.

162. Escobar, S., Lintner, K., Closs, B., and Kerob, D. (2020). 16832 Peptides: An effective ingredient to complement vitamin C for skin aging, *J Am Acad Dermatol*, **6, Supplement**, p. AB187.

163. Swaminathan, J. and Ehrhardt, C. (2012). Liposomal delivery of proteins and peptides, *Expert Opin Drug Deliv*, **12**, pp. 1489–1503.

164. Patravale, V. B. and Mandawgade, S. D. (2008). Novel cosmetic delivery systems: An application update, *Int J Cosmet Sci*, **1**, pp. 19–33.

165. Castangia, I., Caddeo, C., Manca, M. L., Casu, L., Latorre, A. C., Díez-Sales, O., Ruiz-Saurí, A., Bacchetta, G., Fadda, A. M., and Manconi, M. (2015). Delivery of liquorice extract by liposomes and hyalurosomes to protect the skin against oxidative stress injuries, *Carbohydr Polym*, pp. 657–663.

166. Foco, A., Gasperlin, M., and Kristl, J. (2005). Investigation of liposomes as carriers of sodium ascorbyl phosphate for cutaneous photoprotection, *Int J Pharm*, **1–2**, pp. 21–29.

167. Padamwar, M. N. and Pokharkar, V. B. (2006). Development of vitamin loaded topical liposomal formulation using factorial design approach: Drug deposition and stability, *Int J Pharm*, **1–2**, pp. 37–44.

168. Fang, J. Y., Hwang, T. L., Huang, Y. L., and Fang, C. L. (2006). Enhancement of the transdermal delivery of catechins by liposomes incorporating anionic surfactants and ethanol, *Int J Pharm*, **1–2**, pp. 131–138.

169. Manconia, M., Pendás, J., Ledón, N., Moreira, T., Sinico, C., Saso, L., and Fadda, A. M. (2009). Phycocyanin liposomes for topical anti-inflammatory activity: In-vitro in-vivo studies, *J Pharm Pharmacol*, **4**, pp. 423–430.

170. Yarosh, D. B., Both, D., and Brown, D. (2000). Liposomal ursolic acid (merotaine) increases ceramides and collagen in human skin, *Horm Res*, **5–6**, pp. 318–321.

171. Takahashi, M., Kitamoto, D., Asikin, Y., Takara, K., and Wada, K. (2009). Liposomes encapsulating Aloe vera leaf gel extract significantly enhance proliferation and collagen synthesis in human skin cell lines, *J Oleo Sci*, **12**, pp. 643–650.

172. Vanaja, K., Wahl, M. A., Bukarica, L., and Heinle, H. (2013). Liposomes as carriers of the lipid soluble antioxidant resveratrol: Evaluation of

amelioration of oxidative stress by additional antioxidant vitamin, *Life Sci*, **24**, pp. 917–923.

173. Caddeo, C., Teskac, K., Sinico, C., and Kristl, J. (2008). Effect of resveratrol incorporated in liposomes on proliferation and UV-B protection of cells, *Int J Pharm*, **1–2**, pp. 183–191.

174. R., S. S. A. (2010). Development and evaluation of cosmeceutical nanolipogel. *Res J Top Cosmet Sci*, pp. 18–24.

175. Shigeta, Y., Imanaka, H., Ando, H., Ryu, A., Oku, N., Baba, N., and Makino, T. (2004). Skin whitening effect of linoleic acid is enhanced by liposomal formulations, *Biol Pharm Bull*, **4**, pp. 591–594.

176. Wen, A. H., Choi, M. K., and Kim, D. D. (2006). Formulation of liposome for topical delivery of arbutin, *Arch Pharm Res*, **12**, pp. 1187–1192.

177. Liu, D., Hu, H., Lin, Z., Chen, D., Zhu, Y., Hou, S., and Shi, X. (2013). Quercetin deformable liposome: Preparation and efficacy against ultraviolet B induced skin damages in vitro and in vivo, *J Photochem Photobiol B, Biol*, pp. 8–17.

178. Di Marzio, L., Ventura, C. A., Cosco, D., Paolino, D., Di Stefano, A., Stancanelli, R., Tommasini, S., Cannavà, C., Celia, C., and Fresta, M. (2016). Nanotherapeutics for anti-inflammatory delivery, *J Drug Deliv Sci Technol*, pp. 174–191.

179. Golmohammadzadeh, S., Jaafarixx, M. R., and Khalili, N. (2008). Evaluation of liposomal and conventional formulations of octyl methoxycinnamate on human percutaneous absorption using the stripping method, *J Cosmet Sci*, **5**, pp. 385–398.

180. Perugini, P., Genta, I., Pavanetto, F., Conti, B., Scalia, S., and Baruffini, A. (2000). Study on glycolic acid delivery by liposomes and microspheres, *Int J Pharm*, **1**, pp. 51–61.

181. Park, S., Jo, N. R., and Jeon, S. H. (2014). Chitosan-coated liposomes for enhanced skin permeation of resveratrol, *J Ind Eng Chem*, pp. 1481–1485.

182. Yücel, Ç., Şeker Karatoprak, G., and Değim İ, T. (2019). Anti-aging formulation of rosmarinic acid-loaded ethosomes and liposomes, *J Microencapsul*, **2**, pp. 180–191.

183. Kaur, C. D. and Saraf, S. (2011). Topical vesicular formulations of *Curcuma longa* extract on recuperating the ultraviolet radiation-damaged skin, *J Cosmet Dermatol*, **4**, pp. 260–265.

184. Shen, L. N., Zhang, Y. T., Wang, Q., Xu, L., and Feng, N. P. (2014). Enhanced in vitro and in vivo skin deposition of apigenin delivered using ethosomes, *Int J Pharm*, **1–2**, pp. 280–288.

185. Wu, P. S., Li, Y. S., Kuo, Y. C., Tsai, S. J., and Lin, C. C. (2019). Preparation and evaluation of novel transfersomes combined with the natural antioxidant resveratrol, *Molecules*, **3**.

186. Avadhani, K. S., Manikkath, J., Tiwari, M., Chandrasekhar, M., Godavarthi, A., Vidya, S. M., Hariharapura, R. C., Kalthur, G., Udupa, N., and Mutalik, S. (2017). Skin delivery of epigallocatechin-3-gallate (EGCG) and hyaluronic acid loaded nano-transfersomes for antioxidant and anti-aging effects in UV radiation induced skin damage, *Drug Deliv*, **1**, pp. 61–74.

187. Jeswani, G. (2014). Topical delivery of *Curcuma longa* extract loaded nanosized ethosomes to combat facial wrinkles, *J Pharm Drug Deliv Res.*

188. Heydari, S., Ghanbarzadeh, S., Anoush, B., Ranjkesh, M., Javadzadeh, Y., Kouhsoltani, M., and Hamishehkar, H. (2017). Nanoethosomal formulation of gammaoryzanol for skin-aging protection and wrinkle improvement: A histopathological study, *Drug Dev Ind Pharm*, **7**, pp. 1154–1162.

189. Manosroi, A., Jantrawut, P., Akazawa, H., Akihisa, T., Manosroi, W., and Manosroi, J. (2011). Transdermal absorption enhancement of gel containing elastic niosomes loaded with gallic acid from *Terminalia chebula* galls, *Pharm Biol*, **6**, pp. 553–562.

190. Priprem, A., Janpim, K., Nualkaew, S., and Mahakunakorn, P. (2016). Topical niosome gel of *Zingiber cassumunar* Roxb. extract for anti-inflammatory activity enhanced skin permeation and stability of compound D, *AAPS PharmSciTech*, **3**, pp. 631–639.

191. Junyaprasert, V. B., Singhsa, P., Suksiriworapong, J., and Chantasart, D. (2012). Physicochemical properties and skin permeation of Span 60/Tween 60 niosomes of ellagic acid, *Int J Pharm*, **2**, pp. 303–311.

192. Yadav, N. K., Nanda, S., Sharma, G., and Katare, O. P. (2016). Systematically optimized coenzyme q10-loaded novel proniosomal formulation for treatment of photo-induced aging in mice: Characterization, biocompatibility studies, biochemical estimations and anti-aging evaluation, *J Drug Target*, **3**, pp. 257–271.

193. Maramaldi, G., Togni, S., Pagin, I., Giacomelli, L., Cattaneo, R., Eggenhöffner, R., and Burastero, S. E. (2016). Soothing and anti-itch effect of quercetin phytosome in human subjects: A single-blind study, *Clin Cosmet Investig Dermatol*, pp. 55–62.

Chapter 11

Nanoparticle Dispersions for Engineering Application

Maje A. Haruna[a,b] **and Saminu M. Magami**[c,d]
[a]*School of Chemical and Process Engineering, University of Leeds, LS2 9JT, Leeds, United Kingdom*
[b]*Federal University Dutse, 720101, Dutse, Jigawa State, Nigeria*
[c]*Wigan, Greater Manchester, WN3 6GJ, United Kingdom*
[d]*Yusuf Maitama Sule University, P.M.B. 3220, Kano, Nigeria*
majeharuna@gmail.com, m.a.haruna@leeds.ac.uk

In the recent years, nanoparticle suspensions have shown many interesting properties. Their distinctive features offer unprecedented potential for many applications. This chapter provides an overview of the current research on nanoparticle dispersions. Nanoparticle production techniques are described, covering *in situ* and *ex situ* methods, and paying particular attention to how nanoparticle dispersions in polymer matrices are assembled. An overview of the key concepts such as the dispersibility of nanoparticles, the stability of the nanoparticles in the dispersed systems, and the use of dispersant additives to enhance performance properties is given. An account is given of key techniques that involve analysis

Science and Applications of Nanoparticles
Edited by Waqar Ahmed and Ehsan Nourafkan

ISBN 978-981-4267-34-2 (Hardcover), 978-1-003-28029-3 (eBook)
www.jennystanford.com

of the particles contained in the dispersions or the analysis of the dispersion systems as a whole. These include zeta potential methods of analyses, common particle morphology test procedures, spectral/spectroscopic measurements, analyses of the specific surface area of the particles, light-scattering techniques, and rheological analyses methods. Beyond the general application of nanoparticle dispersions, the processes and mechanisms associated with their use in enhanced oil recovery and in fire-resistant glass assemblies are described.

11.1 Introduction

Nanoparticle dispersions are compositions where sub-micron particles are dispersed in a continuous medium. This medium can be aqueous, nonaqueous, or polymeric. There is a wide variety of applications for nanoparticle dispersions in the preparation of materials such as coatings [1], paints, inks [2], drug-delivery systems, ceramics, and nanocomposites. Nanoparticle dispersions are characterized by having a harmonized composition where the overall particles have overcome a low energy barrier that is characterized by reduced electrostatic repulsive forces that exist between the primary particles, hence having a low probability for significant particle–particle collisions to occur. This phenomenon is correlated with the particle size, the particle size distribution, the zeta potential characteristic, and the present and the amount of stabilizing additives in the formulation(s).

Nanoparticle dispersions are typically assembled by dispersing powders in a fluid medium. High-intensity energy is often required in order to breakdown clusters of particles contained in the powder [3]. These could include aggregates of particles, held together by physical, chemical, or mechanical forces. Agglomerates of particles also need to be dispersed, and these are typically held by physical forces. In some cases, high-pressure homogenization techniques can be used, providing an effective approach to break down any agglomerated nanoparticles that are suspended in the continuous medium [4]. Table 11.1 provides some examples of nanoparticle dispersions, their continuous medium, characteristic particle size distribution, and associated stabilization method(s).

Table 11.1 Some examples of nanoparticle dispersions, their continuous medium, particle size and stabilization

Nanoparticle Dispersions	Continuous Medium	Stabilization	Particle Size	Reference
Ag nanoparticles	Silicon oil-based dispersion	Stabilized by a surfactant system	3 nm average particle diameter	[4]
Al_2O_3 nanoparticles	Aqueous dispersion	Stabilized by electrostatic forces and the enhancement of colloidal stability using electrostatic repulsion	13 nm primary particles, 150 nm aggregates	[3]
Al_2O_3 nanoparticles	Organic solvent-based dispersion	Stabilized by steric effect	Large agglomerates ~1700 nm and smaller components ~220 nm	[3]
$Ca(OH)_2$ nanoparticles	Ethanol or 2-propanol dispersions	Alcohol-induced stabilization	—	[5]
CeO_2 nanoparticles	Aqueous dispersion	Stabilized by the activity of poly (acrylic acid)	—	[6]
Ta_2O_5 nanoparticles	Aqueous salt dispersion	Surfactant-free stabilization	2 nm hydrodynamic size	[7]
TiO_2 nanoparticles	Polymer-based dispersion	Stabilized by surface treatment of particles	310–355 nm	[8]
α-Fe_2O_3 nanoparticles	Ethylene glycol-based dispersion		18 nm	[9]
BN	Aqueous dispersion	Surfactant-based electrostatic effect	<10 nm bundle size	[10]
ZrO_2 nanoparticles	Aqueous dispersion	Non-ionic polyvinyl pyrrolidone surfactant	59.9 ± 13.5 nm	[11]

The characteristics of the particles and the nature of their continuous medium often govern the application of the dispersions. These properties and the methods of determining them are outlined in designated sections. As Table 11.1 shows, properties such as particle size and size distribution of nanoparticles in dispersions are key to the dispersion's stability. These also affect the opacity or transparency of the dispersions. There is typically a relationship between particle size and light scattering, as supported by the Mie theory. This theory describes the relationship between the amount of incident light scattered by a spherical medium, having a refractive index that differs from that of the dispersing medium and the size and size distribution of the spherical particles. The larger the size of the particles, the greater the scattering of light and, therefore, the more opaque the dispersed system appears.

11.2 Nanoparticle Production Methods

11.2.1 "Bottom-Up" Process

The "bottom-up" approach comprises the chemical synthesis and/or the highly guarded deposition and material growth. The chemical synthesis may be accomplished in the liquid, vapor, or solid phase. Attaining a nanoscale system in a solid state is difficult. However, it is more common for vapor and liquid fabrication to be achieved using this method [12]. The nanofabrication using the "bottom-up" method depends on building the nanostructure atom-by-atom using either the manipulation of atoms or self-assembly techniques. These can be monitored by applying scanning probing microscopy [13].

11.2.2 "Top-Down" Process

The "top-down" method is more elementary and relies either on the division or the removal of bulk materials in order to produce the desired structure with suited properties, by minimizing bulk fabrication processes [12]. The method is based on microlithographic and physical philosophy, contrary to the "bottom-up" process where molecular or atomic units are used to gather molecular structures, in the "top-down" approach. Milling is a common fabrication method and is often referred to as mechanical alloying or mechanical attrition. This technique is highly desirable in industrial processes because it can be operated on a large scale. However, the lithographic process is the most common method. This process uses ultraviolet light or X-rays or ions or electrons to project image consisting of a liable pattern onto a photo-resisting surface. The advantage of the "top-down" approach is that it does not require an assembly step because the parts are both patterned and build in place [12].

11.3 Nanoparticle Dispersions in Polymer Matrices

Polymer systems are used extensively in different industrial applications due to their unique characteristics. These characteristics include light weight, ductility, and ease of production. However, polymers can have lower strength and modulus as compared to ceramics and metals. To enhance their properties, such polymers have to be reinforced with inclusions (particles, whiskers, or platelets) [14]. Using this approach, properties of the polymers can be improved and at the same time maintaining their ductility and light weight [14]. These improvements can be achieved even at low filler content. Depending on the particle shape, size, chemical nature, and specific surface area, the properties of the polymers are improved [15].

The mixing of polymers with nanoparticles provides a pathway for flexible engineering composites. The physical presence of the nanoparticles, the polymer–particle interaction, and the state of the dispersion lead to enhanced properties in the compositions [16]. Polymer nanocomposites are at the lead of advanced materials applications, where the incorporation of dispersed nanoparticles into a polymer matrix renders materials with tailored and controlled properties without compromising the processing ease of the parent polymer [17, 18].

Polymer nanocomposites can be fabricated by either a mechanical or a chemical process. One of the major problems in polymer nanocomposite preparation is achieving homogenous and uniform dispersion of the nanoparticles in the polymer matrix [16]. The nanofillers tend to form micro-size and aggregate clusters.

However, this can limit the nanoparticle dispersions since it can cause deterioration of some properties of the nanocomposites [15, 16, 19]. Many attempts have been made to homogeneously and uniformly fabricate polymer matrices by polymerization reactions or by surface modification of the filler materials [19, 20]. The following methods are used for dispersing nanoparticles in polymer solutions.

11.3.1 Direct Mixing or *ex situ* Process

Direct mixing involves the breakdown of aggregated particles while mixing [19, 20]. This method consists of two main approaches: *melt-compounding* and *solvent method.*

Melt-compounding method: This process involves the mixing of the polymer without the use of any solvents followed by the addition of the nanofillers into the polymer solution [19, 20]. In this method, the hydrodynamic force (shear stress) is induced in the melt polymer by viscous drag. Thus, the shear stress is used to break down the particle aggregates, leading to the homogeneous dispersion of the polymer nanocomposites. Figure 11.1 shows the SEM micrographs of selected areas of the fracture surface of a negative control composite sample fabricated by melt-compounding [21].

Solvent/solution method: This method involves mixing nanoparticles and the polymer with solvents. The nanoparticles are dispersed in a solvent, and the polymer is dissolved in a co-solvent. The nanocomposites are then collected via the solvent coagulation or the solvent evaporation method. The shear stress in this method is lower than that in the melt-compounding method. To break down the particle aggregates, the nanoparticles are pre-dispersed in the solvent, followed by sonication [20]. The process involves using a magnetic stirrer or an ultra-sonicator. The sonication could be either high-power sonication, using a tip or horn, or a mild sonication in a bath. An early study conducted in order to prepare polymer-nanotube composite using the direct mixing resulted in a good dispersion [22–25]. Hu et al. [26] prepared the hybrid suspension of SiO_2 and HPAM via direct mechanical mixing, and a product with good dispersion was obtained under the influence of 8 wt.% NaCl. The solution of different acrylamide-based polymers with MWCNTs has also been prepared via the direct mixing technique [27].

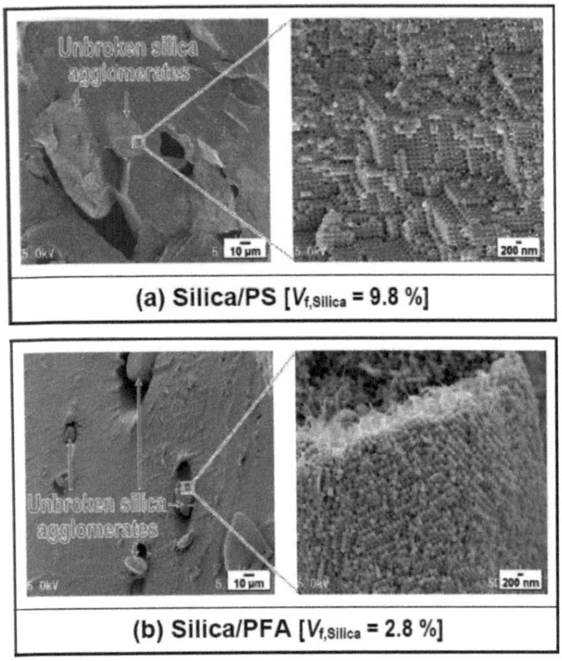

Figure 11.1 SEM micrographs showing dispersion states of silica nanoparticles in silica/polymer composites fabricated by melt-compounding with dense silica agglomerates (a) silica/PS composite (Vf, Silica = 9.8%) and (b) silica/PFA composite (Vf, Silica = 2.8%) [21].

11.3.2 *In situ* Polymerization Method

In situ polymerization methods consist of the nanoparticles dispersed into a monomer, followed by polymerization under the influence of a free-radical initiator [20]. The monomer is polymerized between interlayers to form either an intercalated or an exfoliated nanocomposite. In a similar technique called *in situ* template synthesis, the nanoparticles are synthesized in the presence of polymer chains. The nanoparticle layers and the polymer chains are diffused in an aqueous phase, and the resulting gel undergoes reflux under high temperature. The nucleation and growth of the nanoparticles occur after the polymer chains are trapped inside the particle layers at high temperatures. The major limitation of this method is that polymer decomposition occurs due to high temperature [13]. Figure 11.2 shows the preparation of transparent nanocomposites of *N,N*-dimethyl-acrylamide, styrene, and divinylbenzene with ZnS nanoparticles via free-radical polymerization using 2,2-azobisisobutyronitrile as initiator [28].

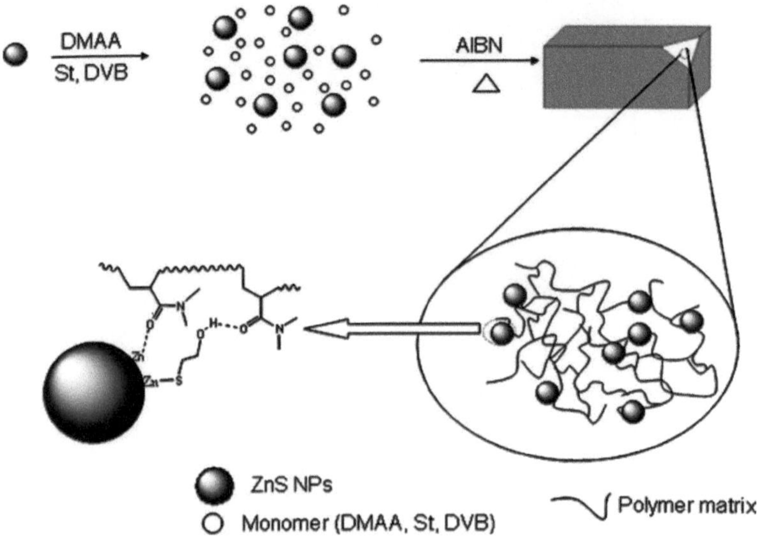

Figure 11.2 Schematic illustration of the preparation process and the structure of some nanocomposites prepared via the *in situ* polymerization method [28].

11.4 Concept of the Dispersibility of Nanoparticles

The dispersibility of nanoparticles in a given medium can be defined as the uniformity or quality of the dispersion or the state that the dispersed particles are in a given medium. The condition of the nanoparticles in the dispersion can be looked at from the point of view of the polydispersity index (PDI). Often the polydispersity index is given in a scale between 0 and 1, where 0 represents an ideal monodispersed system and 1 represents a typical polydispersed system. Therefore, the PDI can be used in order to define the homogeneity of dispersion systems. Figure 11.3 shows that a PDI value of 0.3 in an Al_2O_3 nanoparticle dispersion could be achieved with consistent ultrasonic treatment for a period of 30 min. It can be seen that a monodispersed dispersion of the nanoparticles could not be achieved with ultrasonic treatment of less than 30 min. In this, clearly, the extent of the ultrasonic treatment had greater effect on the PDI of the dispersion than the loading/concentration of the nanoparticles [29].

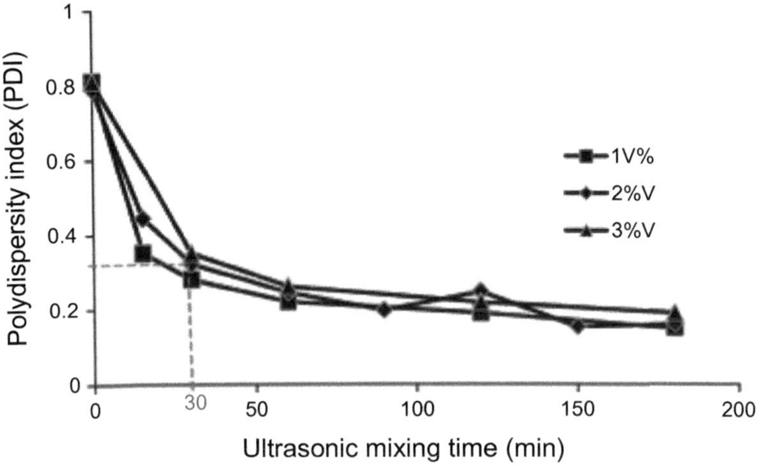

Figure 11.3 Effect of ultrasonic treatment on the PDI of Al_2O_3 nanoparticle dispersions at various concentrations [29].

11.5 Concept of the Stability of the Dispersions

The physical stability of nanoparticle dispersions is often governed by two factors: steric and electrostatic repulsion. In the absence of stabilizing agents, electrostatic repulsion becomes a dominant force. Electrostatic repulsion arises from charges that are present at the nanoparticle's surface; with this hampering the collisions of particles as they move due to the Brownian motion. The weaker the electrostatic repulsion, the greater the chance of nanoparticle aggregation. The zeta potential, therefore, represents the electric potential at the diffuse shear plane of a dispersed nanoparticle and becomes lower when particles are repelled from one another. The thickness of the ionic cloud around the nanoparticle is often described as the Debye screening length, where the greater the absolute value of the zeta potential, the greater the suspension stability. The stronger the surface charge, the better the dispersion stability. Figure 11.4 shows an image in which the extent of a surface modification of TiO_2 nanoparticles using 3-(trihydroxysilyl)-1-propanesulfonic acid can be seen. This treatment directly led to electrostatically and sterically stabilized dispersion of the water-based TiO_2 nanofluids.

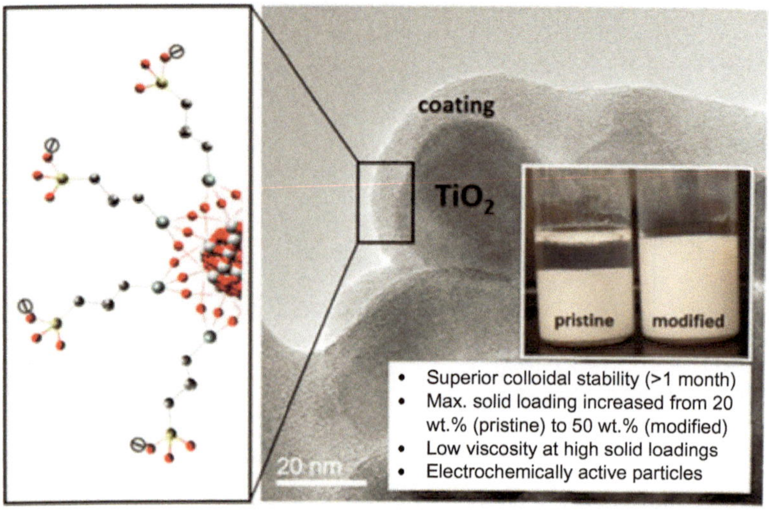

Figure 11.4 Surface modification of TiO_2 nanoparticles with 3-(trihydroxysilyl)-1-propanesulfonic acid to electrostatically and sterically stabilize dispersion of water-based nanofluids [30].

The loading of nanoparticles in a given dispersion can be determined by measuring the solid content of the dispersion. Gravimetric methods are often used, where a comparison is made of the weight of a sample before and after complete removal of the continuous medium (typically by drying methods). Ceramic crucibles containing a re-weighed dispersion can be placed in an oven and heated at a temperature that will ensure removal of all components except the nanoparticles. Following this, the sample is then re-weighted and the residual solid amount is determined.

11.6 Concept of Additives that Promote Dispersibility and Stability

Additive dispersants are widely used in order to promote the stability of nanoparticle dispersions. In a given dispersion of nanoparticles, the aggregation of particles becomes preferential when the van der Waals forces of attraction are greater than the repulsive forces. In order to aid or maintain stability, additive surfactants are, therefore, designed to anchor, adsorb, or bond to the surface of the nanoparticles. Additive substances can also be in the form of common acidic or common alkaline substances. The addition of acidic or basic substances into nanoparticle dispersions often has a profound effect on the pH of the dispersed systems. This then ensures a relationship between the pH and the stability of the nanoparticle dispersions. In the case of alkali-based additives, when the pH is greater than the isoelectric point, the condition at which nanoparticle surfaces are neutral or carry very low charges, the higher pH leads to negative charges that build up at the nanoparticles' surfaces, providing electrostatic repulsive forces necessary to overcome van der Waals forces. This, therefore, ensures that the probability of nanoparticles coming together is minimal. The electrostatic stabilization of nanoparticles in a continuous media is illustrated in Fig. 11.5.

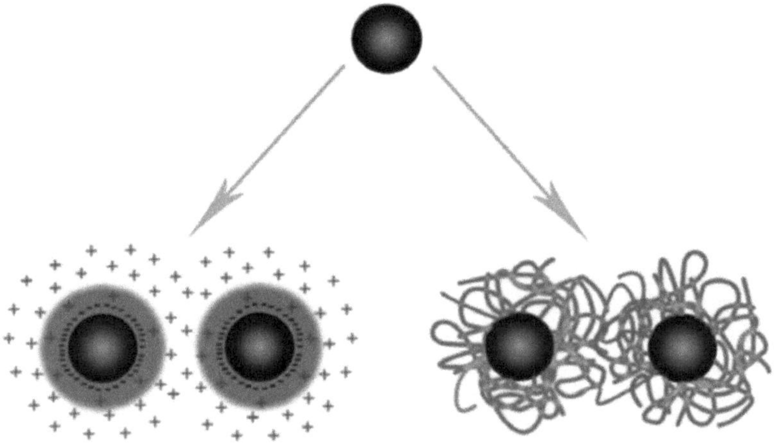

Figure 11.5 Schematic illustration of the electrostatic and the steric stabilization of nanoparticles [31].

Surfactant molecules are also commonly used in the stabilization of nanoparticle dispersions, in a phenomenon commonly referred to as the steric stabilization, illustrated in Fig. 11.6. Surfactants are typically low-molecular-weight organic molecules that are added into the dispersions or are incorporated during the synthesis of the nanoparticles [32, 33]. Commonly used surfactants include the cationic surfactant cetrimonium bromide and the anionic surfactant sodium dodecyl sulfate, both having significant applications in many nanoparticle dispersions [34]. An anion surfactant comprising a hydrophilic poly (ethylene glycol) chain, a hydrophobic alkyl chain, and a polymerizable vinyl group has been reported to act successfully as a capping agent of nanoparticles. This surfactant is highly effective against TiO_2 nanoparticles, ensuring that they become completely dispersible in various organic solvents, ranging from nitriles, alcohols, ketones, and acetates. Consequently, stable dispersions of TiO_2 nanoparticles can be achieved in various polymers with different properties, including in thermosetting epoxy resins and in poly (methylmethacrylate) [35]. Figure 11.6 shows how the surface of the TiO_2 nanoparticles becomes functionalized, following treatment with the PEG-based surfactant.

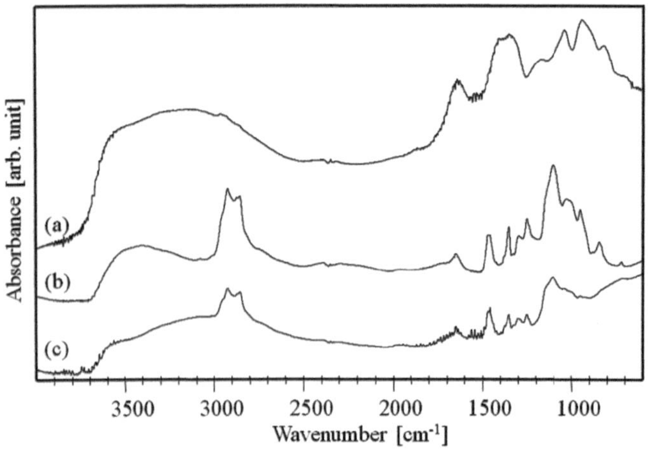

Figure 11.6 Infrared spectrum of (a) untreated TiO_2 nanoparticles, (b) surfactant (1), and (c) surface-modified TiO_2 nanoparticles [35].

11.7 Key Analyses Methods and Performance Tests

Many specialized analyses methods have been developed in order to determine the performance characteristics of nanoparticle dispersions. Typically, these involve either an analysis of the particles contained in the dispersions or an analysis of the dispersion compositions as a whole. The analysis of particles already dispersed in a medium can include particle size analysis, specific surface area analysis, or zeta potential measurement [36]. Other methods of analysis are used in order to decide the stability of nanoparticles in the dispersions, such as by allowing the sample dispersion to stand in a container for a particular period, followed by visual or instrumental measurements. There are techniques like Turbiscan analyses in which backscattering intensity/transmission profiles versus the sample height are used in determining the dispersion stability. These techniques, in addition to zeta potential methods of analyses, common particle morphology test procedures, spectral/spectroscopic measurements, analyses of the particles' specific surface area, light-scattering techniques, and rheological analyses methods, are described in the next sub-sections.

11.7.1 Sedimentation and Centrifugation Methods

The sediment weight or the sediment volume of nanoparticles in a dispersion under an external force field is an indication of the stability of the characterized nanofluid. The variation in the concentration or the particle size of the supernatant particle with the sedimentation time can be determined. Dispersions are considered to be stable when the concentration or the particle size of the supernatant particles remains constant. Sedimentation photographs of the dispersions in test tubes can be taken by a camera, and this is already used by many as a method for observing the stability of nanofluids [7, 37]. A sedimentation balance method can be used to measure the stability of the dispersions, and in this, the tray of sedimentation balance is immersed in the dispersion. This is then followed by measurements of the weight of the sediment nanoparticles at the desirable time.

11.7.2 Zeta Potential Technique

Zeta potential is the electric potential in the interfacial double layer at the location of the slipping plane versus a point in the bulk fluid away from the interface. It shows the potential difference between the dispersion medium and the stationary layer of the fluid attached to the dispersed particle. The significance of the zeta potential is that its value can be related to the stability of colloidal dispersions. Therefore, colloids with high zeta potential (negative or positive) are electrically stabilized, while colloids with low zeta potentials tend to coagulate or flocculate. Zeta potential values well below the threshold of −30mV are typically considered for stable water-based dispersions. This indicates that the nanoparticle dispersion compositions are well stabilized by particle surface charges and that any flocculates or aggregates that may be present do not have a dominant effect. In nanoparticle dispersions where electrostatic stabilization is achieved by using acidic or alkaline species, the value of the zeta potential varies with a variation in the pH, since this translates directly into the concentration of the ions in the dispersion. Figure 11.7 shows an illustration of the relationship between pH, zeta potential, and stability in surfactant-free tantalum oxide nanoparticle dispersions.

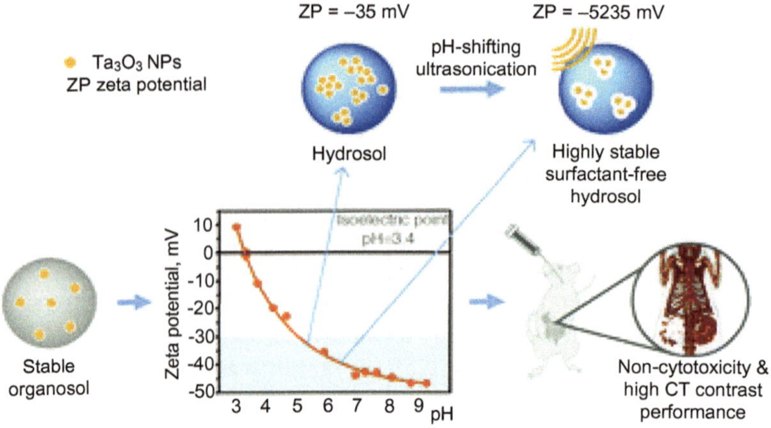

Figure 11.7 Illustration showing the relationship between pH, zeta potential, and stability in surfactant-free tantalum oxide nanoparticle dispersions [7].

11.7.3 Spectral Absorbency Techniques

Spectral absorbency analysis is another efficient way to evaluate the stability of nanofluids. In general, there is a linear relationship between the absorbency intensity and the concentration of nanoparticles in a fluid. The dispersion characteristics of nanoparticles such as alumina, copper, and others in suspensions can, therefore, be evaluated. The sedimentation kinetics can also be determined by examining the absorbency of the particles in the continuous medium [37]. If the nanomaterials dispersed in the fluids have characteristic absorption bands in the wavelength region of 190–1100 nm, it is an easy and reliable method to evaluate the stability of the nanofluids using UV-vis spectral analysis. The variation in the supernatant particle concentration of the nanofluids with the sediment time can be obtained by the measurements of the absorption of the nanofluids. This is because there is a linear relation between the supernatant nanoparticle concentration and the absorbance of the suspended particles.

11.7.4 Centrifugal Techniques

Centrifugal techniques, such as the LUMiSize, can also be used to detect particle stability. The multi-sample analytical centrifuge displays an instantaneous *in situ* change in the particle concentration by NIR light transmitted across the full length of the samples. The analytical centrifuges allow measuring the extinction profiles during centrifugation over the whole sample length. Real-time transmission profiles across the sample are then recorded at certain time intervals during the centrifugation. The sedimentation velocity is then determined. An example of the measurement scheme of the multi-sample analytical centrifuge with photometric detection is shown in Fig. 11.8. In this, a parallel NIR light is passed through the sample cells and the distribution of the local transmission is recorded at pre-set time intervals over the entire sample length [38].

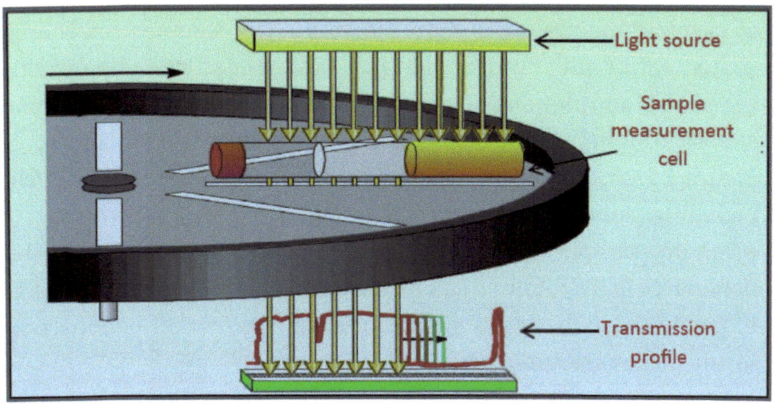

Figure 11.8 Measurement scheme of the multi-sample analytical centrifuge [38].

11.7.5 Analysis of Specific Surface Area

The specific surface area of nanoparticle dispersions can be determined by the means of accelerated surface area and porosimetry techniques. The nanoparticles need to be separated from the dispersed medium through techniques such as filtration or evaporation. This can then lead to nanoparticle powders, which can be treated with liquid nitrogen, in order to promote particle adsorption. The test procedure then involves situating the nanoparticle powdered sample in a flask and degassing the powder considerably. When high-vacuum is reached, a controlled supply of nitrogen gas is then fed to the powder until the surface area was measured by the adsorption of the gas on the surface of the nanoparticles [39]. Figure 11.9a shows an isotherm, where the hysteresis loop represents a mesoporous structure in the MnO_2 nano-sheets. From this particular figure, the BET surface area was determined to be 337.01 m^2/g [40].

11.7.6 Particle Morphology Analyses

In order to study the morphology (shape and size) of nanoparticles dispersed in a colloidal system or to determine the size distribution of the particles, it may not be necessary to remove the particles from the continuous medium. For example, dilute aqueous nanoparticle dispersions can be applied directly on stubs, when carrying out SEM analysis. These may need to be allowed to dry out before strips of carbon tape are applied on the stubs. The samples can then be coated with a thin layer of gold [2]. SEM images can help establish the shape and dimensions of the nanoparticles, with these being spherical, tubular, rod like, star like, irregular shaped, and so on. Figure 11.10 shows some micrographs of spherical nanoparticles, including cobalt cores and silver nanoparticles. In contrast to using dry samples (or drying samples out) in SEM analyses, dilute nanoparticle dispersions can be used directly for the determination of particle size distribution and the zeta potential of nanoparticles in any given medium.

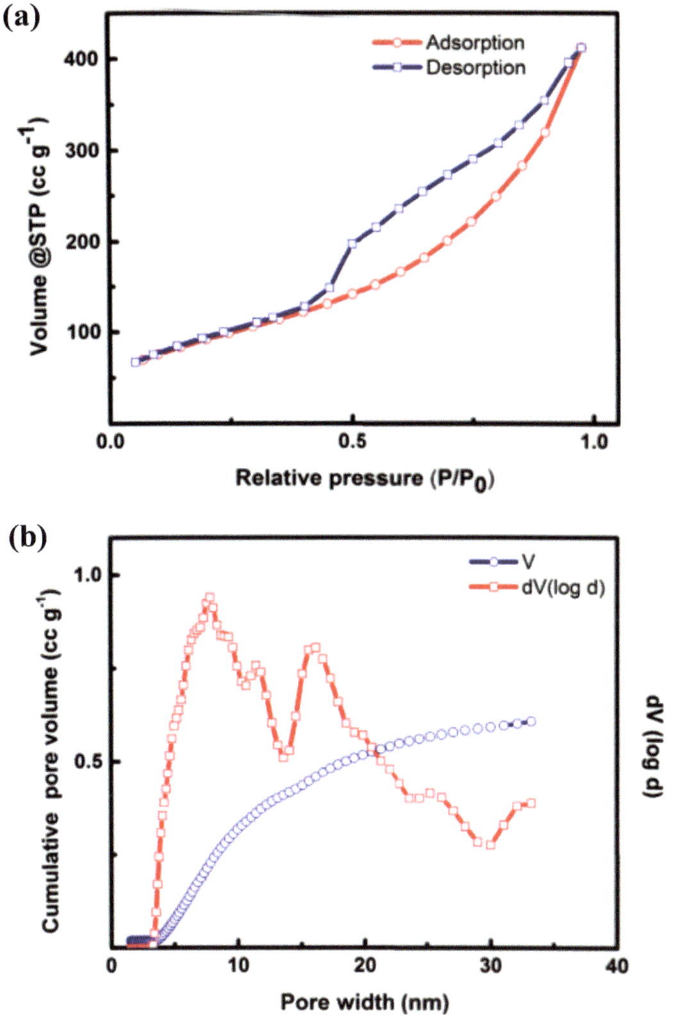

Figure 11.9 Nitrogen gas isotherm (a) and a corresponding pore size distribution (b) of the as-prepared MnO_2 [40].

11.7.7 Light Scattering in Dispersions

Dynamic light scattering (DLS) can be used to study nanoparticle dispersions at different concentrations in aqueous media, organic solvents, or polymer solutions. Such a technique can reveal the direct effect of the presence and size of aggregates, the effect of the solid concentration, and the effect of the presence and amount of surfactant(s) or polymers in the dispersion [3]. An illustration of the technique involved in DLS is given in Fig. 11.11 [42].

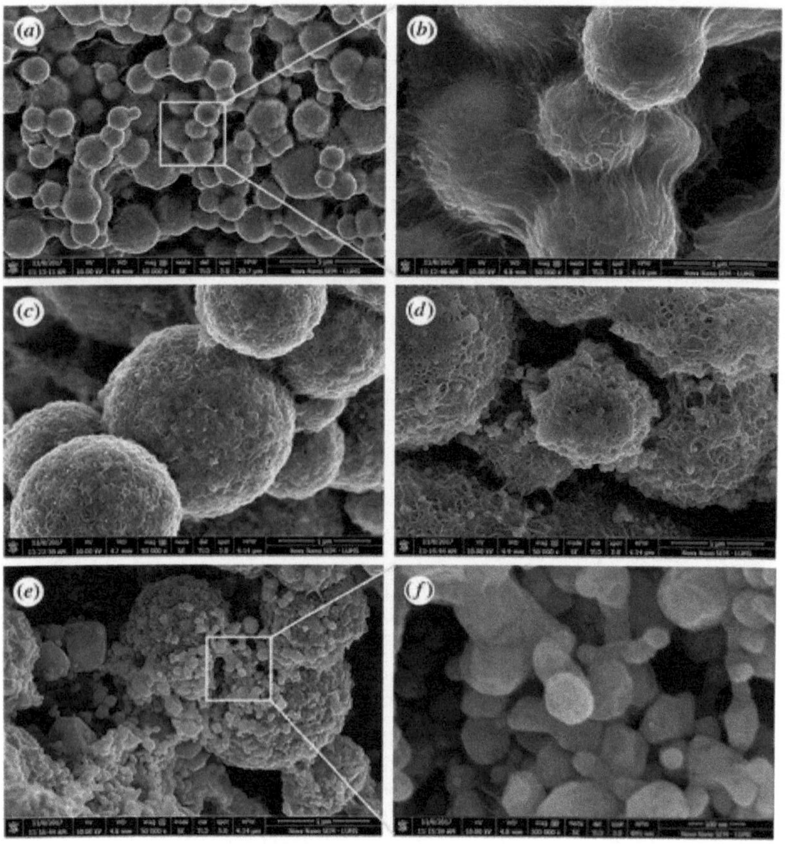

Figure 11.10 SEM micrographs of spherical nanoparticles, including cobalt cores and silver nanoparticles [41].

In the DLS analysis of dilute nanoparticle dispersions, particles in the sample that are in a state of Brownian motions are illuminated by a laser beam. The light scattered by the particles is collected and measured with a photomultiplier. Light fluctuations in the detector are also caused by the moving particles. The extent of the fluctuation depends on the size of the individual particles, with larger particles causing a more rapid fluctuation at the detector than smaller particles. The intensity of the fluctuations is computed to provide an intensity correlation function, whose analysis provides the diffusion coefficient of the particles. This is also known as the diffusion constant, as represented by Eq. 11.1. This is known as *the Stokes–Einstein equation.* In this equation, D is the diffusion coefficient, R is the radius of the particle, k is the Boltzmann constant, T is the temperature, and η is the viscosity.

$$D = \frac{kT}{6\pi R\eta} \tag{11.1}$$

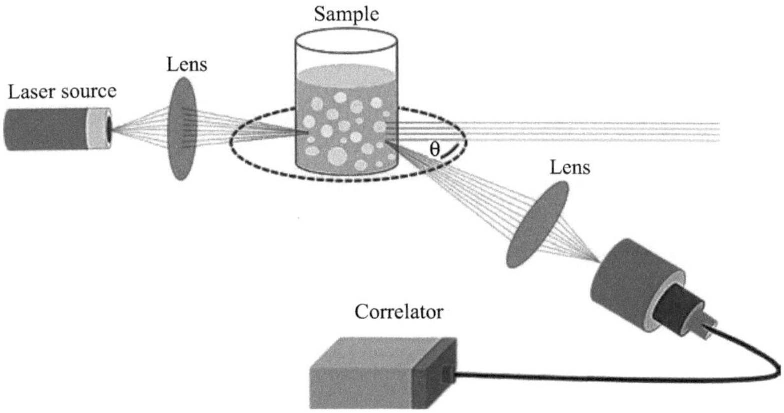

Figure 11.11 Representation of dynamic light-scattering principle [42].

11.7.8 Rheology and Flow Behavior

Rheological analyses of nanoparticle dispersions could provide important information on the flow and deformation behavior of the materials. Nanoparticle dispersions that exhibit sheer thickening behavior, by virtue of viscosity increase with shear rate and shear stress, could reveal interactions such as particle–particle interactions, particle–polymer interactions, or polymer–polymer interactions [43]. This behavior can be advantageous in applications such as the designing of stimuli-responsive sensors or actuators, and in the development of protective materials [43]. On the other hand, nanoparticle dispersions can exhibit shear thinning behavior, where their viscosity decreases with the shear rate and the shear stress. Shear thinning can arise when nanoparticles orient themselves in the direction of flow and its gradient. Agglomerates are then broken down, and as a consequence, the amount of fluid rendered inactive by the particles is decreased. The interaction forces may then decrease and lead to a decrease in the overall resistance to the flow of the dispersion [9]. This behavior can be advantageous in applications such as ink and coating manufacture [44]. The flow characteristics of many materials, including shear thinning and shear thickening, are represented in Fig. 11.12.

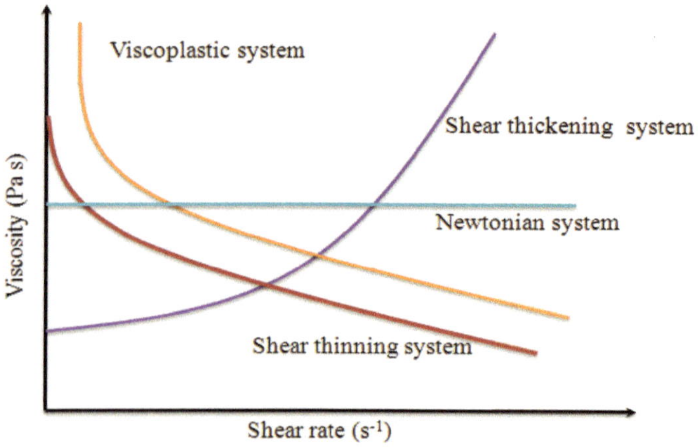

Figure 11.12 Typical flow profiles of materials [45].

Nanoparticle dispersions fall into a broad range of particles contained in a continuous medium. The solid fraction of the dispersion directly affects the viscosity of the whole composition. Particle–particle interactions can occur in dispersions with high solid fractions with the overall rheological behavior of the dispersion being dictated by the Newtonian flow behavior of the continuous phase. When the solid fraction of the dispersion increases, the probability of particle–particle collision increases. In a high solid fraction system, considerable shear force is required when friction occurs during particle–particle collision. Figure 11.13 illustrates dispersion systems where the solid fraction is varied and how this affects the rheological behavior of the materials.

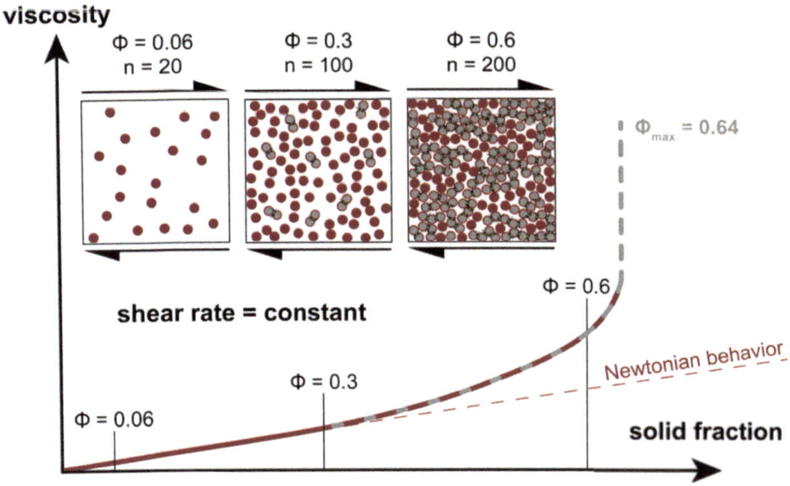

Figure 11.13 Illustration of the viscosity of suspensions of particles in different kinds of solid fraction systems, for a constant shear rate and constant sphere size [46].

A number of rheological test approaches can be used in order to study and understand the performance of nanoparticle dispersions. These could include shear ramp measurements, where evaluations are typically carried out in a continuous shear ramp mode at a desired temperature. Since typical nanoparticles are of low viscosity, parallel plates with wide diameters and low gap settings should be used. These will ensure that inertial effects are minimal and that the effects of surface tension on artifact shear thinning under low torque are minimized. Alternatively, concentric cylinders can be used in the rheological evaluation of nanoparticle dispersions. These allow high shear rates to be used in the measurements and are designed to minimize the effect(s) of sample evaporation. The advantage of using parallel plates over concentric cylinders is the small sample volume that is required in the former. Common geometries used for the testing low viscosity fluids, dispersions, and solutions are represented in Fig. 11.14.

Gradient temperature ramp tests can also be carried out using parallel plates with wide diameters. A nanoparticle dispersion can then be evaluated at a constant shear rate, while changing the temperatures from a set point. In each case, the change in viscosity (flow behavior) can be observed as the temperature rises. The data obtained from such tests can be given a kinetic treatment, leading to the determination of the yield point/stress of the nanoparticle dispersion, the flow activation energy of the dispersion, and its infinite shear viscosity. This treatment of data was found to be particularly useful in understanding the dispersion/composite of carbon quantum dots in poly (acrylamide). In this, the extent of the sensitivity of the yield stress to the shear history of the nanofluid was assessed. A low value of the yield stress obtained for the composites shows that the material had higher shear sensitivity in comparison to a corresponding pure poly (acrylamide) solution. This effect resulted due to the rapid movement and subsequent breakdown of structural linkages that priory exist between the nanoparticles and the poly (acrylamide) macromolecular chains [48].

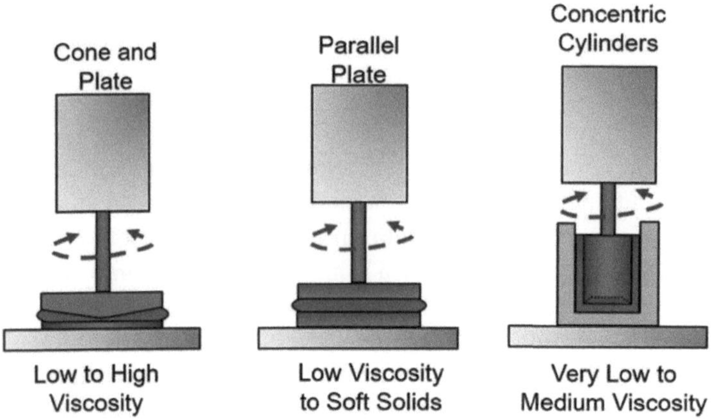

Figure 11.14 Illustration of common geometries commonly used for the testing of low viscosity fluids [47].

11.8 Selected Applications

11.8.1 Nanoparticle Dispersions in Enhanced Oil Recovery

The use of nanotechnology for enhanced oil recovery is a recent development in which many studies have shown that adding nanoparticles into a flooding fluid could reinforce the polymer properties, alter the oil/water interfacial tension (IFT) and substrate wettability, and subsequently increase the oil recovery rate by means that are similar to those for chemical EOR methods [49–51].

Many nanoparticles can be injected into the porous medium with a micron-sized pore throat, but it is better to control the stability of the nanofluids in nanometer size because nanoparticles aggregate can easily form once they are hydrated [27, 52]. Nanoparticle aggregates are characterized by having large surface-area-to-volume ratio. Nanoparticles retain high surface energies and, therefore, form aggregates in order to reduce their surface energies [52]. In oil recovery applications, the formation of nanoparticle aggregates may lead to pore throat plugging/locking and can result in rock properties alteration. Many different EOR mechanisms using nanofluids have been reported and proposed in order to increase the recovery of oil. Applications of some of the mechanisms involved are reviewed in this section.

11.8.1.1 Interfacial tension

Oil and water are two immiscible liquids that have high interfacial tension between them. Introducing nanoparticles into the system will alter the interfacial tension and lead to oil recovery enhancement. The layers of nanoparticles produce lower IFT than produced by surfactants. Consequently, IFT is sensitive to nanofluid concentration because as particle concentration increases, the IFT decreases [53, 54]. Nanoparticle adsorption mechanisms tend to affect the IFT in comparison to the surfactants because adsorption at the interface arises due to their hydrophilic head and hydrophobic tail. In contrast, neutral nanoparticle adsorption arises due to the less energy required for maintaining the particle–fluid interface. Neutral nanoparticles change the IFT by applying a cheaper particle–fluid interface to replace the energetically fluid–fluid expensive interface. Experimental and theoretical studies have demonstrated the induced effect of adding nano-silica in natural surfactant adsorption on sandstone rock. Hydrophilic nano-silica is less efficient than its hydrophobic counterpart from the point of view of obstructing adsorption damage on shale-sandstone [53].

11.8.1.2 Wettability

The actual oil reservoir can be oil, intermediate, or water wet [55]. Research has shown that oil production by the water-wet formation is more desirable than production by the oil-wet formation [56]. In addition, production by the intermediate-wet formation is better than production by the water-wet formation [56]. Change in the rock wettability effected by the use of nanoparticles has been extensively studied [57]. Most oil production occurs before water breakthrough in the water-wet system, after which the rate will decline. Certainly, in an intermediate oil-wet system, the water breakthrough will occur earlier than in a water-wet system. The relative permeability controls the flow and spatial fluid distribution in a porous medium, and it is, therefore, affected by the wettability effect [57, 58]. It is necessary to have a reliable wettability modifier with improved mobility properties. Therefore, nanoparticles can be applied to alter the wettability because it is possible for applications under broader oil reservoir conditions [58].

11.8.1.3 Structural disjoining pressure

The mechanism of oil displacement by nanoparticles is still under investigation, although Chengara et al. have mentioned the importance of disjoining pressure [59]. An experiment conducted revealed that nanoparticles that are contained in the three-phase region between oil, water, and rock tend to force themselves to be placed between the solid rock surfaces and the discontinues phase. A wedge-like structure is formed by the particles, resulting in structural disjoining pressure [59]. When liquid/oil drop dispersed in nanofluids approach the hydrophilic solid surface, the nanoparticles will arrange themselves orderly forming solid-like structure at the three-phase contact region. When this arrangement is toward the bulk phase, the structure becomes more disordered and attains a fluid-like structure. Studies have shown that in the confined region, the pressure rising from such an ordering manner will enhance the spreading of nanoparticle behavior [60]. The particles that are in the three-phase contact region will contribute to generating a wedge-like structure and push themselves to stay in between the solid rock and the discontinuous phase. The available particles in the bulk fluid will employ a pressure that forces the particles forward, toward the wedge structure. This kind of resulting force is known as the structural disjoining pressure or the film tension gradient [61]. The forces that drive this phenomenon are electrostatic force, van der Waals force, and Brownian motion [62].

11.8.1.4 Retention in porous media

The size and shape, the flow rate, the concentration of suspended particles, and the mechanism of carrying the fluids are the factors that control the migration and retention of solid particles in the porous media [63, 64]. Skauge et al. have demonstrated that silica nanoparticles can propagate through a pore system so quickly, and pose no damage to the environment, due to their natural existence in the reservoir. They cannot also strain or block pores because they are too small, making them preferable for EOR applications [65]. The transport of nanoparticles to the porous media is governed by hydrodynamics, convection, and diffusion [65].

11.8.1.5 Mobility control

In EOR, oil viscosity is considerably higher than the injected fluid's viscosity. Water-soluble polymers are widely used in EOR processes in order to boost the fluid's viscosity, leading to cost-effectiveness and good rapport with other chemicals [66–68]. Consequently, the application is still limited to certain conditions of the reservoir. However, adding nanoparticles into the fluid will improve its efficiency and increase the viscosity up to an optimum level with net effect targeting improved mobility, and hence better oil recovery efficiency. SiO_2 dispersions are widely used in order to improve the rheological performance of polymers used in EOR [18, 62, 69]. For example, Saito et al. [70] have demonstrated the effect of adding 3 wt.% SiO_2 nanoparticles into a polymer solution, leading to a shear thinning behavior and a change in the overall flow profile. Dingwe Zhu et al. [71] have demonstrated the rheology and the flooding of HAPAM/SiO_2 hybrids. These exhibit good viscosity and are more heat and salt tolerant than HAPAM alone. This improvement is as a result of hydrogen bond formation between the hybrids. Similarly, HAHPAM/SiO_2 dispersions with increased viscosity and high elastic modulus lead to more oil recovery compared to a pure HAPAM solution [R]. In this, a bridge between the SiO_2 nanoparticles and the

HPAM is formed, consequently enhancing the rheological properties of the HPAM, as reported by Hu et al. [26]. Table 11.2 lists some other nanoparticles that are used to this effect.

Table 11.2 Some nanoparticles used in various dispersed systems with the view of improving rheological properties

Nanoparticle	Application	Result	Reference
TiO_2	Used alongside polyisobutylene in decalin solution in order to improve rheological properties.	Viscosity increase was observed with less effect at a high shear rate, with Newtonian plateau followed by a shear thinning region.	[54, 72]
CNTs	Used in order to improve polymer rheological properties.	Rheological properties were improved due to the establishment of a state of the dispersion of the CNTs in the mixture.	[62, 73, 74]
Nanoclay	Used in order to improve polymer rheological properties.		[67, 75–79]
Layered double hydroxide	Used in order to improve the tendency of the polymer to deformation.		[80–82]
Fe_2O_3	Used in order to promote better flow characteristics.	Addition of only 0.1 wt.% polymer and surfactant significantly improves the mobility of the nanoparticles in the porous medium.	[83]

Nanoparticle	Application	Result	Reference
SiO_2	Used as a thickening additive for high molecular weight anionic PAM and HPAM.		[26, 50, 84]
Laponite XLG	To evaluate the fracture-plugging and the mobility control of polymer in a mature reservoir.	Improvement in the polymer viscosity effectively leading to better sweep efficiency, thereby improving the recovery efficiency.	[67, 75]

11.8.2 Nanoparticle Dispersions in Fire-Resistant and Fire-Retardant Applications

Colloidal silica nanoparticle dispersions, in particular, have found applications in the field of fire-resistant and fire-retardant materials. Typically, fumed silica nano-powders are used in order to make silica dispersions. The nano-silica dispersions are then used to prepare interlayers, which can be used in fire-resistant glazing applications. The fire resistance of the nano-silica-based interlayer(s) contained in the fire-resistant glazing arises due to the reaction of the interlayer to high temperatures or fire situations, leading to the formation of an intumesced barrier that helps prevent the spread of fire, thermal energy, and smoke [85]. The mechanism involved in the intumescent process is the same when nano-silica is used in transparent fire-retardant coatings. In addition to the nanoparticles helping in maintaining the high degree of transparency, by forming chemical bonds with the other coating components, they increase the thermal stability of the coatings, confer lame retardancy and smoke suppression properties to the coatings, and form an intumesced layer when in contact with fire. Figure 11.15 shows compact and intumescent char layers after the combustion of some intumescent fire-retardant coatings [30]. Examples of nanoparticle-based fire-resistant assemblies are summarised in Table 11.3.

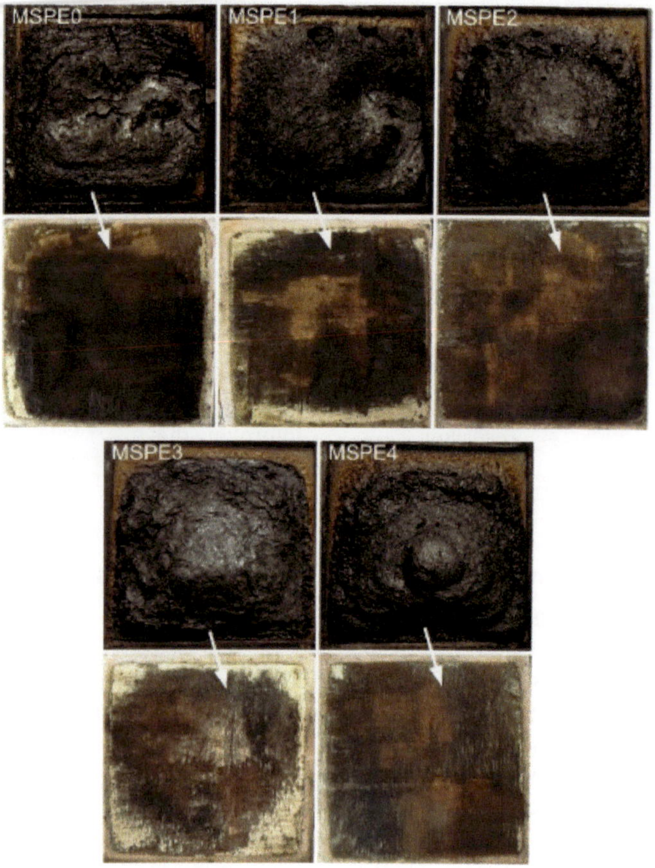

Figure 11.15 Images showing the intumesced properties of some nano-silica, relevant to their influence on the flame retardancy and smoke suppression properties of transparent fire-retardant coatings [30].

Many fire-resistant interlayers can be prepared by mixing nano-silica dispersions with a polyhydric alcohol and an aqueous metal hydroxide solution, in desirable proportions. These components are then mixed under reduced pressure, while controlling the temperature. The mixture is heated for a desired duration at boiling temperatures until a specific reaction is achieved. After the mixture had cooled down, it is then injected into a cavity between arranged glass sheets and cured at elevated temperatures. The curing process transforms the translucent, aqueous composition into a transparent, fire-resistant assembly.

Table 11.3 Examples of nanoparticle-based fire-resistant assemblies

Technology	Key Constituent for Fire Resistance	Performance	Reference
Fireproof foam compositions	Carbon nanotube, up to 15 wt%	Fire resistance of polymer foams	[86]
Heat insulating coatings	90 ~ 99 wt% silicate substrate	Fire-retardant and heat-proof insulation	[87]
Multi-layer fire-barrier systems	Nano-silica particles	Fire barrier, insulation, and intumescence systems	[88]
Fire-resistant glazing laminate	Nano-silica dispersion, with at least 50% by weight of silica	Thermal insulation and intumescence of interlayer	[89]
Fire-resistant glazing laminate	Nano-silica, particle size in the range 7–50 nm	Thermal insulation and intumescence of interlayer	[90]
Fire-resistant glazing laminate	Nano-silica from a water-based dispersion	Intumescent fire protection layer	[91]

11.9 Conclusion

There are so many different applications of nanoparticle dispersions, including enhanced oil recovery; fire-resistant glazing systems; assembly of coatings, paints, and inks; drug-delivery systems; ceramics; and nanocomposites. These varied applications are ensured and underlined by having dispersible nanoparticles and by such nanoparticles being stable in the assembled or applied media. Often, the dispersibility and the stability of the nanoparticles are required to be controlled by the use of additive stabilizers/ dispersants. There are clearly various analyses and performance test methods that are applied to the study of the nanoparticles contained in the dispersions, and there are those applied directly to the dispersion systems. These methods are key in ensuring that applications of the nanoparticle dispersions, such as those detailed under the enhanced oil recovery and the fire-resistant glazing subsections, are possible. Despite this, one area for further development and application of nanoparticle dispersions is the generation of blended, multi-nanoparticle dispersed systems where the synergetic effects in the systems can pave the way for wider applications.

References

1. Magami, S.M., et al., The physical–chemical behaviour of amino cross-linkers and the effect of their chemistry on selected epoxy can coatings' hydrolysis to melamine and to formaldehyde into aqueous food simulants. *Progress in Organic Coatings*, 2015. **78**: pp. 325–333.
2. Meng, Q. and Magami, S. M., Metallic powders for composite inks: Formulation, application, and characterization. *Research Journal of Nanoscience and Engineering*, 2017. **1**(1): pp. 14–23.
3. Nguyen, V. S., Rouxel, D., and Vincent, B., Dispersion of nanoparticles: From organic solvents to polymer solutions. *Ultrasonics Sonochemistry*, 2014. **21**(1): pp. 149–153.
4. Hwang, Y., et al., Production and dispersion stability of nanoparticles in nanofluids. *Powder Technology*, 2008. **186**(2): pp. 145–153.
5. Rodriguez-Navarro, C., Suzuki, A., and Ruiz-Agudo, E. J. L., Alcohol dispersions of calcium hydroxide nanoparticles for stone conservation. *Langmuir*, 2013. **29**(36): pp. 11457–11470.
6. Booth, A., et al., Freshwater dispersion stability of PAA-stabilised cerium oxide nanoparticles and toxicity towards *Pseudokirchneriella subcapitata*. *Science of the Total Environment*, 2015. **505**: pp. 596–605.
7. Koshevaya, E., et al., Surfactant-free tantalum oxide nanoparticles: Synthesis, colloidal properties, and application as a contrast agent for computed tomography. *Journal of Material Chemistry*, 2020. **8**(36): pp. 8337–8345.
8. Magami, S. M., et al., The effect of TiO_2, pigmentation on the hydrolysis of amino resin crosslinked epoxy can coatings. *Journal of Coatings Technology and Research*, 2014. **11**(6): pp. 1013–1022.
9. Pastoriza-Gallego, M. J., et al., Rheological non-Newtonian behaviour of ethylene glycol-based Fe_2O_3 nanofluids. *Nanoscale Research Letters*, 2011. **6**(1): pp. 1–7.

10. McWilliams, A. D. S., et al., Surfactant-assisted individualization and dispersion of boron nitride nanotubes. *Nanoscale Advances*, 2019. **1**(3): pp. 1096–1103.
11. Ordóñez, F., et al., Synthesis of ZrO_2 nanoparticles and effect of surfactant on dispersion and stability. *Ceramics International*, 2020. **46**(8): pp. 11970–11977.
12. Kelsall, R. W., Hamley, I.W., and Geoghegan, M. *Nanoscale Science and Technology*. 2005: Wiley Online Library.
13. Mongillo, J. F., *Nanotechnology 101*. 2007: ABC-CLIO.
14. Jordan, J., et al., Experimental trends in polymer nanocomposites: A review. *Materials Science and Engineering A*, 2005. **393**(1): pp. 1–11.
15. Hanemann, T. and Szabó, D. V., Polymer-nanoparticle composites: From synthesis to modern applications. *Materials*, 2010. **3**(6): pp. 3468–3517.
16. Schmidt, G. and Malwitz, M. M., Properties of polymer–nanoparticle composites. *Current Opinion in Colloid & Interface Science*, 2003. **8**(1): pp. 103–108.
17. Chatterjee, T. and Krishnamoorti, R., Rheology of polymer carbon nanotubes composites. *Soft Matter*, 2013. **9**(40): pp. 9515–9529.
18. Kumar, S. K. and Krishnamoorti, R., Nanocomposites: Structure, phase behavior, and properties. *Annual Review of Chemical and Biomolecular Engineering*, 2010. **1**: pp. 37–58.
19. Tanahashi, M., Development of fabrication methods of filler/polymer nanocomposites: With focus on simple melt-compounding-based approach without surface modification of nanofillers. *Materials*, 2010. **3**(3): pp. 1593–1619.
20. Khan, W. S., Hamadneh, N. N., and Khan, W. A., Polymer nanocomposites–synthesis techniques, classification and properties. In *Science and Applications of Tailored Nanostructures*. 2016: One Central Press (OCP).
21. Tanahashi, M. J. M., Development of fabrication methods of filler/polymer nanocomposites: With focus on simple melt-compounding-based approach without surface modification of nanofillers. *Materials (Basel)*, 2010. **3**(3): pp. 1593–1619.
22. Jin, L., Bower, C., and Zhou, O., Alignment of carbon nanotubes in a polymer matrix by mechanical stretching. *Applied Physics Letters*, 1998. **73**(9): pp. 1197–1199.
23. Coleman, J. N., et al., High performance nanotube-reinforced plastics: Understanding the mechanism of strength increase. *Advanced Functional Materials*, 2004. **14**(8): pp. 791–798.

24. Cadek, M., et al., Reinforcement of polymers with carbon nanotubes: The role of nanotube surface area. *Nano Letters*, 2004. **4**(2): pp. 353–356.
25. Cadek, M., et al., Morphological and mechanical properties of carbon-nanotube-reinforced semicrystalline and amorphous polymer composites. *Applied Physics Letters*, 2002. **81**(27): pp. 5123–5125.
26. Hu, Z., et al., Rheological properties of partially hydrolyzed polyacrylamide seeded by nanoparticles. *Industrial & Engineering Chemistry Research*, 2017. **56**(12): pp. 3456–3463.
27. Haruna, M. A., et al., Improved rheology and high-temperature stability of hydrolyzed polyacrylamide using graphene oxide nanosheet. *Journal of Applied Polymer Science*, 2019: p. 47582.
28. Cheng, Y., et al., Preparation and properties of transparent bulk polymer nanocomposites with high nanophase contents. *Journal of Material Chemistry*, 2008. **18**(34): pp. 4062–4068.
29. Sadeghi, R., et al., Investigation of alumina nanofluid stability by UV–vis spectrum. *Microfluidics and Nanofluidics*, 2015. **18**(5): pp. 1023–1030.
30. Yan, L., Xu, Z., and Wang, X., Influence of nano-silica on the flame retardancy and smoke suppression properties of transparent intumescent fire-retardant coatings. *Progress in Organic Coatings*, 2017. **112**: pp. 319–329.
31. Tchalala, M. R., Anjum, D. H., and Chaieb, S., Effect of ionic liquid (emim BF4) on the dispersion of gold nanoparticles. *Journal of Physics: Conference Series*. 2016: IOP Publishing.
32. Kamiya, H. and Iijima, M., Surface modification and characterization for dispersion stability of inorganic nanometer-scaled particles in liquid media. *Science and Technology of Advanced Materials*, 2010. **11**: p. 044304.
33. Amjad, M., Haruna, M. A., and Gardy, J., Nanomaterials for solar energy capture and steam generation. In *Emerging Nanotechnologies for Renewable Energy*. 2021, Elsevier. pp. 37–48.
34. Alam, M. S., et al., The self-association and mixed micellization of an anionic surfactant, sodium dodecyl sulfate, and a cationic surfactant, cetyltrimethylammonium bromide: conductometric, dye solubilization, and surface tension studies. *Journal of Dispersion Science and Technology*, 2016. **37**(11): pp. 1645–1654.
35. Iijima, M., et al., Anionic surfactant with hydrophobic and hydrophilic chains for nanoparticle dispersion and shape memory polymer

nanocomposites. *Journal of the American Chemical Society*, 2009. **131**(45): pp. 16342–16343.

36. Mourdikoudis, S., Pallares, R. M., and Thanh, N. T. K., Characterization techniques for nanoparticles: Comparison and complementarity upon studying nanoparticle properties. *Nanoscale,* 2018. **10**(27): pp. 12871–12934.

37. Liu, L.-H., et al., *Advanced Nanohybrid Materials: Surface Modification and Applications*. 2012, Hindawi.

38. Haruna, M.A., *Novel Polymer-Nanocomposites for Enhanced Oil Recovery*. 2019, University of Leeds.

39. Shaji, A. and Zachariah, A. K., Surface area analysis of nanomaterials. In *Thermal and Rheological Measurement Techniques for Nanomaterials Characterization*. 2017, Elsevier. pp. 197–231.

40. Li, Y. L., et al., Facile preparation of MnO_2 with large surface area in a rotor-stator reactor for supercapacitors. *International Journal of Electrochemical Science*, 2016. **11**: pp. 9644–9655.

41. Kanwal, Z., et al., Synthesis and characterization of silver nanoparticle-decorated cobalt nanocomposites (Co@ AgNPs) and their density-dependent antibacterial activity. *Royal Society Open Science*, 2019. **6**(5): pp. 182135.

42. Choudhary, R. C., et al., Synthesis, characterization, and application of chitosan nanomaterials loaded with zinc and copper for plant growth and protection. In *Nanotechnology*. 2017, Springer. pp. 227–247.

43. Sha, X., et al., Shear thickening behavior of nanoparticle suspensions with carbon nanofillers. *Journal of Nanoparticle Research*, 2013. **15**(7): pp. 1–11.

44. Magami, S. M., Functional can coatings–Part 1: Underlying basic physical and chemical principles. *Surface Coatings International*, 2013. **96**(2): pp. 88–93.

45. Magami, S. M., Functional can coatings Part 2: Composition, attributes, applications and performance. *Surface Coatings International*, 2013. **96**(3): pp. 148–155.

46. Anton Parr, *The influence of particles on suspension rheology,"* Anton Parr.

47. Whaley, J. K., Templeton, C., and Anvari, M. Rheological testing for semisolid foods: Traditional rheometry. In *Rheology of Semisolid Foods*. 2019, Springer. pp. 63–96.

48. Haruna, M. A., et al., Influence of carbon quantum dots on the viscosity reduction of polyacrylamide solution. *Fuel*, 2019. **248**: pp. 205–214.

49. Haruna, M. A., Amjad, M., and Magami, S. M., Nanocomposites for enhanced oil recovery. In *Emerging Nanotechnologies for Renewable Energy*. 2021, Elsevier. pp. 81–113.

50. Haruna, M. A., et al., Nanoparticle modified polyacrylamide for enhanced oil recovery at harsh conditions. *Fuel*, 2020. **268**: p. 117186.

51. Haruna, M. A. and Wen, D., Stabilization of polymer nanocomposites in high-temperature and high-salinity brines. *ACS Omega*, 2019. **4**(7): pp. 11631–11641.

52. Oberdisse, J., Aggregation of colloidal nanoparticles in polymer matrices. *Soft Matter*, 2006. **2**(1): pp. 29–36.

53. Sharma, T., Iglauer, S., and Sangwai, J. S., Silica nanofluids in an oilfield polymer polyacrylamide: Interfacial properties, wettability alteration, and applications for chemical enhanced oil recovery. *Industrial & Engineering Chemistry Research*, 2016. **55**(48): pp. 12387–12397.

54. Suleimanov, B., Ismailov, F., and Veliyev, E., Nanofluid for enhanced oil recovery. *Journal of Petroleum Science and Engineering*, 2011. **78**(2): pp. 431–437.

55. Morrow, N. R., Wettability and its effect on oil recovery. *Journal of Petroleum Technology*, 1990. **42**(12): pp. 1476–1484.

56. Ju, B., Fan, T., and Ma, M., Enhanced oil recovery by flooding with hydrophilic nanoparticles. *China Particuology*, 2006. **4**(01): pp. 41–46.

57. Han, D., et al. The effect of wettability on oil recovery by alkaline/surfactant/polymer flooding. In *SPE Annual Technical Conference and Exhibition*. 2006. Society of Petroleum Engineers.

58. Karimi, A., et al., Wettability alteration in carbonates using zirconium oxide nanofluids: EOR implications. *Energy & Fuels*, 2012. **26**(2): pp. 1028–1036.

59. Chengara, A., et al., Spreading of nanofluids driven by the structural disjoining pressure gradient. *Journal of Colloid and Interface Science*, 2004. **280**(1): pp. 192–201.

60. Wasan, D. T. and Nikolov, A. D., Spreading of nanofluids on solids. *Nature*, 2003. **423**(6936): pp. 156–159.

61. Mcelfresh, P. M., Holcomb, D. L., and Ector, D., Application of nanofluid technology to improve recovery in oil and gas wells. In *SPE International Oilfield Nanotechnology Conference and Exhibition*. 2012. Society of Petroleum Engineers.

62. El-Diasty, A. I. and Aly, A. M. Understanding the mechanism of nanoparticles applications in enhanced oil recovery. In *SPE North*

Africa Technical Conference and Exhibition. 2015. Society of Petroleum Engineers.

63. Yu, J., et al. Transport study of nanoparticles for oilfield application. In *SPE International Conference on Oilfield Scale*. 2010. Society of Petroleum Engineers.
64. Kanj, M. Y., Funk, J. J., and Al-Yousif, Z., Nanofluid coreflood experiments in the ARAB-D. In *SPE Saudi Arabia Section Technical Symposium*. 2009. Society of Petroleum Engineers.
65. Skauge, T., Spildo, K., and Skauge, A., Nano-sized particles for EOR. In *SPE Improved Oil Recovery Symposium*. 2010. Society of Petroleum Engineers.
66. Ye, Z., et al., Synthesis and performance of an acrylamide copolymer containing nano-SiO_2 as enhanced oil recovery chemical. *Journal of Chemistry*, 2013.
67. Thomas, A., Gaillard, N., and Favero, C. Some key features to consider when studying acrylamide-based polymers for chemical enhanced oil recovery. *Oil & Gas Science and Technology–Revue d'IFP Energies nouvelles*, 2012. **67**(6): pp. 887–902.
68. Heller, H. and Keren, R., Anionic polyacrylamide polymers effect on rheological behavior of sodium-montmorillonite suspensions. *Soil Science Society of America Journal*, 2002. **66**(1): pp. 19–25.
69. Nair, N., et al., Coarse-grained simulations of polymer-grafted nanoparticles: Structural stability and interfacial behavior. *The Journal of Physical Chemistry B*, 2016. **120**(35): pp. 9523–9539.
70. Saito, Y., Ogura, H., and Otsubo, Y., Rheological behavior of silica suspensions in aqueous solutions of associating polymer. *Colloid and Polymer Science*, 2008. **286**(13): pp. 1537–1544.
71. Zhu, D., et al., Aqueous hybrids of silica nanoparticles and hydrophobically associating hydrolyzed polyacrylamide used for EOR in high-temperature and high-salinity reservoirs. *Energies*, 2014. **7**(6): pp. 3858–3871.
72. Ehtesabi, H., Ahadian, M. M., and Taghikhani, V., Enhanced heavy oil recovery using TiO_2 nanoparticles: Investigation of deposition during transport in core plug. *Energy & Fuels*, 2014. **29**(1): pp. 1–8.
73. Nourafkan, E., et al., Improved rheological properties and stability of multiwalled carbon nanotubes/polymer in harsh environment. *Journal of Applied Polymer Science*, 2018: p. 47205.
74. Zhang, L., Tao, T., and Li, C. Formation of polymer/carbon nanotubes nano-hybrid shish–kebab via non-isothermal crystallization. *Polymer*, 2009. **50**(15): pp. 3835–3840.

75. Tongwa, P. and Bai, B., Degradable nanocomposite preformed particle gel for chemical enhanced oil recovery applications. *Journal of Petroleum Science and Engineering*, 2014. **124**: pp. 35–45.
76. Tongwa, P., Nygaard, R., and Bai, B., Evaluation of a nanocomposite hydrogel for water shut-off in enhanced oil recovery applications: Design, synthesis, and characterization. *Journal of Applied Polymer Science*, 2013. **128**(1): pp. 787–794.
77. Nezhad, S. S. K. and Cheraghian, G., Mechanisms behind injecting the combination of nano-clay particles and polymer solution for enhanced oil recovery. *Applied Nanoscience*, 2016. **6**(6): pp. 923–931.
78. Chen, G., et al., An attenuated total reflection FT-IR spectroscopic study of polyamide 6/clay nanocomposite fibers. *Macromolecular Rapid Communications*, 2004. **25**(11): pp. 1121–1124.
79. Tjong, S. C., Structural and mechanical properties of polymer nanocomposites. *Materials Science and Engineering: R: Reports*, 2006. **53**(3): pp. 73–197.
80. Fu, P., et al., Preparation, stability and rheology of polyacrylamide/ pristine layered double hydroxide nanocomposites. *Journal of Materials Chemistry*, 2010. **20**(19): pp. 3869–3876.
81. Hu, Z. and Chen, G. Aqueous dispersions of layered double hydroxide/ polyacrylamide nanocomposites: preparation and rheology. *Journal of Materials Chemistry A*, 2014. **2**(33): pp. 13593–13601.
82. Hu, Z. and Chen, G. Novel nanocomposite hydrogels consisting of layered double hydroxide with ultrahigh tensibility and hierarchical porous structure at low inorganic content. *Advanced Materials*, 2014. **26**(34): pp. 5950–5956.
83. Nourafkan, E., et al., Synthesis of stable iron oxide nanoparticle dispersions in high ionic media. *Journal of Industrial and Engineering Chemistry*, 2017. **50**: pp. 57–71.
84. Zeyghami, M., Kharrat, R., and Ghazanfari, M., Investigation of the applicability of nano silica particles as a thickening additive for polymer solutions applied in EOR processes. *Energy Sources, Part A: Recovery, Utilization, and Environmental Effects*, 2014. **36**(12): pp. 1315–1324.
85. Magami, S. M., Functional crosslinked hydrogels. In *Reactive and Functional Polymers*, Gutiérrez, T. J. (Eds), Volume Two. Springer, Cham., 2020.
86. Mayeres, J.-P., et al., *Fireproof foam compositions*. 2011, Google Patents.

87. 曹发斌, 孙., 李辽沙, 王平毕影,, "*A kind of silicate powder of ATO nanoparticle parcel and application thereof,*" 2015. CN104004387B, 2015.

88. Robinson, J., Mazany, A., and Cartwright, C., *Multi-layer fire barrier systems*. 2005, Google Patents.

89. Richard, C. N., Ainsworth, V. C., Holland, J., and Magami, S., "*UV stable fire-resistant glazing laminates,*" 2021. GB201916952D0, 2021.

90. J. R. H. Karikath Su Varma, D. H., "*Method for the Production of Fire Resistant Glazing,*" 2006. US20120061008A1, 2006.

91. Gelderie, N. S. U., "*Protective element and method for the production thereof,*" 2009. US20120321901A1, 2009.

Chapter 12

Effect of Ultrahigh Pressure on Nanomaterials Structure

Ehsan Nourafkan[a] and Yaser Kashcooli[b]
[a]*School of Engineering and Materials Science,*
Queen Mary University of London, London, UK
[b]*Department of Chemical and Biological Engineering,*
Drexel University, USA
e.nourafkan@qmul.ac.uk, enorafkan@gmail.com

Ultrahigh pressure can change the nanomaterial structure, which is interesting for future research on the design of new materials. The chemical and physical properties of soft and hard nanomaterials (e.g., structure, shape, and electron configuration) could be changed at high and ultrahigh pressure. The external pressure, as an additional processing parameter, enables the formation of new nanomaterial structures via manipulating system energy. This chapter aims to review the effect of high and ultrahigh pressure on the chemical and physical properties of both soft nanostructures, including microemulsions and lyotropic liquid crystals, and hard nanomaterials, including solid nanoparticles. The effect of

Science and Applications of Nanoparticles
Edited by Waqar Ahmed and Ehsan Nourafkan

ISBN 978-981-4267-34-2 (Hardcover), 978-1-003-28029-3 (eBook)
www.jennystanford.com

ultrahigh pressure on the size, structure, morphology, and electron configuration and stability of nanostructures has been be reviewed.

12.1 Introduction

High-pressure synthesis is one of the frontiers of searching new materials. For instance, a large elastic strain field and an electronic state transition made by an isotropic pressure field can make new crystal structures and textures. Unlike heating or cooling, the pressure propagates through the sample homogeneously without forming any gradients. High pressure in the GPa scale can alter the basic bonding of compounds (e.g., carbon based and organic) and reduce the atomic volume of bulk materials by 10% or more. Applications of high pressure have greatly enhanced our understanding of the electronic, phonon, and doping effects on the newly emerged nanomaterials. Pressure also affects the fundamental activity and structure in biological assemblies. Pressures in the range encountered in the biosphere are known to have many effects on biomacromolecules, including alteration of ligand binding, membrane permeability, bacterial motility, and molecular association and aggregation. Thus, high pressure is a useful tool both for the synthesis of new matter and for probing the existing phases of scientific or technological significance. High pressure can modify nano-objects in two different ways:

1. Modification of the crystalline lattice structure of nanoparticles by changing the properties without any aggregation or agglomeration.
2. Formation of a network between nanoparticles (e.g., clathrate, 3D architecture). Joining nanoscale carbon materials such as carbon nanotubes (CNTs), graphene, and nanodiamonds with covalent bonding has promising super-hard and ultra-strong properties.

In this chapter, we tried to cover both path by reviewing the available knowledge and propose future opportunities for the design of new nano-objects size and geometry under high-pressure condition.

12.2 High-Pressure Instrumentation

To carry out pressure tuning, a high-pressure apparatus is necessary. The techniques for the generation of high pressure have been shown in Fig. 12.1. Large-volume press (LVP) and diamond anvil cell (DAC) are the most common techniques for ultrahigh pressure generation.

12.2.1 Large-Volume Press

Static pressures as high as 400 GPa are now achievable inside a diamond anvil cell (DAC). However, the sample obtained in a DAC is too small to characterize when the pressure is released. Compared to a DAC, the sample volume in a large-volume press (LVP) apparatus is large, and it can vary between the mm^3 and cm^3 scales.

12.2.2 Diamond Anvil Cell

A diamond anvil cell (DAC) is a standard and powerful tool for studying materials under extreme conditions of pressures (P) and temperatures (T), which was initially introduced by Ming and Basset [1]. Two cut diamonds are used as the anvil to squeeze the sample, which exert a lot of force over a tiny area. The area doing the pressing is a tiny facet at the bottom of each diamond called the culet. Hundreds of gigapascal pressure could be produced using conventional DACs by pushing the diamonds together really hard. The main advantage of DACs is that the optical properties of diamonds and their transparency in a wide range of wavelengths help the scientists to study the sandwich between the two culets suspended in a bit of fluid. Visible light, infrared light, Raman, and X-ray and γ-ray spectroscopy could be applied to study the atomic, photonic, electronic, and magnetic structures of nanomaterials at ultrahigh pressure. The culet size of diamonds has a great influence on the generated pressure since the pressure is defined by the force applied per surface area. The choice of the diamond anvil type (Fig. 12.1) mainly depends on the P-T range, which is intended to reach during the experiment. Basically, flat, single-bevel, and double-bevel diamonds are used for the generation of pressure up to 400 GPa, while toroidal and double-stage diamonds are applied for the generation of pressure over 400 GPa [2].

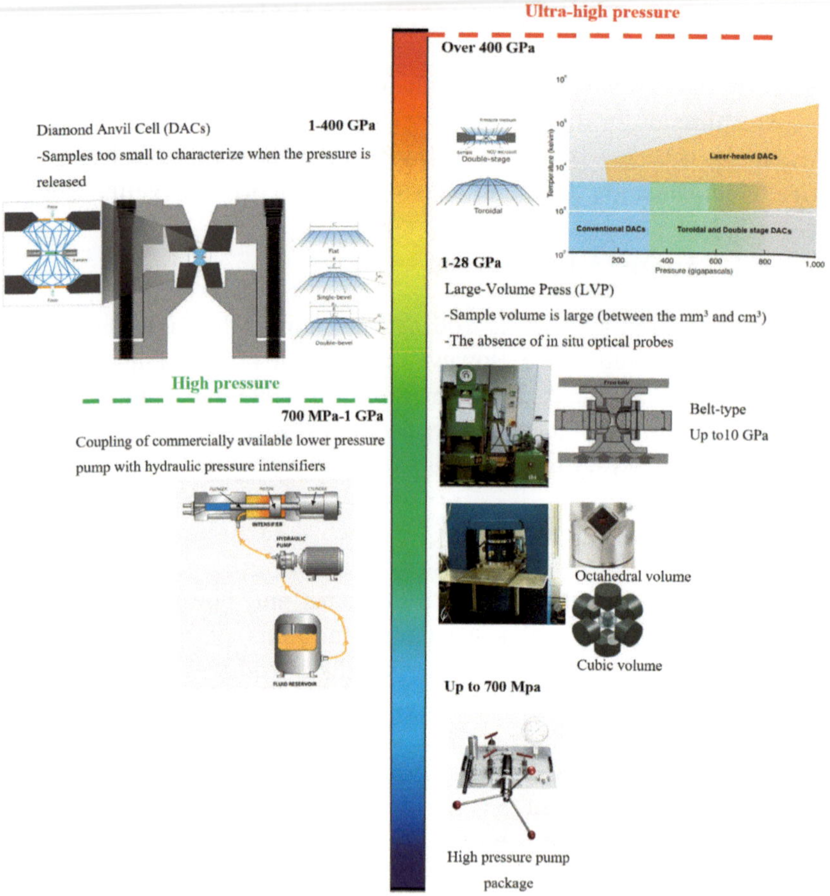

Figure 12.1 Different techniques for high-pressure generation.

12.3 Effect of Pressure on Soft Matter Nanostructures

Soft materials such as emulsions, liquid crystals, and colloidal assemblies have wide engineering and medical applications. For instance, emulsion-based systems are widely used for encapsulating, protecting, and delivering active chemical ingredients and drugs. Investigating the effect of pressure on the structure and phase behavior of emulsions is crucial for the improvement in their ability as carrier for chemical ingredients. In many cases, the concept of soft matter can be imported from physics and chemistry, which have traditionally focused on materials such as colloids, polymers, surfactants, and liquid crystals. However, biological soft materials are categorized as organic materials (e.g., lipids and bio-membrane) that are present in biological tissues. Different life forms exist in the deepest ocean trenches at high pressures. Understanding the intrinsic effect of pressure on membrane structures is critical to understanding the adaptation and evolution of deep-sea organisms as well as soft material processing. Hydrostatic pressure can be used to drive structural changes in soft matters, which offer significant advantage over other structure-change triggers such as temperature. Here we review the recent progress in the investigation of the effect of pressure on the structure of emulsions, liquid crystals, and biological soft matters.

The nature of pressure tends to reduce the volume of materials in any structure, and surfactant molecules are not an exception. As Fig. 12.2 shows, the hydrophobic tail of the surfactant/lipid is stretched at high pressure, leading to a change in the curvature of the surfactant layer, which may lead to phase transitions of microemulsions [3–5].

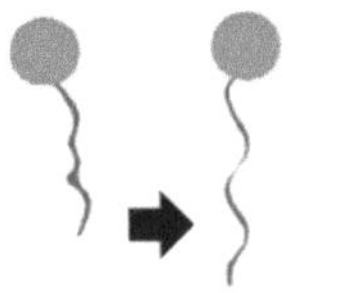

Figure 12.2 Effect of pressure on the stretching of the hydrophobic tail.

Berghaus et al. studied the effect of pressure up to 400 MPa on the structure of bicontinuous microemulsion of tetraethyleneglycol decanoyl ether (C10E4)-water-n-octane by small-angle X-ray scattering (SAXS) (Fig. 12.3) [6]. They reported that high pressure leads to an increase in the stiffness and hence the bending of the O/W interface. However, the bending was not enough to result in a phase transition to a more ordered lamellar phase.

Nagao et al. and Kawabata et al., in two different studies, reported similar observations for the effect of pressure (up to 60 MPa) and temperature on the interaction of surfactant molecules in the microemulsion system of the AOT–water–decane ternary system using SANS analysis. They reported that the surfactant monolayer is more flexible at higher temperature and more rigid at higher pressure [3, 4]. The reason was that the attractive interaction between the hydrophobic tails of the surfactant molecules is enhanced at high pressure, while the head–head repulsion is increased at high temperature. In fact, the rigidity of the interface increases with increasing pressure, while the interface becomes more flexible with increasing temperature.

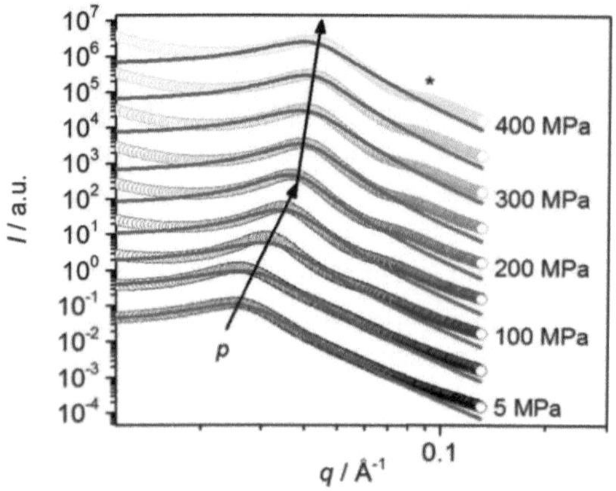

Figure 12.3 Bulk scattering intensity profiles, $I(q)$, of the bicontinuous microemulsion at different pressures in double-logarithmic representation. Reprinted from Ref. [6].

The development of new technologies in recent years lets us to study the curvature properties of soft materials at high pressure. Purushothaman et al. developed a microscope cell to study the membrane structure, rigidity, and fluctuation at high pressure up to 80 MPa [5]. As Fig. 12.4 shows, both membrane (dioleoyl-phosphatidylcholine) bending rigidity and membrane tension significantly increased at higher pressures. The SAXS data also confirm that increasing pressure leads to decreasing of the area per hydrophobic chain of dioleoyl-phosphatidylcholine (an increase in the bilayer thickness).

Liquid crystals are one of the fundamental forms of soft matters and conventionally divided into thermotropic or lyotropic [7]. Thermotropic liquid crystals are those that occur in a certain temperature range and are not in the scope of this chapter. Briefly, the effect of pressure and temperature on the phase diagram of a thermotropic liquid crystal systems has been shown in Fig. 12.5. According to Fig. 12.5, liquid crystal transfers from isotropic liquid to nematic phase and to solid state on increasing pressure.

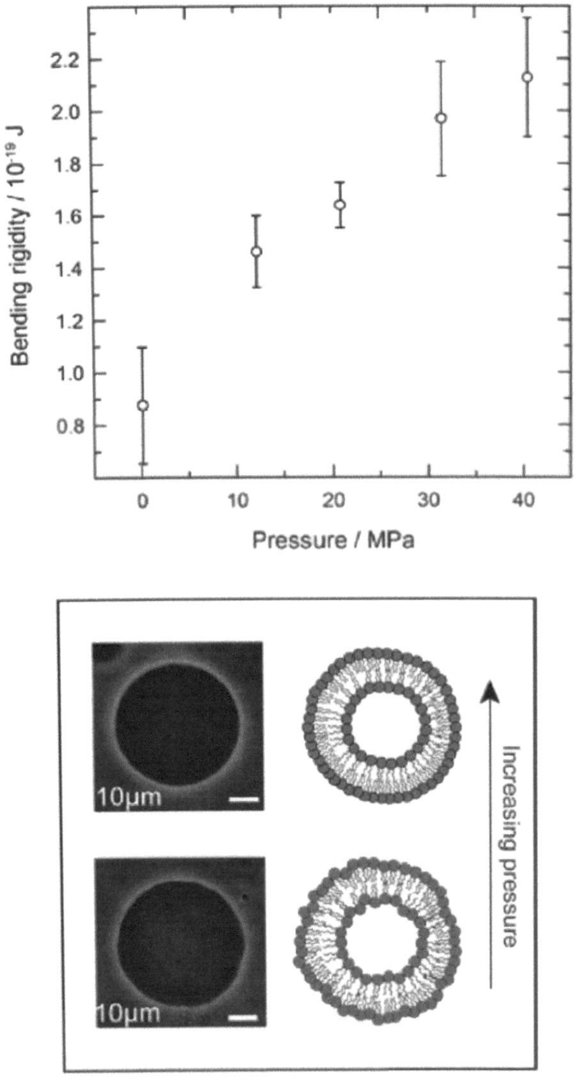

Figure 12.4 Average bending rigidity of dioleoyl-phosphatidylcholine vesicle membrane and schematic showing the pressure-induced reduction in membrane fluctuation. Reprinted from Ref. [5].

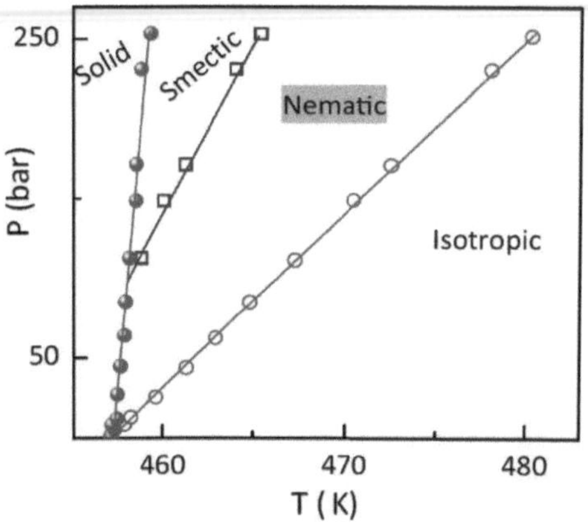

Figure 12.5 Pressure–temperature phase diagram of p-methoxybenzoic acid showing the formation of pressure-induced nematic and smectic phases. Reprinted from Ref. [8].

Lyotropic liquid crystals (LLCs) consist of two or more components and generally formed by amphiphilic molecules (e.g., surfactants). Therefore, LLCs could form by dispersing self-organized structures, including micelles (spherical, cylindrical), membranes, or vesicles in an isotropic liquid. The effects of temperature, concentration of amphiphilic fraction as well as their properties on LLCs were topics of research in numerous studies and review papers [7, 9, 10]. For instance, the critical packing parameter (CPP) of amphiphilic molecules could have a great influence on the LLC structure by changing the interface curvature between components [10] (Fig. 12.6). The CPP is defined as follows:

$$\text{CPP} = v/a_0 l_\text{c} \tag{12.1}$$

where a_0 is the surface area of the headgroup, l_c is the maximum length of the chain, and v is the molecular volume of the chains.

As mentioned earlier, the interface curvature could change by increasing pressure. Investigation of the impact of pressure on the LLC phase diagram has been the topic of a few researches, which particularly focused on the self-assembly of lipid membranes composition in natural organs. Salvador-Castell et al. presented the effect of high static pressure (up to 100 MPa) on lipid membrane containing archaeal-like phospholipids 1,2-di-O-phytanyl-sn-glycero-3-phosphocholine (DoPhPC) and 1,2-di-Ophytanyl-sn-glycero-3-phosphoethanolamine (DoPhPE)-squalane fatty component using small-angle X-ray scattering (SAXS) [11]. Figure 12.7 shows the inverse effect of temperature and pressure on the phase transformation of LLC from lamellar and hexagonal phases verified by SAXS data. At elevated temperature, the hexagonal (HII) phase appeared due to an increase in the membrane's negative curvature, while higher pressure (above 350 bar) promoted the lamellar phase formation.

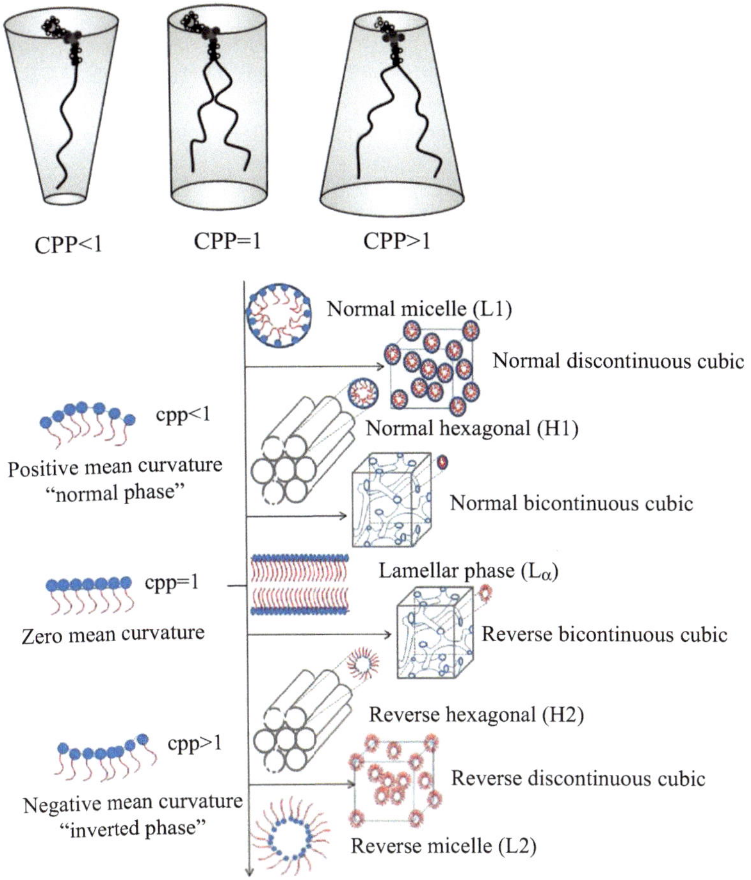

Figure 12.6 The shape of amphiphilic molecule versus critical packing parameter values: CPP < 1 (cone), CPP = 1 (cylindrical), and CPP > 1 (inverted cone). Reprinted from Refs. [9, 10].

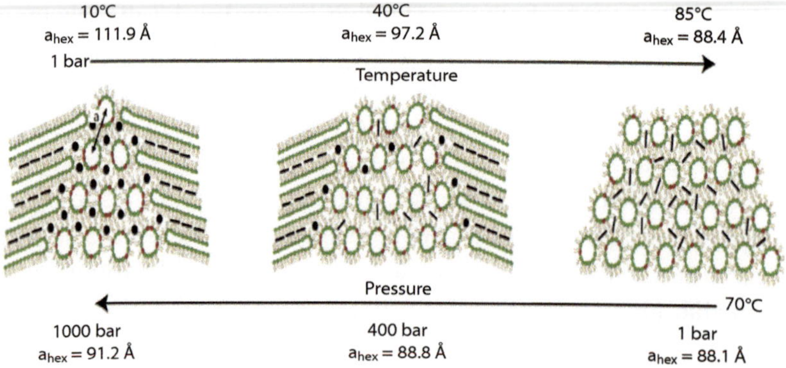

Figure 12.7 Schemes of phase coexistence and lattice parameter evolution as a function of pressure and temperature (squalane is represented in black; black spheres represent squalane aggregates; lipid hydrocarbon chains are in brown lipid headgroups are shown in green and in red). Reprinted from Ref. [11].

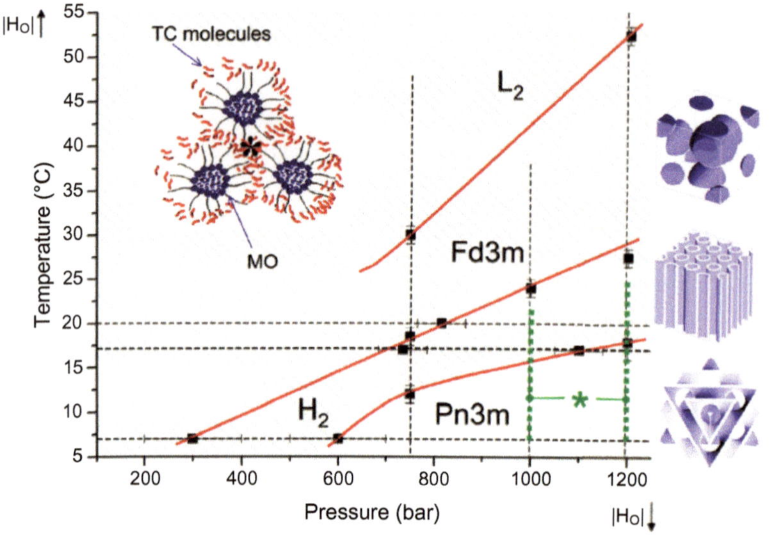

Figure 12.8 LCC phase transformation of tetradecane-monoolein-water system at different temperature–pressure. Reprinted from Ref. [12].

In another research, the effect of pressure (up to 120 MPa) on the phase transformation of hydrated tetradecane-monoolein-water system was studied by Yaghmur et al. [12]. At ambient pressure, the lamellar (Lα) phase was observed in a wide range of temperature;

however, LLC structural transitions to inverted hexagonal (H2), 3D cubic micellar (Fd3m), and bicontinuous cubic-Pn3m were observed at different temperatures (Fig. 12.8).

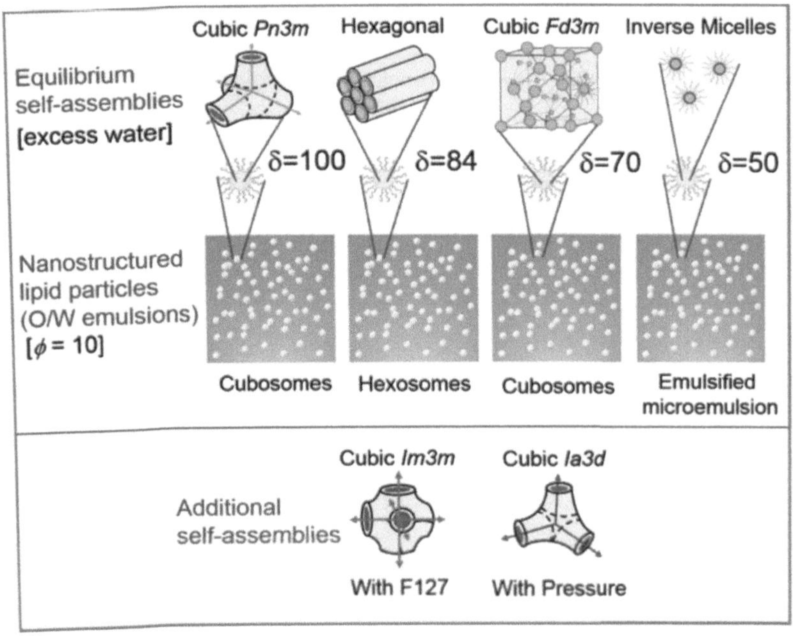

Figure 12.9 Different types of self-assembly controlling by tuning the δ value (i.e., the composition of the lipid phase); bicontinuous cubic Im3m and Ia3d phases were formed at high pressure. Reprinted from Ref. [12].

Kulkarni et al. (2016) investigated the pressure-dependent phase behavior of nanostructured lipid particles using SAXS measurements in the range of 1–1200 bar at 25°C. Nondispersed self-assembly structures were prepared in an aqueous medium using two different types of lipids [13]. They successfully synthesized various structures of bicontinuous cubic (Pn3m and Im3m), micellar cubic (Fd3m), hexagonal (H2), and inverse micellar (L2) phases by tuning the lipid-to-oil ratio (δ and ϕ value) and hydrostatic pressure (Fig. 12.9).

$$\delta = \frac{m_{\text{lipid}}}{m_{\text{lipid}} + m_{\text{oil}}} \times 100$$

$$\phi = \frac{m_{\text{lipid}} + m_{\text{oil}}}{m_{\text{water}} + m_{\text{lipid}} + m_{\text{oil}}} \times 100 \qquad (12.2)$$

where m_{lipid}, m_{oil}, and m_{water} are the masses of lipid, solubilized oil, and water, respectively.

12.4 Effect of Pressure on Hard Nanomaterials

Ultrahigh pressure can provide access to solid-state compounds that cannot be formed with traditional techniques. Some of the interesting applications of NPs particularly for electronic, magnetic, and optical devices are related to ordering or densely packing NPs in structures of two and three superlattices (SLs). Therefore, the synthesis of the building blocks of NPs and nanocrystals (NCs), referred to as "supercrystals" (SCs), and studies of their properties have gained great interest in recent years [14]. The high pressure is able to not only change the crystalline structure of NPs at the atomic scale but also regulate the alignment of particles at the nano and meso scales. At high pressure, the amount of Gibbs free energy of the system changes due to the change in the PV term in the Gibbs energy equation of $G = E + PV - TS$, which provides the opportunity of formation of a new phase or tune configuration. The oriented alignment or attachment involves arrangement of NPs, which might be followed by the structural transformation of NPs to develop a specific crystallographic phase. Basically, during the structure transformation, the crystalline facets with higher surface are eliminated, which finally results in the coalescence of NPs into the SC structure. Table 12.1 reviews the most interesting research for the effect of static pressure on the structural transformations of nanomaterials and emerging of new phase structures. Walsh and Freedman also reviewed the effect of ultrahigh pressure on the formation of novel intermetallic compounds, including bismuth element [15]. Here our focus is to review the recent progress on the effect of ultra-pressure on gold, quantum dots, and carbon-based nanomaterials as heart of the next technological revolution in solid-state electronics.

Table 12.1 Review of the most interesting research for the effect of static pressure on emergence of new nanostructures

Type of Nanoparticles	Type of Property	Max Pressure and Method	Pressure Generation	Ref.
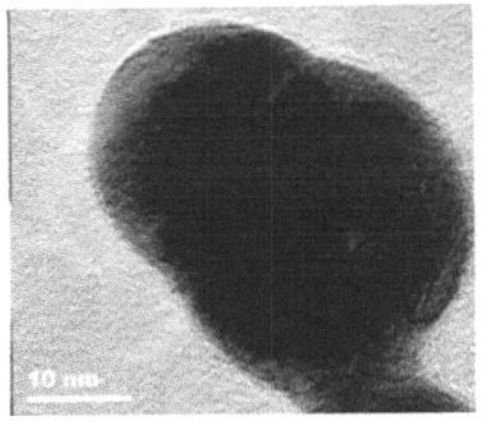 Gold (30 nm) and silver (10 nm)	Stiffness	DAC (30 GPa)	Polysynthetic domain twinning and lamellar defects as the main origin for the strong decrease in compressibility	[16]
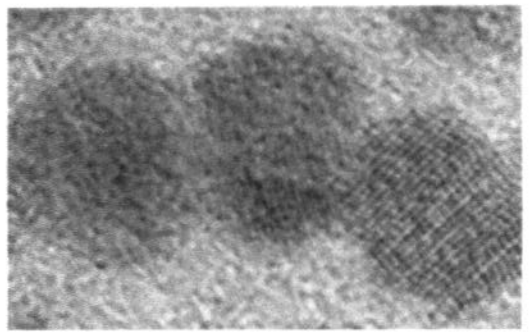 QDs of CdSe (6–8 nm)	Optical properties Crystalline and phase transfer	DAC (15 GPa)	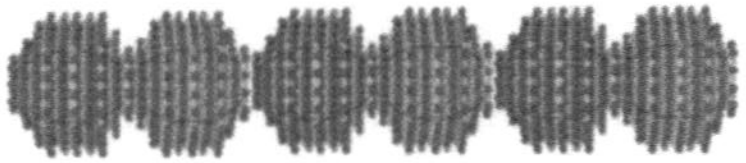	[17]

(*Continued*)

Table 12.1 (*Continued*)

Type of Nanoparticles	Type of Property	Max Pressure and Method	Pressure Generation	Ref.
 Single gold NPs (22 nm ± 3 nm) coated with PVP	Crystalline structure	DAC (12 GPa)	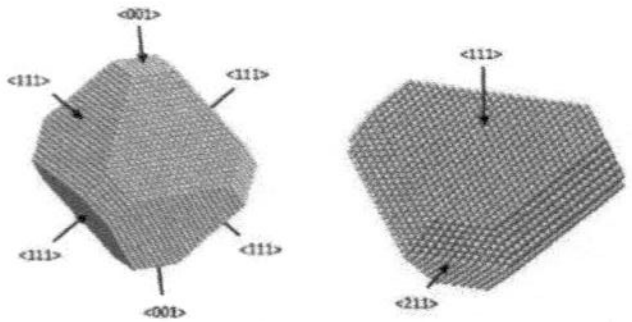 • No lattice distortion was observed for triangle morphology after imposing to the same pressure. • An asymmetrical lattice distortion of ~3% that led to a transition from a cubic to an orthorhombic structure for octahedral morphology of gold NPs.	[18]
Gold (5.2 nm)	Crystalline structure	DAC (6 GPa)	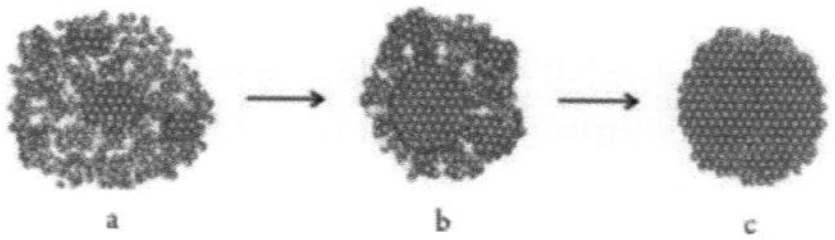 Pressure-induced super crystallization of nanoparticle arrays.	[19]

Type of Nanoparticles	Type of Property	Max Pressure and Method	Pressure Generation	Ref.
 Supercrystals of PbS NCs (7 nm) stabilized with oleic acid	Crystalline preferential orientation Structural stability Bulk modulus	DAC (55.5 GPa)	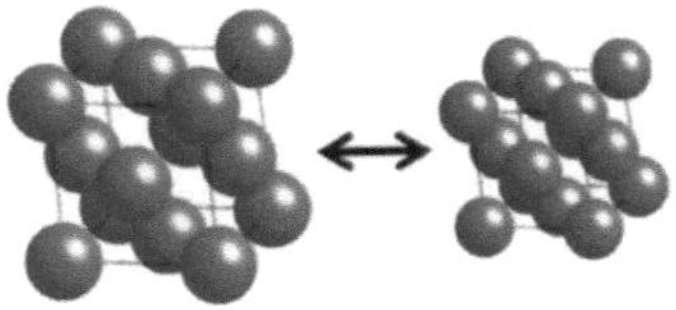 • Single unit cell of fcc-organized NPs depicting that the structure was transformed to a faceted 3D supercrystal (SC) self-assembled structure. • The ambient pressure bulk modulus of the SCs was calculated to be ~5 GPa for compression and ~14.5 GPa for decompression from fitting of Vinet and Birch-Murnaghan equations of state.	[14]
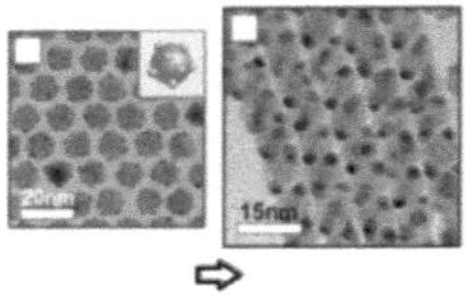 Quantum-dot-gold (1.6 ± 0.2 nm)	Self-assembled superlattices formation	DAC (15.8 GPa)	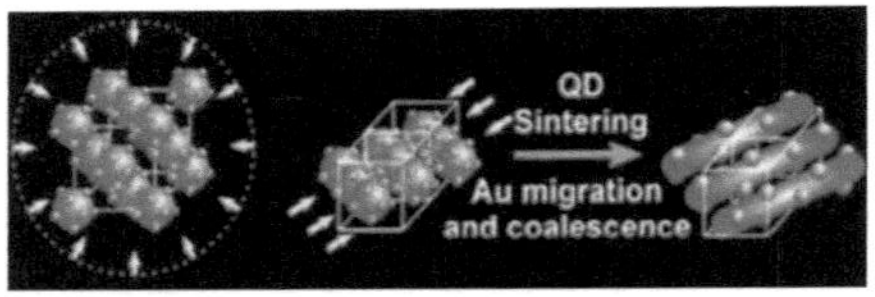	[20]

(Continued)

Table 12.1 (*Continued*)

Type of Nanoparticles	Type of Property	Max Pressure and Method	Pressure Generation	Ref.
PbS nanocubes (13 nm) coated by oleic acid	Structural transformation and stability	DAC (15.4 GPa)	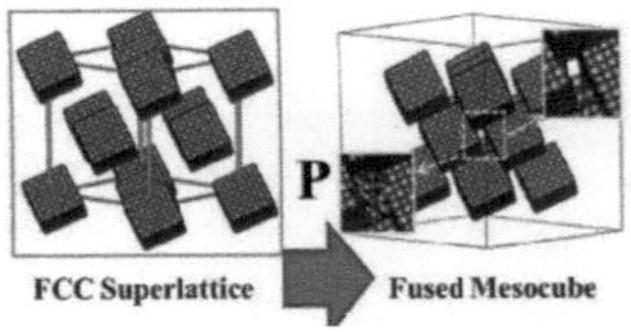 • Transformation of PbS fcc structure to pressure-driven 3D porous architecture. • Larger compressibility of oleic acids was irreversible after the release of pressure, which maintains the 3D structure stability under ambient conditions.	[21]
 PbS NPs (3.5 nm)	Structural transformation Transformation mechanism	DAC (14.5 GPa)	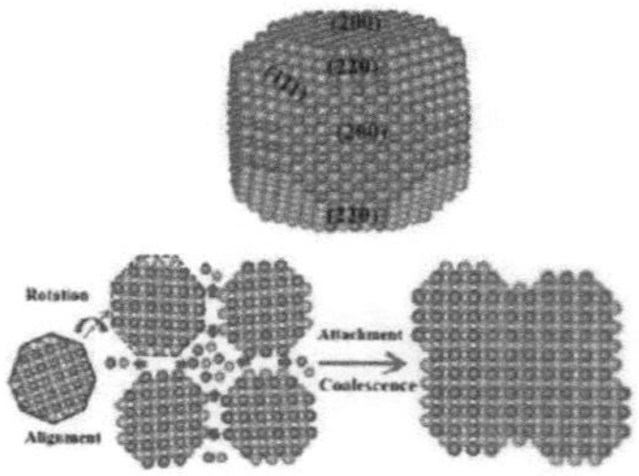 PbS NPs converted to two-dimensional single-crystal PbS nanosheets with a lamellar mesostructured.	[22]

Type of Nanoparticles	Type of Property	Max Pressure and Method	Pressure Generation	Ref.
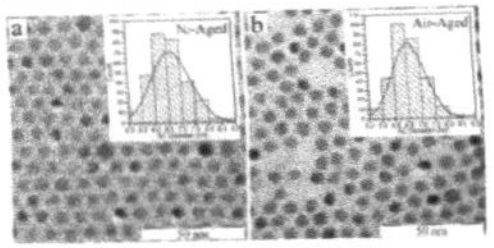 PbS (6.6 nm)	Structural transformation and stability	DAC (16.5 GPa)	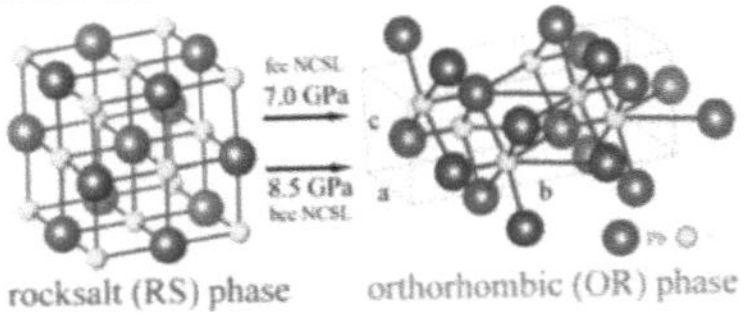 Colloidal PbS nanocrystals (NCs) self-assembled into superlattice (SL) allotropes of either face-centered cubic (fcc) or body-centered cubic (bcc) symmetry.	[23]
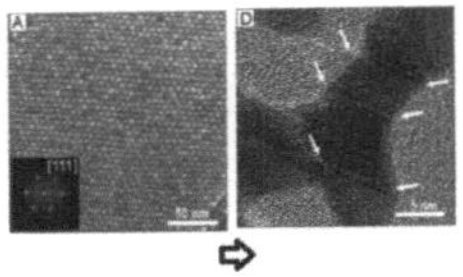 Gold (30 nm)	Structural transformation and stability	DAC (12.4 GPa)	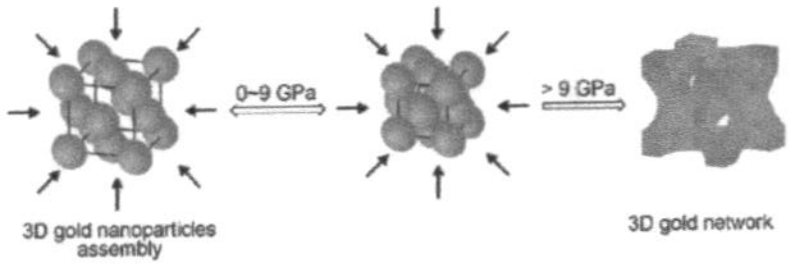 • A film of nanoparticles with face-centered cubic (fcc) structure converted to a 3D nanostructured gold architecture. • The 3D gold nanostructure was maintained under ambient conditions after the pressure releasing from 12.4 GPa to ambient pressure.	[24]

(Continued)

Table 12.1 (*Continued*)

Type of Nanoparticles	Type of Property	Max Pressure and Method	Pressure Generation	Ref.
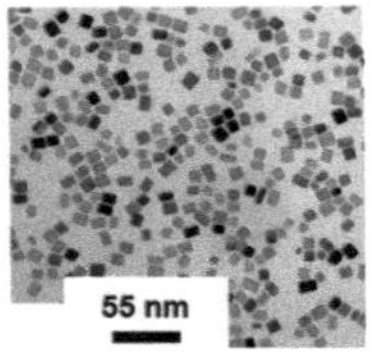 Pd nanocubes (10 nm)	Structural transformation and stability	DAC (24.8 GPa)	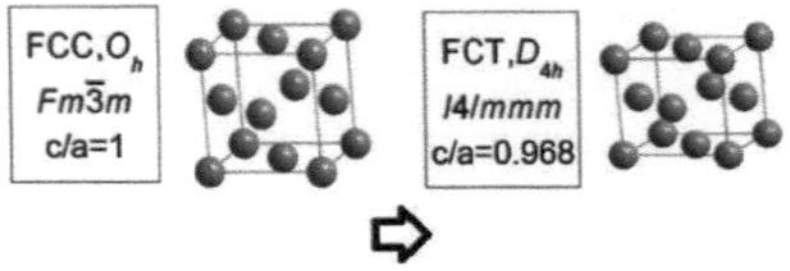 Face-centered cubic fcc to ace-centered-tetragonal (fct) transformation was observed for the cold-compressed Pd nanocubes	[25]
Film of gold NPs (5.2 nm)	Structural transformation and MD simulation	Pulsed power generator (12 GPa)	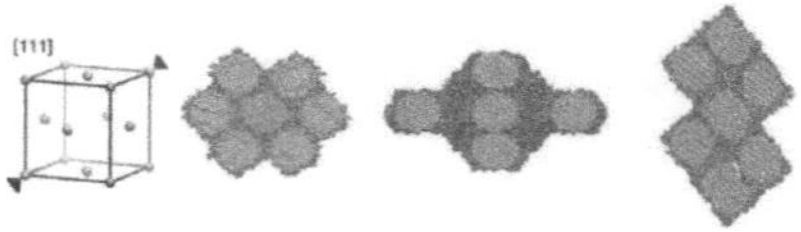 Various gold architectures are generated through compressive mesoscale coalescences of spherical NPs	[26]
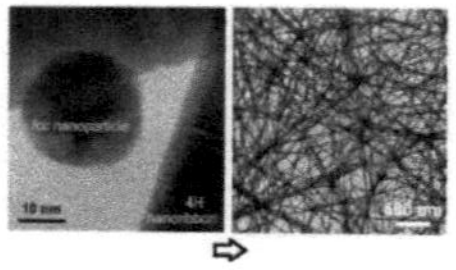 Gold NPs	Structural transformation and stability	DACs (14.3 GPa)	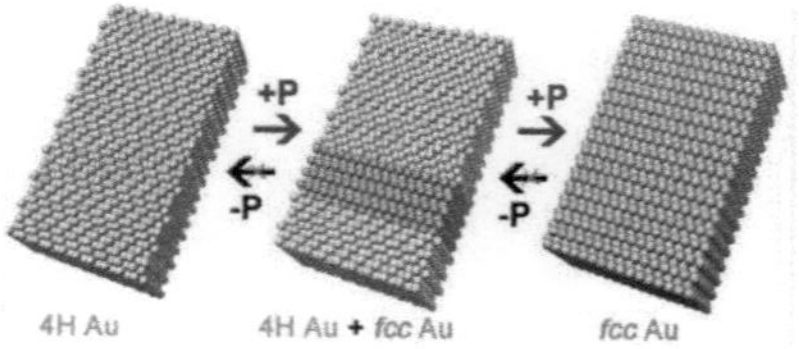 Phase transformation of hexagonal 4H to face-centered cubic (fcc) phase in Au nanoribbons (NRBs)	[27]

Type of Nanoparticles	Type of Property	Max Pressure and Method	Pressure Generation	Ref.
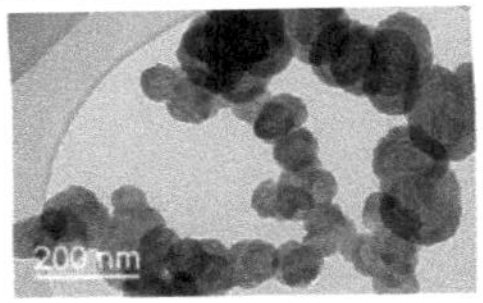 Cubic boron nitride (average diameter of 130 nm)	Structural transformation	Multianvil system (15 GPa)	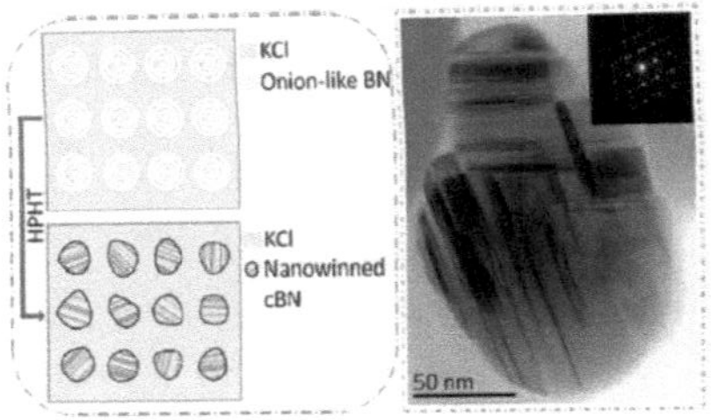 The highly wrinkled and defective oBN prefers to phase transition into nanotwinned cBN under HPHT conditions.	[28]
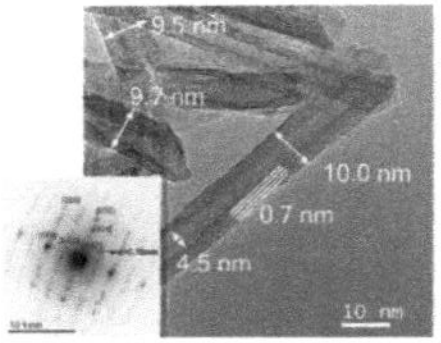 Nano-sized titania nanotube (anatase)	Structural transformation Optical properties	Merrill-Bassett DAC (6GPa)	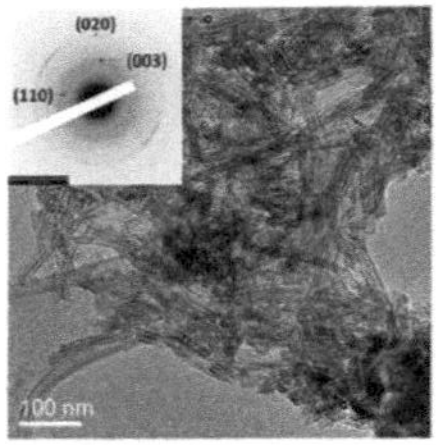 A significant shift in the Raman band structure of nanotube was observed due to distortions and a 5% contraction in the interlaminar spacing.	[29]

(Continued)

Table 12.1 (*Continued*)

Type of Nanoparticles	Type of Property	Max Pressure and Method	Pressure Generation	Ref.
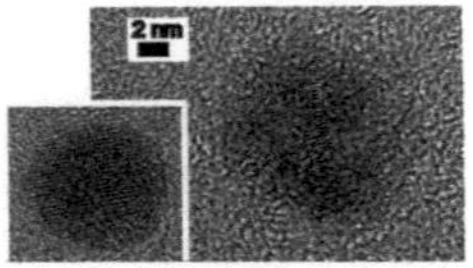 5-10 nm silver nanoparticles	Structural transformation	DAC (10.2GPa)	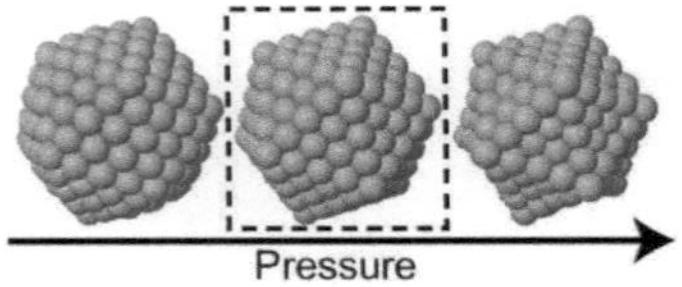 Silver nanoparticles undergo a reversible noncubic distortion with pressure	[30]
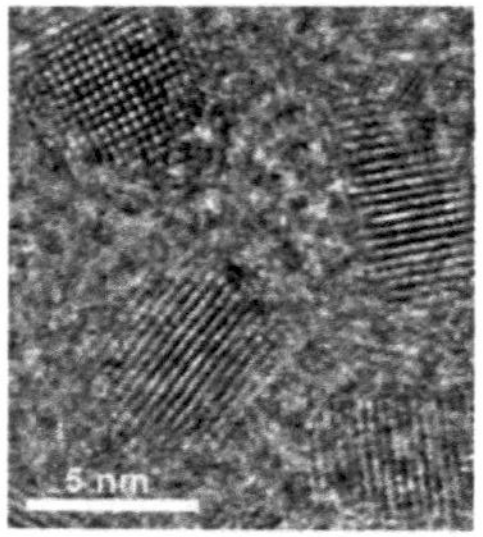 PbTe nanoparticles (6 nm)	Structural transformation	DAC (19.7 GPa)	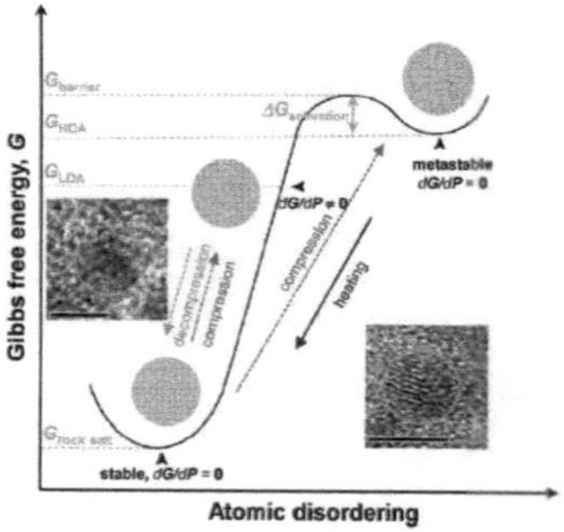 • The PbTe nanocrystal started to become amorphous above 10 GPa and then underwent a low-to-high density amorphous phase transformation at pressures over 15 GPa. • Proposing a structural mechanism for pressure-induced switching between amorphous and crystalline phases as well as their associated growth processes.	[31]

Type of Nanoparticles	Type of Property	Max Pressure and Method	Pressure Generation	Ref.
 Gold (5.2 nm)	Structural transformation	DAC (13 GPa)	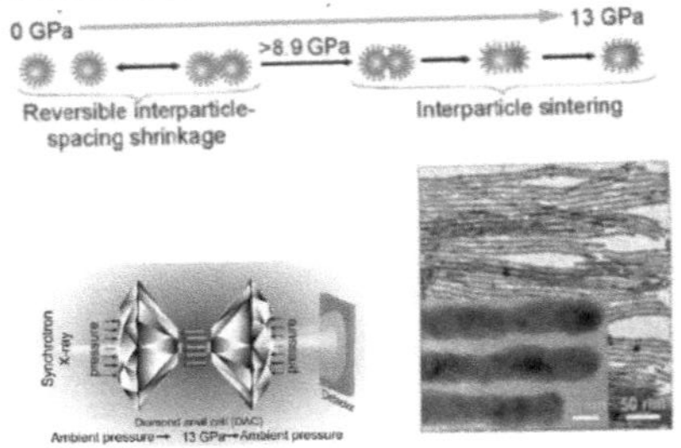 The ordered fcc spherical-nanoparticle superlattice had been converted into an ordered nanowire array after release of the pressure from 13 GPa to ambient pressure.	[32]
Nickel (Ni) and carbon nanotube (CNT)	Hardness	Cold pressed (4 GPa)	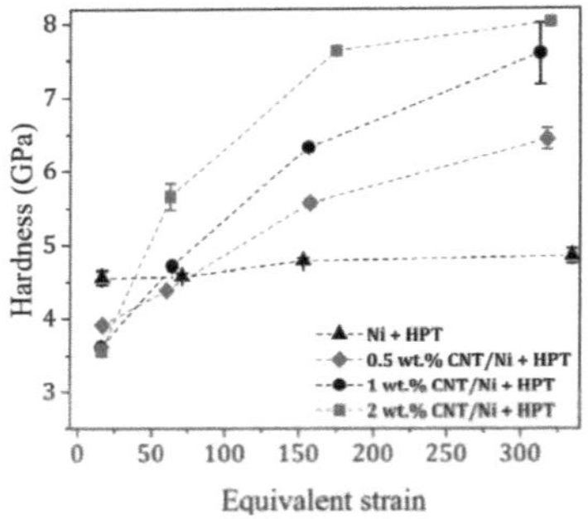 Reinforced Ni-matrix composites were processed by high-pressure torsion (HPT).	[33]

12.4.1 Gold Nanoparticles

Gold nanostructures, including nanorods, nanowires, and nanosheets, exhibit promising properties that enable a wide variety of applications in nanoelectronics, biosensors, and medical phototherapy [34]. Recently, several research studies show that ultrahigh pressure can produce new atomic modes of gold microstructures. Schroer et al. studied the effect of pressure up to 0.4 GPa on the phase transition of gold nanoparticle from disordered liquid state to ordered crystalline structure [35]. The 2D-SAXS analysis at 0.4 GPa shows the appearance of Debye–Scherrer rings and confirms the formation of ordered supercrystals of face-centered cubic (Fig. 12.10a,b). However, the reflections disappeared after the release of pressure, demonstrating a reversible transition.

Mendoza-Cruz et al. reported an orthorhombic lattice distortion in single-crystal truncated-octahedral gold nanoparticles induced by applying a high pressure up to 12 GPa [18]. By the analysis of the HR-TEM results, they showed that the crystalline deformation of NPs depends on the morphology of particles. In this way, the imposed high pressure led to a transition from a cubic to an orthorhombic structure for truncated-octahedral gold NPs, while any lattice distortion was not observed for triangular NPs at the same condition. Figure 12.11 illustrates the HAADF-STEM and fast Fourier transform (FFT) images of a single truncated-octahedral and single triangular NPs subjected to high pressure. Figures 12.11 (left) c and g show the magnified areas indicated in Fig. 12.11 (left) a and e. The intensity profiles in Fig. 12.11 (left) d and h relevant to the magnified areas in Fig. 12.11 (left) c and g highlight the difference in [1] the distances of octahedral NPs. Unlike the octahedral particles, no significant lattice distortion was observed for triangular morphology in Fig. 12.11 (right) d.

Li et al. investigated the effect of tens of 10 GPa pressure on gold NPs using a pulsed power generator (Fig. 12.12a). Different d-spacing values were observed at different pressures and also the formation of nanosheets and hexagonal bundles of gold nanowires was reported (Fig. 12.12b).

Medeghini et al. studied the optical properties of a single bipyramid gold nanoparticle (Fig 12.13a) up to 10 GPa by coupling a DAC and high-sensitivity spatial modulation spectroscopy (SMS) technique [36]. The optical extinction spectrum of the nanoparticle was measured by SMS at high pressure, which showed a further distinct red-shift (Fig. 12.13b). The localized surface plasmon resonances (LSPR) and optical absorption and scattering of nano-object are among important properties for the development of nanoelectromechanical systems, or nanosensing devices. The study shows the tuning of the plasmonic response of metal nanoparticles by controlling the surrounding pressure (Fig. 12.13b).

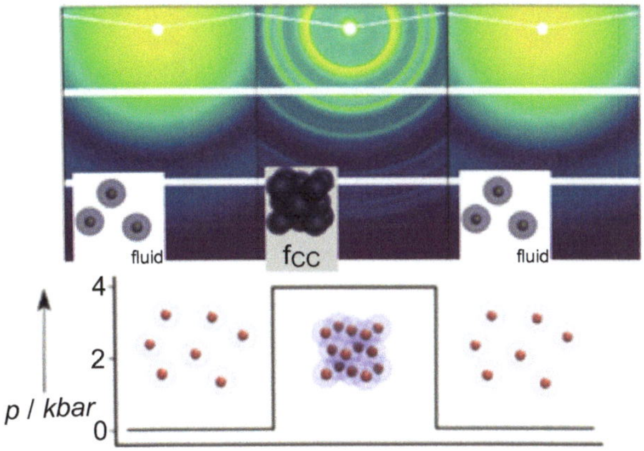

(a)

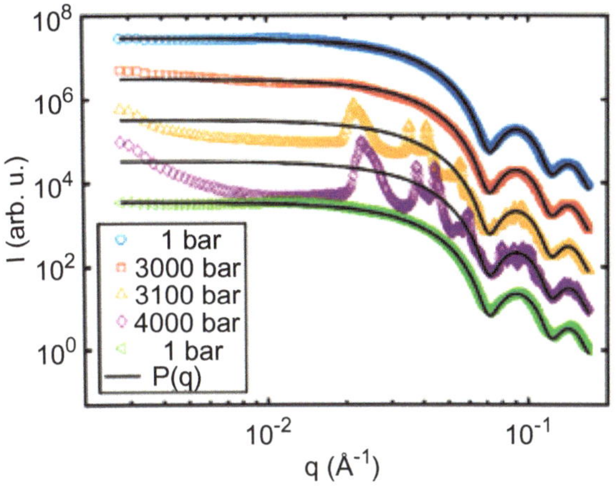

(b)

Figure 12.10 (a) Debye–Scherrer rings formation at high pressure (2D SAXS); (b) SAXS curves at different pressures.

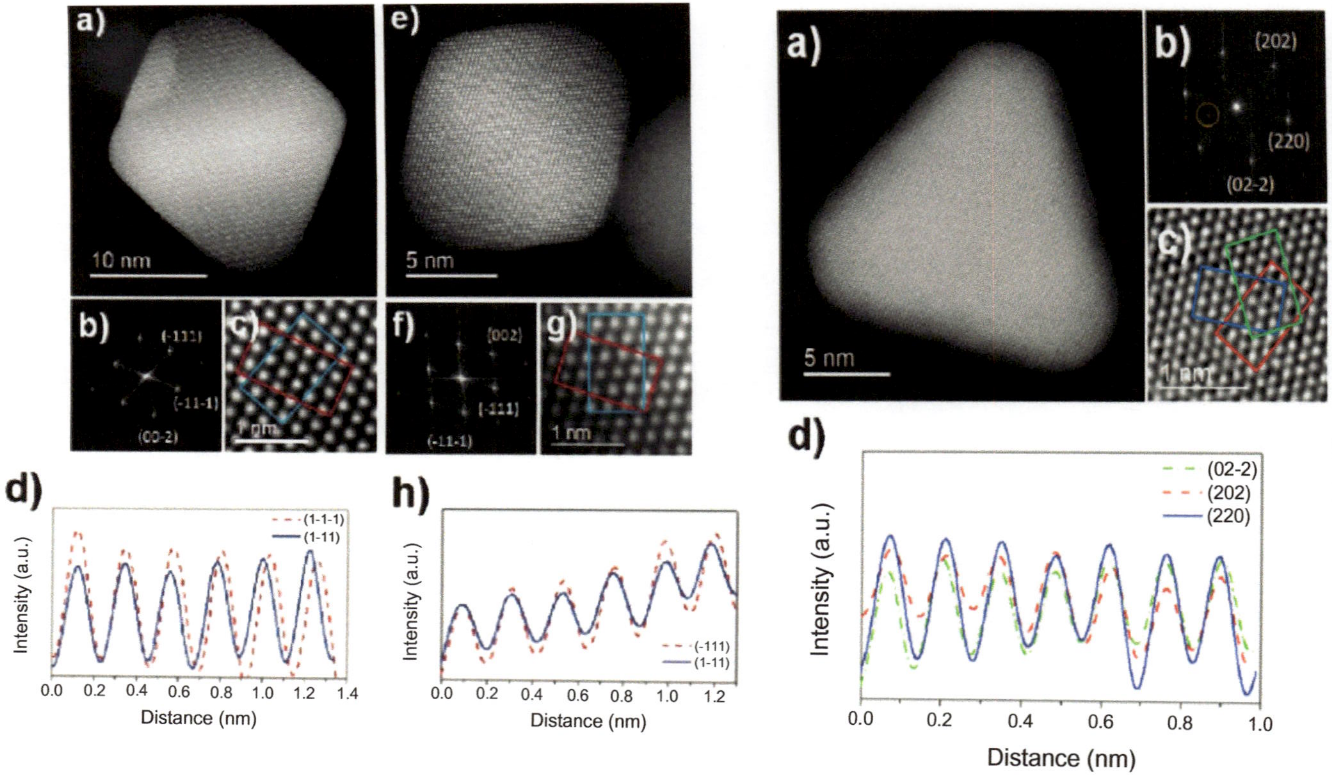

Figure 12.11 HAADF-STEM high-magnification images of (left) a *truncated-octahedral* and *(right)* a *triangular NP subjected to high* pressure. Reprinted from Ref. [18].

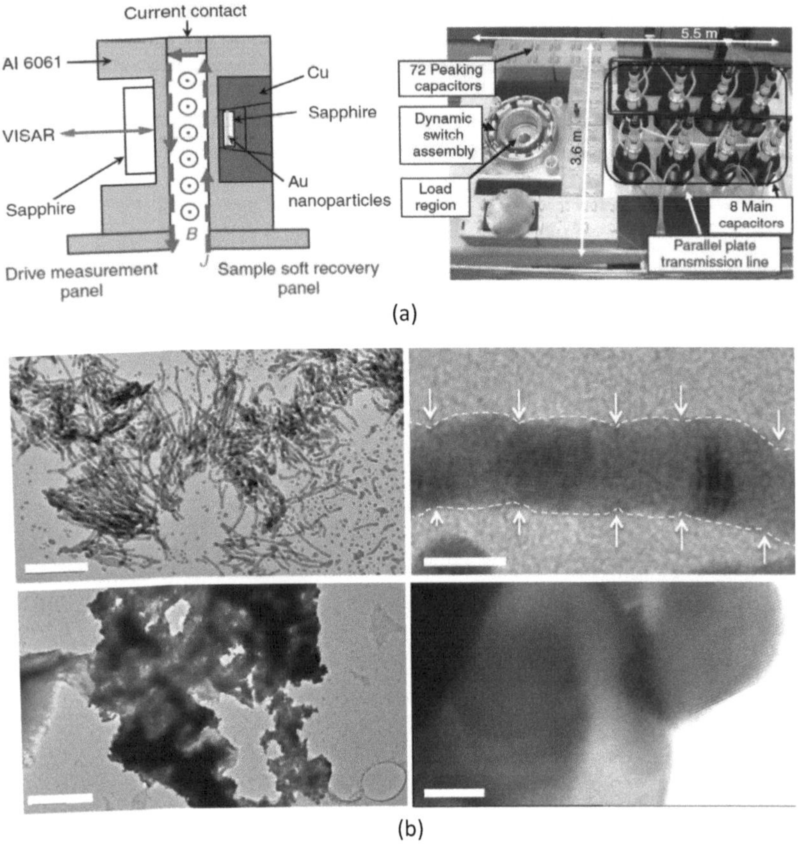

Figure 12.12 (a) Pulsed power generator, (b) formation of nanosheets and nanowires from spherical gold NPs imposing the high pressure. Reprinted from Ref. [34].

12.4.2 Quantum Dots

Li et al. formed CdSe nanowires (lengths roughly of 600 nm, 5 nm) after high-pressure compression of the 6–8 nm spherical nanocrystal arrays in a DAC [17]. The result of high-pressure small-angle scattering (HP-SAXS) confirmed the phase transitions of CdSe from Wurtzite (WZ) CdSe to rock salt (RS) started at 7 GPa. Below 7 GPa, the nanocrystals remained as Wurtzite (WZ). The mechanism of conversion of CdSe from WZ to RS under high pressure was proposed by Bealing et al., using molecular dynamic (MD) simulation as shown in Fig. 12.14 [37]. The HP-SAXS also shows peak shifting to higher angles (lower d-spacing), which conform the shrinking of crystal facets distances. After releasing the pressure from 15 GPa to ambient, the RS phase is a dominant crystalline structure until 0.7 GPa, and then it was converted to zinc blende (ZB).

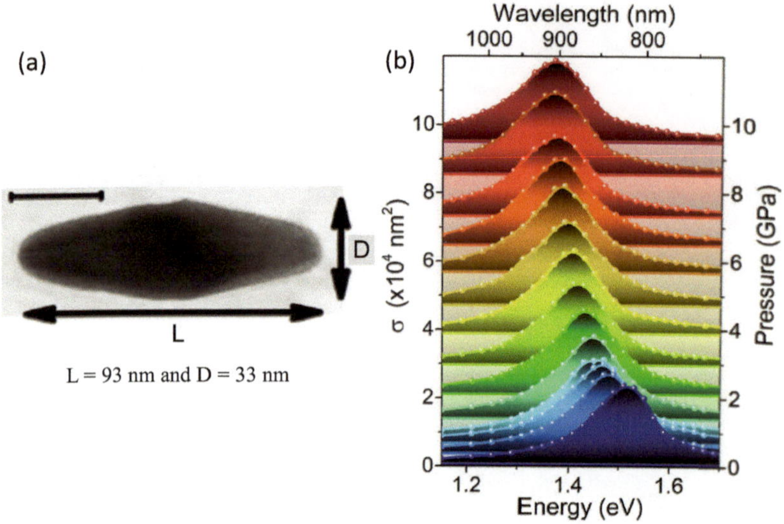

Figure 12.13 (a) TEM image of single gold nanoparticle, (b) the optical extinction spectrum of gold particle extinction, measured at increasing pressures. Reprinted from Ref. [36].

In a similar research on quantum dots, Zh et al. investigated the effect of pressure (up to 15.8 GPa) on the nanocrystal structure of the quantum dots of CdSe-CdS QD hosts (9.7 ± 0.5 nm)-Au satellite NCs (Fig. 12.15a) [20]. The SAXS confirmed the fcc structure at the ambient pressure with a lattice parameter of 16 nm (Fig. 12.15b). By increasing the pressure up to 6 GPa, the d-spacing of the (111) facet of gold was decreased from 2.76 to 2.68 nm. A higher pressure than 6 GPa leads to the overlap of (200) QD's peak with the (111) Au's peak. Further compression caused a gradual transformation of the fcc structure to the lamellar superstructure (Fig. 12.15a). At 15.8 GPa, the fcc structure was completely transformed to the lamellar structure and retained even after releasing the pressure to the ambient condition. The transformation process has been evaluated by SAXS and HR-TEM. The SAXS and HR-TEM results as well as schematic of the proposed pressure phase transformation process were provided in Fig. 12.15.

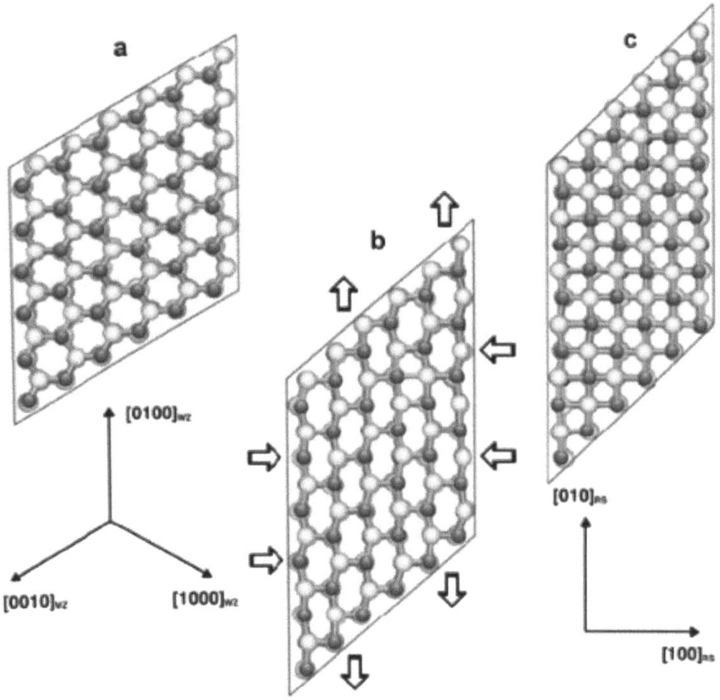

Figure 12.14 Mechanism of the WZ to RS transformation raised to a pressure of P = 10.5 GPa. Reprinted from Ref. [37].

12.4.3 Carbon-Based Nanotubes

Carbon has different allotropes, such as graphite, diamond, fullerenes, nanotubes, and graphene, due to its ability to form sp-, sp^2-, and sp^3-bonds, each one with unique physical and chemical properties. High pressure can induce a change from sp^2- to sp^3-bond in carbon, and the produced elusive new carbon phases such as cubic and hexagonal diamonds. Two-dimensional graphene nanosheets exhibit extremely high electron mobility, current density [15], and thermal conductivity [16], making them a promising candidate for applications in micro-electromechanical systems (MEMS), solar cells, and new capacitors. Ke et al. studied the bandgap of a

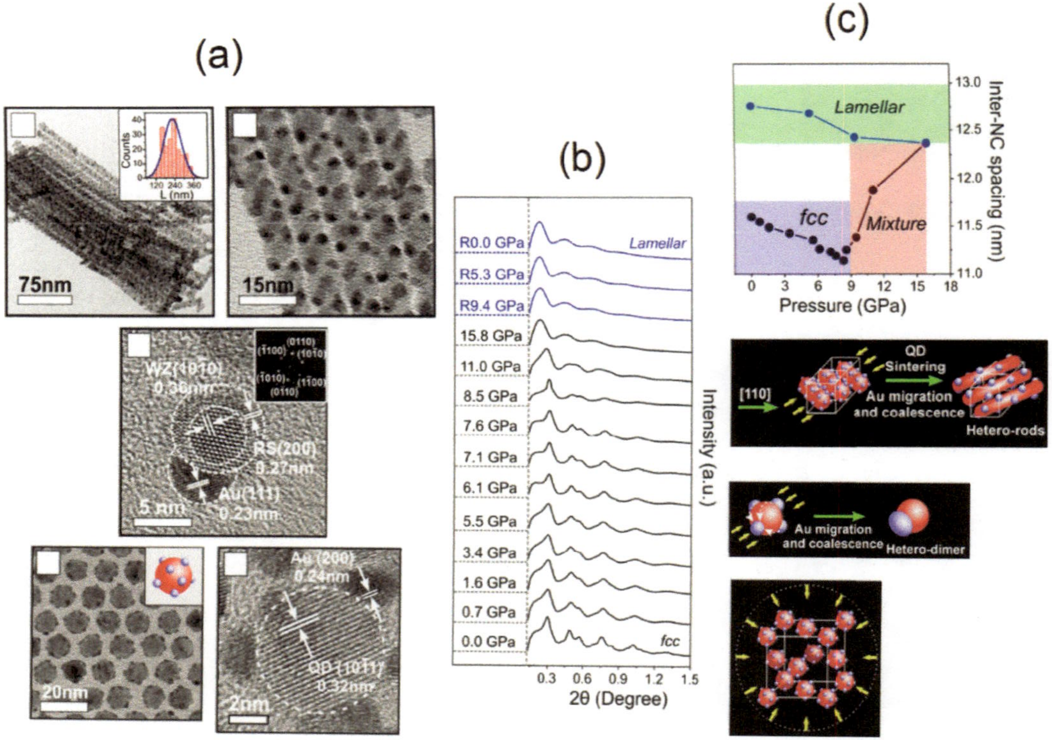

Figure 12.15 (a) HR-TEM images of QD-Au (fcc structure and lamellar), (b) SAXS patterns during compression (black) and decompression (blue), (c) proposed pressure phase transformation process. Reprinted from Ref. [20].

trilayer graphene by tuning the interlayer interaction with high pressure using a DAC (Fig. 12.16a) [38]. Trilayer graphene flakes were produced mechanically by the exfoliation of graphite on Si/SiO_2 substrate. They reported that an intrinsic bandgap of 2.5 eV is achievable in compressed trilayer graphene, which is important for post-silicon electronics application. The resistances of trilayer graphene were measured by two- and four-terminal electrodes and a sudden increase in the resistance was observed at pressure higher than 33.0 GPa, which indicated bandgap opening (Fig 12.16b). The spectroscopic techniques of IR and vis-UV absorption were used to confirm the bandgap of 2.5 ± 0.3 eV in pressurized trilayer graphene due to tuning the interlayer coupling. During decompression, the semiconducting phase remains unchanged down to a few GPa, but almost returns to its original state again at ambient pressure. The bandgap of 2.5 ± 0.3 eV is still larger than that of silicon (1.1 eV) under ambient conditions, but the high-pressure approach could facilitate the practical graphene-based electronic devices.

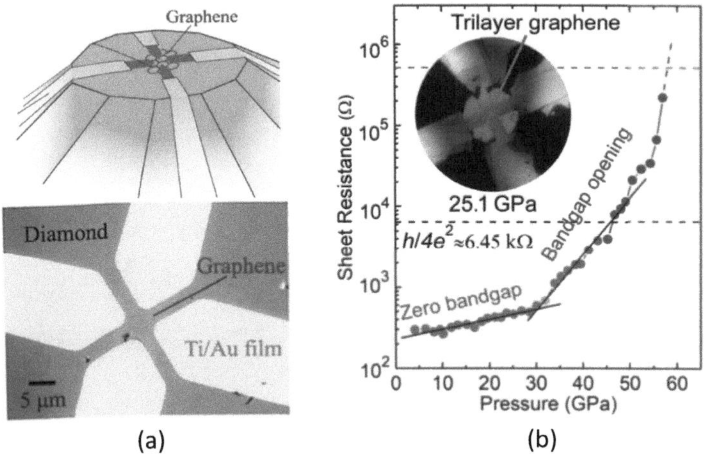

Figure 12.16 (a) Optical microscope image of trilayer graphene and Ti/Au film electrodes on the diamond surface; (b) resistance–pressure curves of trilayer graphene measured at room temperature. Reprinted from Ref. [38].

Smith et al. generated fully hydrogenated graphane by heating graphene at high pressure up to 5 GPa and temperature up to 200°C in molecular hydrogen using DAC [39]. The high pressure led to a reaction of sp^2-bonded graphene with hydrogen, which was investigated by Raman spectroscopy. Figure 12.17 shows a sharp peak at 1350 cm^{-1} in the Raman spectrum of graphene after exposing to hydrogen at combined high pressure and high temperature. The ID/IG ratio is an indicator for the level of hydrogenation under employing atomized hydrogen to react with graphene [39, 40].

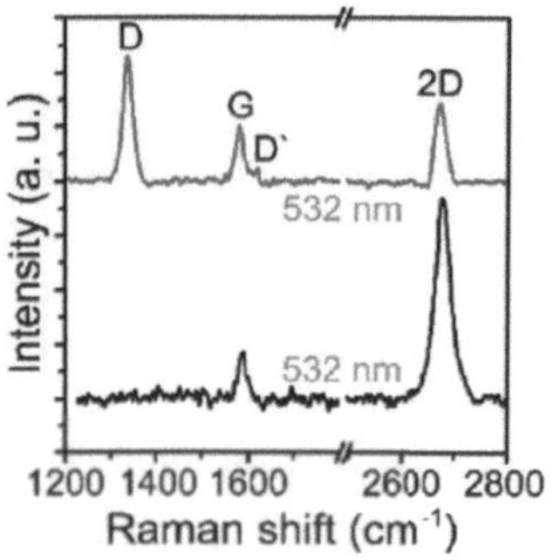

Figure 12.17 (a) Raman spectrum of (black) graphene before and (red) hydrogenated graphene after treatment at 4.0 GPa and 200°C. Reprinted from Ref. [39].

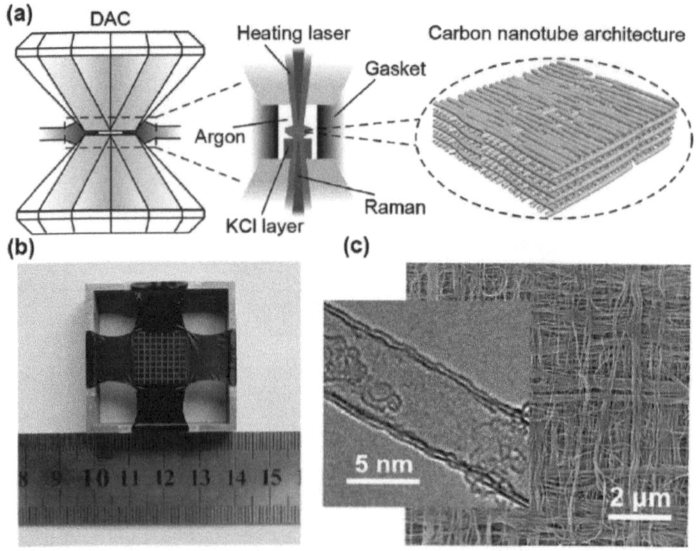

Figure 12.18 (a) Schematic illustration of laser-heated DAC. (b) The red lines indicate the sample area. (c) SEM characterization of samples consisting of entirely cross-stacked CNT bundles. Reprinted from Ref. [41].

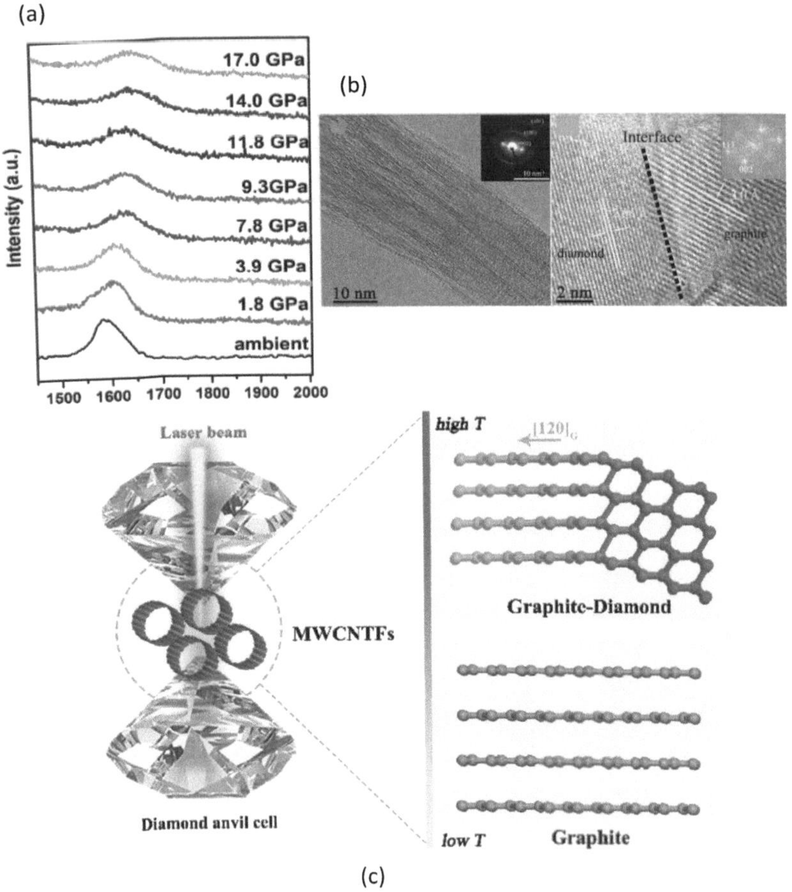

Figure 12.19 (a) Raman spectra of the G-band of pristine MWCNTFs at different pressures, (b) HR-TEM image of the diamond-graphite nanocomposite, and (c) possible transformation pathway of MWCNTFs to diamond-graphite hybrid structure. Reprinted from Ref. [42].

Carbon nanotubes (CNTs) and fullerenes are two classes of fully sp^2 bonding carbons. Fullerenes have hollow cage structures, and the compression C70 fullerenes can produce interesting 1D, 2D, and 3D fullerene network. Li et al. synthesized a cross-stacked CNT architecture by forming sp^3 covalent cross-stacked CNTs under high pressure up to 30 GPa and temperature of 765–1345 K using a laser-heated DAC [41]. The Raman spectrum was used to confirm the bonding formation. A cross-stacked architecture of CNT film (Fig. 12.18) was formed by placing a CNT film over another one with an angle of 90°. This process was repeated until the thickness of the cross-stack reached 10 μm.

Yang et al. proposed the synthesis of a diamond-graphite hybrid structure from multi-walled carbon nanotubes at high pressure–high temperature conditions (up to 17.0 ± 1 GPa and 2000 ± 200 K) [42]. The Raman spectra for samples prepared at higher temperature and pressure showed that the G-band broaden and shifts to higher frequencies that are similar to graphite (Fig. 12.19a). The HR-TEM images of the materials before and after the test confirmed the formation of tilted graphitic carbon layers, which are bonded with diamond crystals (Fig. 12.19b). In a similar research, an irreversible mechanical failure of carbon nanotubes was reported at 72 GPa and the formation of ribbon-like structures was confirmed by Raman and HR-TEM analyses (Fig. 12.20) [43].

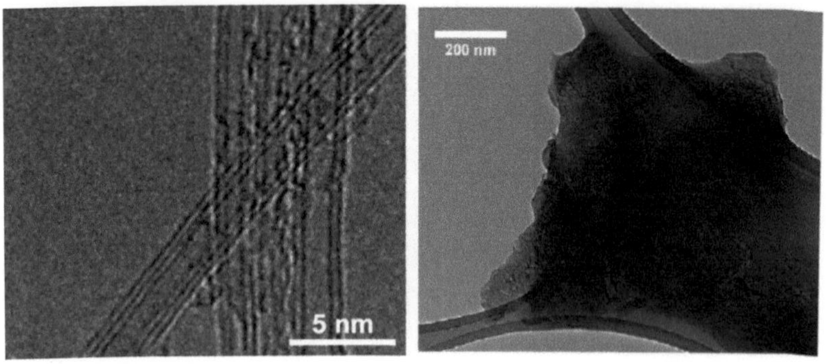

Figure 12.20 TEM images of initial triple-wall carbon nanotubes (TWCNTs) and the recovered sample after 71.9 GPa pressure cycle. Reprinted from Ref. [43].

12.5 Conclusion

Despite the tremendous progress in the investigation of the effect of high-pressure on the structure of nanomaterials, scientists are at the beginning of a long journey and lots of discoveries are still remaining. There is a gap in the study of the effect of ultrahigh pressure on more complex systems, including ternary and quaternary nanomaterials particularly for pressure <100 GPa. This chapter provides a summary of how the chemical and physical properties of the soft and hard nanomaterials could be changed and altered under high pressure. The chapter would help for future studies that will probe the chemistry at higher pressures and explore more complex nanomaterials than those that have been studied to date.

References

1. Ming, L. C. and W. A. Bassett, Laser heating in the diamond anvil press up to 2000 C sustained and 3000 C pulsed at pressures up to 260 kilobars. *Review of Scientific Instruments*, 1974. **45**(9): pp. 1115–1118.
2. O'Bannon III, E. F., et al., Contributed review: Culet diameter and the achievable pressure of a diamond anvil cell: Implications for the upper pressure limit of a diamond anvil cell. *Review of Scientific Instruments*, 2018. **89**(11): p. 111501.
3. Kawabata, Y., et al., Temperature and pressure effects on the bending modulus of monolayers in a ternary microemulsion. *Physical Review Letters*, 2004. **92**(5): p. 056103.
4. Nagao, M., et al., Effects of temperature and pressure on phase transitions in a ternary microemulsion system. *The Journal of Chemical Physics*, 2001. **115**(21): pp. 10036–10044.
5. Purushothaman, S., et al., Influence of high pressure on the bending rigidity of model membranes. *The Journal of Physical Chemistry B*, 2015. **119**(30): pp. 9805–9810.
6. Berghaus, M., et al., Near-surface and bulk behavior of bicontinuous microemulsions under high-pressure conditions. *The Journal of Physical Chemistry B*, 2016. **120**(29): pp. 7148–7153.
7. Dierking, I., et al., *Polymer-Modified Liquid Crystals*. 2019: Royal Society of Chemistry.
8. Chandrasekhar, S., et al., *Pressure-Induced Mesomorphism*. 1975.
9. Huang, Y. and S. Gui, Factors affecting the structure of lyotropic liquid crystals and the correlation between structure and drug diffusion. *RSC Advances*, 2018. **8**(13): pp. 6978–6987.
10. Shearman, G., et al., Inverse lyotropic phases of lipids and membrane curvature. *Journal of Physics: Condensed Matter*, 2006. **18**(28): p. S1105.
11. Salvador-Castell, M., et al., Induction of non-lamellar phases in archaeal lipids at high temperature and high hydrostatic pressure by apolar

polyisoprenoids. *Biochimica et Biophysica Acta (BBA)-Biomembranes*, 2020. **1862**(2): p. 183130.

12. Yaghmur, A., et al., Effects of pressure and temperature on the self-assembled fully hydrated nanostructures of monoolein–oil systems. *Langmuir*, 2010. **26**(2): pp. 1177–1185.
13. Kulkarni, C. V., et al., Effects of high pressure on internally self-assembled lipid nanoparticles: A synchrotron small-angle X-ray scattering (SAXS) study. *Langmuir*, 2016. **32**(45): pp. 11907–11917.
14. Podsiadlo, P., et al., High-pressure structural stability and elasticity of supercrystals self-assembled from nanocrystals. *Nano Letters*, 2011. **11**(2): pp. 579–588.
15. Walsh, J. P. and D. E. Freedman, High-pressure synthesis: A new frontier in the search for next-generation intermetallic compounds. *Accounts of Chemical Research*, 2018. **51**(6): pp. 1315–1323.
16. Gu, Q., et al., Unexpected high stiffness of Ag and Au nanoparticles. *Physical Review Letters*, 2008. **100**(4): p. 045502.
17. Li, B., et al., Pressure compression of CdSe nanoparticles into luminescent nanowires. *Science Advances*, 2017. **3**(5): p. e1602916.
18. Mendoza-Cruz, R., et al., Orthorhombic distortion in Au nanoparticles induced by high pressure. *CrystEngComm*, 2019. **21**(22): pp. 3451–3459.
19. Wu, H., Z. Wang, and H. Fan, Stress-induced nanoparticle crystallization. *Journal of the American Chemical Society*, 2014. **136**(21): pp. 7634–7636.
20. Zhu, H., et al., Pressure-enabled synthesis of hetero-dimers and hetero-rods through intraparticle coalescence and interparticle fusion of quantum-dot-Au satellite nanocrystals. *Journal of the American Chemical Society*, 2017. **139**(25): pp. 8408–8411.
21. Wang, T., et al., Pressure processing of nanocube assemblies toward harvesting of a metastable PbS phase. *Advanced Materials*, 2015. **27**(31): pp. 4544–4549.
22. Wang, Z., et al., Deviatoric stress driven formation of large single-crystal PbS nanosheet from nanoparticles and in situ monitoring of oriented attachment. *Journal of the American Chemical Society*, 2011. **133**(37): pp. 14484–14487.
23. Bian, K., Z. Wang, and T. Hanrath, Comparing the structural stability of PbS nanocrystals assembled in fcc and bcc superlattice allotropes. *Journal of the American Chemical Society*, 2012. **134**(26): pp. 10787–10790.

24. Wu, H., et al., Nanostructured gold architectures formed through high pressure-driven sintering of spherical nanoparticle arrays. *Journal of the American Chemical Society*, 2010. **132**(37): pp. 12826–12828.

25. Guo, Q., et al., Cubic to tetragonal phase transformation in cold-compressed Pd nanocubes. *Nano Letters*, 2008. **8**(3): pp. 972–975.

26. Li, B., et al., Superfast assembly and synthesis of gold nanostructures using nanosecond low-temperature compression via magnetic pulsed power. *Nature Communications*, 2017. **8**(1): pp. 1–9.

27. Li, Q., et al., Pressure-induced phase engineering of gold nanostructures. *Journal of the American Chemical Society*, 2018. **140**(46): pp. 15783–15790.

28. Ying, P., et al., High-pressure synthesis of cBN nanoparticles with high-density nanotwin substructures. *ACS Omega*, 2019. **5**(1): pp. 650–654.

29. Ojeda-Galvan, H., et al., Deformation behavior of titanate nanotubes subjected to high pressure. *Journal of Applied Physics*, 2017. **121**(2): pp. 025902.

30. Koski, K. J., et al., Structural distortions in 5–10 nm silver nanoparticles under high pressure. *Physical Review B*, 2008. **78**(16): p. 165410.

31. Quan, Z., et al., Pressure-induced switching between amorphization and crystallization in PbTe nanoparticles. *Nano Letters*, 2013. **13**(8): pp. 3729–3735.

32. Wu, H., et al., Pressure-driven assembly of spherical nanoparticles and formation of 1D-nanostructure arrays. *Angewandte Chemie*, 2010. **122**(45): pp. 8609–8612.

33. Aristizabal, K., et al., Evolution of the lattice defects and crystalline domain size in carbon nanotube metal matrix composites processed by severe plastic deformation. *Materials Characterization*, 2019. **154**: pp. 344–352.

34. Li, B., et al., Corrigendum: Superfast assembly and synthesis of gold nanostructures using nanosecond low-temperature compression via magnetic pulsed power. *Nature Communications*, 2017. **8**.

35. Schroer, M. A., et al., Pressure-stimulated supercrystal formation in nanoparticle suspensions. *The Journal of Physical Chemistry Letters*, 2018. **9**(16): pp. 4720–4724.

36. Medeghini, F., et al., High-pressure effect on the optical extinction of a single gold nanoparticle. *ACS Nano*, 2018. **12**(10): pp. 10310–10316.

37. Bealing, C., R. Martoňák, and C. Molteni, The wurtzite to rock salt transition in CdSe: A comparison between molecular dynamics and

metadynamics simulations. *Solid State Sciences*, 2010. **12**(2): pp. 157–162.

38. Ke, F., et al., Large bandgap of pressurized trilayer graphene. *Proceedings of the National Academy of Sciences*, 2019. **116**(19): pp. 9186–9190.
39. Smith, D., et al., Hydrogenation of graphene by reaction at high pressure and high temperature. *ACS Nano*, 2015. **9**(8): pp. 8279–8283.
40. Ryu, S., et al., Reversible basal plane hydrogenation of graphene. *Nano Letters*, 2008. **8**(12): pp. 4597–4602.
41. Li, R., et al., Joining cross-stacked carbon nanotube architecture with covalent bonding. *Applied Physics Letters*, 2017. **110**(18): p. 183101.
42. Yang, X., et al., Diamond-graphite nanocomposite synthesized from multi-walled carbon nanotubes fibers. *Carbon*, 2021. **172**: pp. 138–143.
43. Silva-Santos, S., et al., From high pressure radial collapse to graphene ribbon formation in triple-wall carbon nanotubes. *Carbon*, 2019. **141**: pp. 568–579.

Index